WILDLIFE DAMAGE MANAGEMENT

WILDLIFE
Damage
Management

Prevention, Problem Solving
& Conflict Resolution

RUSSELL F. REIDINGER, Jr. & JAMES E. MILLER

THE JOHNS HOPKINS UNIVERSITY PRESS | BALTIMORE

The Johns Hopkins University Press
2715 North Charles Street
Baltimore, Maryland 21218-4363
www.press.jhu.edu

Library of Congress Cataloging-in-Publication Data

Reidinger, Russell F., 1945–
  Wildlife damage management : prevention, problem
solving, and conflict resolution / Russell F. Reidinger, Jr.,
and James E. Miller.
       pages cm
  Includes bibliographical references and index.
  ISBN-13: 978-1-4214-0944-3 (hardcover : acid-free paper)
  ISBN-10: 1-4214-0944-5 (hardcover : acid-free paper)
  ISBN-13: 978-1-4214-0945-0 (electronic)
  ISBN-10: 1-4214-0945-3 (electronic)
 1. Wildlife management.   2. Wildlife pests—Control.
I. Miller, James E. (James Earl), 1941–   II. Title.
  SK355.R45 2013
  639.9—dc23   2012045475

A catalog record for this book is available from the British
Library.

For Carol, Stephanie, and Benjamin
—RFR

For Doris, Kelly, Michael, Brooks, and for those exemplary heroes and mentors who preceded us in this honorable profession and avocation
—JEM

Contents

Preface

Whereas this textbook may have found very limited use even a few decades ago, today it is intended to meet the present and emerging needs of educators, students, and professional practitioners. We divide the book into six major sections. First, we provide terms relating to wildlife management in general and wildlife damage management in particular as well as a summary of the history of damage management and of the resources available for its study. Second, we relate wildlife damage management to its ecological and wildlife management underpinnings. Third, we survey damaging species. Fourth, we explore current practices, organizing them into physical, chemical, and biological methods. Fifth, we explore the human dimensions of wildlife management, including the social and cultural dimensions, the policy and legal dimensions, and the increased influence of public opinion on biologically and ecologically based practices. Sixth, to pull it all together, we look at overall management strategies and use current research directions to propose a vision of the future.

We begin each chapter or section within a chapter with a statement of concepts, principles, and terms. We follow these with an explanation. For each, we then illustrate the concepts with specific examples from wildlife damage management experience or case studies. Throughout, using concepts and examples, we try to maintain a global rather than local or national perspective on a truly worldwide subject. We include, as appropriate, examples from the entire spectrum of damaging species, from unicellular organisms to plants and animals, but emphasize vertebrate pests throughout. A summary and questions are included at the end of each chapter to assist in reviewing and to further deeper discussion of selected issues.

PART I • AN OVERVIEW OF WILDLIFE DAMAGE MANAGEMENT

In this section we introduce wildlife damage, explore its history, and point to resources that help practitioners choose proven solutions that are based on science and experience.

1

Introduction

This chapter defines and describes wildlife damage management, contrasts notions of the value of wildlife with that of pests and problem plants and animals, and relates wildlife damage management to wildlife management, applied ecology, and conservation.

Statement

Wildlife species are highly valued and beneficial resources, but they also injure and kill people, make people sick, threaten their livelihoods, and damage their property, interests, and industry. Preventing or reducing these negative consequences of wildlife is the purpose of wildlife damage management.

Explanation

The term *wildlife* evokes different responses among people. Some of us think of deer (*Cervidae*), pronghorn antelope (*Antilocapra americana*), pheasant (*Phasianidae*), turkey (*Meleagris gallopavo*), or other game species. This probably reflects a long history of hunting or fishing, either directly or vicariously through magazines, television, and movies. Others think of American bald eagles (*Haliaeetus leucocephalus*), lions (*Panthera leo*) on the Serengeti plains of Africa, snow leopards (*Uncia uncia*) in the mountains and agroecosystems of India, or koalas (*Phascolarctos cinereus*), desert tortoises (*Gopherus agassizii*), green sea turtles (*Chelonia mydas*), least terns (*Sternula antillarum*), or other threatened or endangered animals. These images probably reflect a more recent human concern for our fellow creatures on this planet. Both game and nongame threatened and endangered species are examples of wildlife as the term is used in this book.

What about the **commensals**, animals that "come to the table" and associate with humans, such as the house mouse (*Mus musculus*) or the Norway rat (*Rattus norvegicus*), that live behind storehouse walls and are not likely to be the first animals that come to mind when we think about wildlife? Or, for that matter, what about the fleas on the backs of the commensal rodents that may carry a virus or a bacterium? And how about unwanted dandelions (*Taraxacum* spp.) that grow along with cultivated grasses in the lawn? The mouse, the flea, the virus or bacterium, and the dandelion—nondomesticated, living, and at least irritating humans and possibly seriously affecting their health—are all important wild living things and thus a part of this book.

We define **wildlife** as any nondomesticated plant or animal, but this working definition requires some explanation. Some creatures are not clearly domesticated or nondomesticated. Examples include laboratory rats, strains of wild Norway rats bred by humans over many generations to facilitate handling and use in research or as pets. Others are rainbow trout (*Oncorhynchus mykiss*), bluegills (*Lepomis macrochirus*), channel catfish (*Ictalurus punctatus*), and largemouth bass (*Micropterus salmoides*), which were bred and raised commercially in fish farms and often bought and stocked by landowners for personal use. Still others are deer and elk (*Cervus canadensis*) raised on "game farms" using animal husbandry as practiced with domestic livestock. Whether these animals are indeed wildlife is determined by public law, often worked out in a courtroom (e.g., Bolen and Robinson 2002). For our purposes, they are domesticated and not considered in this book.

Management, in the simplest sense, is getting something or somebody to do what the manager wants. The term is nonjudgmental, accommodating both "good" and "bad" managers. When used in wildlife management, it means getting wildlife to do what people want, such as changing behavior, population size, or community structure. The term also refers to getting people to do what the manager wants, such as changing attitudes about wildlife or its damage. While the manager ultimately decides what is to be done, his or her decisions are almost always based on the views of people affected by the management responses or the damage. While the concept of management is itself straightforward, getting people to agree on what they want can be difficult. Achieving consensus on acceptable strategies and methods for managing the damage is a further difficulty.

Wildlife damage is a perceived or actual harm to humans, their property, or something they value. Practicing managers and scientists within the Wildlife Services program (United States Department of Agriculture, USDA) categorize wildlife damage generally as damage to property, agriculture (including forestry and livestock), public health or safety, or other wildlife species (table 1.1). The approach is comprehensive but overlapping. For example, ingestion of Canada geese (*Branta canadensis*) or other birds by engines of a commercial airplane, with consequent

TABLE 1.1   *Types of wildlife damage*

Type of damage	Example
Property	Bird damaging airplane engine
Agricultural, including forestry and livestock	Rodents eating rice
Public health or safety	Ebola virus species-jumping to humans
Affecting other wildlife	Brown cowbird parasitizing warbler nest

damage to the engines, is both damage to property (jet engines and the plane on crashing) and possibly public safety (harm to humans if the plane crashes [fig. 1.1]). Another approach is to classify damage taxonomically—i.e., damage caused by plants, birds, mammals, or other taxa. For example, **vertebrate pest control**, which for many years was considered the whole of wildlife damage management, focuses on damage caused by birds, mammals, reptiles, and amphibians. Alternatively, the classification may be based on an ecological role such as **predation** (attacking and eating another animal) or **herbivory** (eating plants).

Two types of wildlife damage have received particular attention recently: **invasive species** (nondomesticated species that invade new habitat, adversely affecting it) and **wildlife diseases** (abnormal conditions that impair bodily functions, have specific symptoms and signs, and are carried by nondomesticated organisms). Invasive species have probably affected human enterprise and natural ecosystems since early humans first migrated and traded. Wildlife diseases, particularly **zoonoses** (wildlife diseases that are transmitted between wildlife and humans or their livestock or pets) such as **rabies** (a viral neuroinvasive disease that causes acute encephalitis in warm-blooded animals) and **plague** (caused by the bacterium *Yersinia pestis* and manifested as bubonic, septicemic, or pneumonic), have also had a long history with humans. Problems with invasive species and zoonoses have recently resurfaced with increasing frequency and intensity, prompting public concerns for better understanding of the causes and demands for solutions.

Pest is a term used often in wildlife damage management. The term refers to the individual, population, or species that is causing damage. For example, common vampire bats (*Desmodus rotundus*) that bite donkeys (*Equus africanus asinus*) and cattle (*Bos* spp.) in Central America, transmitting rabies and reducing milk production or weight gain, are called pests (fig. 1.2). So are European starlings (*Sturnus vulgaris*)

Figure 1.1 US Airways Flight 1549 crashed into the Hudson River shortly after ingestion of geese on take-off, January 15, 2009. All 155 passengers and crew survived. *Photo by U.S. Army Corps of Engineers.*

Figure 1.2 Vampire bat (*left*, *Desmodus rotundus*) about to feed on blood from the ankle of a cow, perhaps transmitting rabies to the cow. After a bat is captured (*right*), jelly containing anticoagulant can be smeared onto the fur and the bat released to return to its cave. Conspecifics ingest poison when heterogrooming, making the method species specific and selective. *Photos by USDA.*

at cattle feed lots and Eastern gray squirrels (*Sciurus carolinensis*) at some bird feeders. The dandelions mentioned above are likely called pests by most suburban homeowners.

Today when we think of wildlife damage, we don't conjure up pests as much as we do damage caused both by desirable and, to some people, undesirable wildlife that, knowingly or not, have acted against the will of humans. We assume that wildlife lack **conscience** (the sense of right or wrong with which humans are endowed) and, in the course of being, sometimes offend humans. We recognize that unwanted dandelions in lawns are a nuisance, a negative value. Many old-timers, however, attest that leaves of dandelions make great salads and the flowers great wines. And the root sap of the dandelion may someday provide inexpensive rubber (Bland 2008). European starlings, a negative value at cattle feed lots, also support human interests as predators of injurious insects (Conover and McCoy 2003). Thus, all wildlife species have positive and negative aspects. And in that context, **wildlife damage management** can be viewed as the science and art of diminishing the negative aspects of wildlife while maintaining or enhancing their positive ones. That being said, the term *pest* is a convenient one and we use it in this book, always with the understanding that all species may have some positive values to humans.

Wildlife preservation and wildlife conservation also relate to wildlife damage management. **Wildlife preservation** is underlain by the philosophy that humans should not interfere with nature, essentially letting wildlife take its course. Here it is deemed best for humans to not intervene in the affairs of nature. Wildlife preservation has underpinned some approaches to management of wildlife and its damage, including that sometimes used by the United States National Park Service (USNPS). The approach has yielded many successes. However, public pressure has recently encouraged the USNPS to revise its preservationist approach to some wildlife damage problems—e.g., allowing deer hunting at both Gettysburg National Military Park and the Eisenhower National Historic Site. As stated by the park superintendent, "Deer management is an unfortunate necessity of preserving the Gettysburg and Eisenhower parks. Intense browsing by high numbers of deer damages the historic landscapes. We need to protect the historic woodlots and the farm fields in order to tell the story of these two parks" (Latschar 2009). At the time of writing, the USNPS was also considering controlled deer hunts at the Morristown National Historic Park in New Jersey. Unless humans intervene or nature's course eventually readjusts the deer population to an acceptable level, the historic park has no capability to reproduce its primary canopy, because all seedlings are consumed by hungry deer. The deer have moved into surrounding neighborhoods, including the nearby Frelinghuysen Arboretum, prompting a call for deer culling by the Park Commission and the Audubon Society.

Wildlife conservation contrasts with wildlife preservation in that it advocates the wise use and stewardship of wildlife resources. Championed by Aldo Leopold (1933) in the first textbook on wildlife management, this philosophy talks of humans as a part of both the problem of, and the solution to, wildlife damage. Under this philosophy, for example, humans discovered and implemented a solution to the transmission of rabies to cattle by vampire bats in Central America. Humans brought the cattle to Central America, providing easy prey for the vampire bats. Growth of vampire bat populations was directly related to increased cattle production. Thus, humans were clearly part of the problem. A solution required the discovery that vampire bats are susceptible to low doses of **anticoagulants** (compounds that reduce blood coagulation) and that anticoagulants can be delivered to bats through injection of low doses in cattle (Thompson et al. 1972) or by smearing some on the bats' fur and then by their **heterogrooming** (wherein one animal grooms another; fig. 1.2). Thus, humans were part of a solution.

Ecology is the study of plants or animals "at home"— that is, the study of plants or animals interacting with both the living (biotic) and nonliving (abiotic) parts of their environment (Odum 1971). Odum described wildlife management as a part of the broader application of ecology to natural resources management. He pointed to the intensive studies of individual game species as contributing greatly to population ecology and cited as evidence examples from such studies in his textbook. Ecological studies have provided fundamental knowledge of wildlife species on which to base management practices and methods to evaluate effectiveness of the practices. Wildlife management, therefore, is bolstered by ecology and can be viewed as an application of ecology.

Wildlife damage management exists within the broader base of wildlife management and is therefore also undergirded by ecology. Wildlife managers use damage management, for example, to reduce predation, at a refuge whose mission is to produce ducks, or to eliminate predation by the Arctic fox (*Vulpes lagopus*) so that endangered Aleutian Canada goose populations (*B. c. leucopareia*) can recover. In this book we use an overall **autecological** (ecology of individual organisms or species) organization to look at the management of damage by individual wildlife species and a **synecological** (a holistic approach to ecology, involving groups of organisms such as communities or landscapes) organization to approach damage and management at the community level.

We close by emphasizing the **anthropomorphic** (human-oriented) nature of wildlife damage management. It seems unlikely that wildlife have the cognition necessary to make them keenly aware of their own existence (and therefore the certainty of their own death or that of other wildlife) or that they might have concern for the extinction of their own or other species. It seems also that wildlife will continue to interact with each other, regardless of human presence or absence. It therefore seems incumbent on humans to make the best decisions they can about the needs of humans and wildlife and then to proceed with the wise management of wildlife resources.

Examples

DAMAGE TO AGRICULTURE. It is just past sunset as a Filipino farmer walks along a dike on his hectare rice farm, his steps slow and in unison with the rhythmic rattle of a stone within an old tin can, tied to a stick that he shakes. Resting momentarily, he hears the rustle of rice panicles as rats move from plant to plant, carefully inspecting each one and methodically removing and eating the rice. The farmer presses on, fearing that within a few days his rice will be gone, harvested not by him but by uninvited commensal rodents. The farmer will spend the next few nights walking his fields, rattling the tin can, guessing at the futility of his efforts. Perhaps in the morning he will seek the assistance of a local priest. Or perhaps he will negotiate with the rats. The rice is his family's main income and resource. Without its availability and retail value, his family will suffer severely (Canby 1977).

DAMAGE TO FORESTRY. The landowner looked in dismay at the flooded woodland, thinking out loud that the flooded area must encumber at least 60 acres. When he was here two days ago, there had been no flooding. Winding down a dirt road that quickly became impassable, and observing that water had begun eroding the rock surrounding one culvert and was already reaching for the substructure of the County AU road overpass, the landowner hopped from his pickup and got his first glimpse of the source of the problem. "Beaver dam," he said, "Didn't know those little guys could cause me such trouble!"

A short time later, a local control specialist stood beside the landowner looking at the rising water. "You need to decide what you want to do," he said. "If you want to eliminate the flooding, and prevent further damage to your timber, roads, culverts, bridges, and other crops, then we need to remove the dam, the beaver, and the debris. If you want to retain the beaver and the pond, then we can try a device called a pond leveler, although it will only work if you are willing to put considerable time and effort into maintaining it. If not maintained, damage will soon resume, and it may begin again anyway somewhere upstream." The landowner thought it over. He considered the summer heat, the difficulty for him or anyone to work in ponds under these conditions, and the time and effort he would have to commit if he were to try the pond leveler. **Pyrotechnics** or mechanical/physical removal of the dam would be tough work, but he knew experienced trappers such as this agent were ready and able to do that work, and it offered a permanent solution to the problem. But, he liked seeing the beaver, an animal that was a rare sight only 60 years ago. It was a tough choice.

DAMAGE TO PROPERTY AND HUMAN SAFETY. An American military pilot deftly aborts his landing route, banking his B-2 "Stealth" bomber into a holding pattern near the main runway at Whiteman Airbase in Missouri. The tower has instructed him to do so. Meanwhile, on the runway below, a muffled sound is heard and a deer drops, downed expertly by a federal wildlife biologist sharpshooter. The deer was the unfortunate victim of being in the proverbial wrong place at the wrong time. Shortly, a pickup truck emerges onto the runway and the deer is removed. The plane lands safely.

DAMAGE TO OTHER WILDLIFE. A short distance from the edge of a Michigan pine-oak forest but beyond the easy or likely reach of humans, an endangered Kirtland's warbler (*Dendroica kirtlandii*) quietly builds its nest, spurred on by a growing urge to lay eggs. A few days later, the healthy eggs are laid and the warbler incubates them. The incubation itself, aided by male feeding, is uneventful except for the visit of another bird. Its stay is brief. In the ensuing days the warbler cares for the eggs (fig. 1.3), and when the young hatch, the male switches his energies to capturing and bringing insects for the voracious and rapidly growing young. At long last, the single surviving young is feathered and grown and the "warbler" fledges—in the person of a brown cowbird (*Molothrus ater*)! This reproductive effort, commendable though it was, added nothing to the survival of the warbler species. It took the efforts of humans, first to assess the impact of the brown cowbird on populations of Kirtland warblers and then to aggressively trap and remove cowbirds

Figure 1.3   An egg of a parasitic brown-headed cowbird (*Molothrus ater*) rests in the nest of an eastern phoebe (*Sayornis phoebe*). The phoebes will hatch and fledge the young cowbird, sometimes at the expense of their own offspring. *Photo by Galawebdesign.*

from critical habitat, to reverse the warbler's slide toward extinction (Decapita 2000).

DAMAGE TO PUBLIC HEALTH OR SAFETY. Within a tropical rain forest in the Democratic Republic of the Congo (formerly Zaire) is the small village of Yambuku. In August 1976, a 44-year-old schoolteacher returns to Yambuku from a trip to the northern part of the country. He feels ill and is treated in a local clinic for malaria. However, his symptoms worsen and include diarrhea, uncontrolled vomiting, headache, and dizziness. Eventually he begins bleeding severely and dies about fourteen days later, on September 8. Soon, nosocomial transmission results in others in the clinic dying. In all, 284 (88%) of 318 reported cases are fatal (Pourrut et al. 2005). National and soon international assistance is sought to quarantine the village and prevent further spread of the disease. **Ebola hemorrhagic fever** (a viral disease with malaria-like symptoms, often fatal, first recognized in the Democratic Republic of the Congo in 1976, with bats or other wildlife as reservoirs) becomes an international concern, with sporadic outbreaks occurring in Gabon, Africa, as recently as January 2002.

These examples, although separated geographically and in time, underscore a common theme: regardless of type of damage, wildlife was the principal culprit in some form of ensuing human disaster, actual or potential. Most of us can tell tales of some close encounter that left us feeling bad about wildlife. Have you met a wary striped skunk (*Mephitis mephitis*)? Has your car collided with a deer? Have you had to repair holes made by woodpeckers in your redwood siding or by squirrels or raccoons in your attic? Are you concerned about **West Nile virus** (which infects humans, pets, and wildlife and is often transmitted by the bite of a mosquito), **Lyme disease** (caused by bacteria of the genus *Borrelia*, transmitted by ticks of the Genus *Ixodes*, and causing flu-like symptoms in humans, eventually affecting joints, heart, and nervous system if left untreated), or rabies? Wildlife damage can rightly be viewed as the dark side of wildlife. Perhaps because we humans share more and more space with wildlife, or perhaps because of our successes in the management of such species as deer and geese, wildlife damage management is today a subject of great interest to students as well as the general public.

Summary

- Wildlife damage management is the science and art of diminishing the negative consequences of wildlife while maintaining or enhancing their positive aspects. It is a part of wildlife management and an application of ecology.
- Wildlife includes any nondomesticated living organism, including viruses, bacteria, plants, and animals.
- Damage is an affront, real or perceived, to humans. It is a global issue that can be categorized as damage to property; agriculture, including forestry and livestock; public health or safety; and other wildlife species or their habitat.
- Invasive species and wildlife diseases (including zoonoses) are two emerging concerns.
- Management is getting both wildlife and people to follow the learned guidance of a manager.
- Today we see damage caused by both desirable and undesirable wildlife as working against the best interests of humans.
- The principle of wildlife preservation holds that humans should not interfere with nature, leaving little room for proactive management of damage.
- The principle of wildlife conservation holds that humans are part of the cause of damage and need to be part of the solution.

Review and Discussion Questions

1. Relate wildlife damage management to wildlife management and ecology.
2. Are wildlife species good or bad? How do descriptions of wildlife as pests or varmints relate to today's thinking about wildlife damage?

3. We describe wildlife damage management as anthropomorphic. Do you agree or disagree? Defend your view with specific examples.

4. Try categorizing wildlife damage with a system that you concoct, for example by economic loss or intensity of impacts on humans. How does your system compare to the ones used in this book?

5. List ten examples of wildlife damage. Categorize the examples using one or more of the systems you developed in question 4.

2

History

We provide thoughts on how wildlife damage and its management evolved along with people, from early hominids to modern-day humans.

Statement

As humans evolved and reached milestones, such as using predatory weapons, transporting goods, and domesticating wildlife, corresponding dimensions of wildlife damage and its management also emerged, such as massive killings of predators to reduce competition and predation on humans, invasive species and their control, and shepherding, sheep dogs, and **feral** wildlife.

Explanation

Without much evidence, we believe that humans' earliest wildlife damage management practices had the goal of protecting human life and limb. This is based partly on the belief that humans inherited an instinct for survival, hardwired and ingrained, that also occurs in other animals. Perhaps a sabertooth cat (*Smilodon* spp.), a hyena (see, e.g., Rice 2009, for hominid hair in hyena feces), or a giant short-faced bear (*Arctodus simus*) was killed preemptively by prehistorical humans. Although we doubt there will ever be direct evidence to support this view, there is a growing body of circumstantial evidence.

Hart and Sussman (2005) summarized data suggesting that *Australopithecus afarensis,* an early hominid (the same taxonomic family as humans) that lived 5 million to 2.5 million years ago (YA), served ecologically more as prey than predator (table 2.1). The authors attributed this rather demeaning ecological function to three factors: small size (*A. afarensis* were mostly three to five feet tall in a world with much larger predators); ineptitude at making and using tools; and inability to use fire. As evidence, Hart and Sussman (2005) point to teeth marks on bones of *A. afarensis*, talon scrapes on skulls, and holes in crania that just fit the size and shape of the fangs of a sabertooth cat (fig. 2.1). The authors suggest that at least some of the marks were made in the act of predation rather than scavenging and estimate that about 6–10% of *A. afarensis* succumbed to predators, a percentage consistent with efficiencies observed for modern-day predators.

TABLE 2.1   *Wildlife damage management in human history (speculative)*

Period			Wildlife Damage	
Hominid/Human		Role	Activity	Problem
Hominid (*Australopithecus afarensis*)	5–2.9 M*	Prey	Defended self from predators	Small size (3–5 ft.), ineffective tools, can't use fire, fear of predators
(*A. africanus*)	3–2.1 M			
(*Homo habilis*)	2–1.6 M			
Hominid/ Human		Predator	Massive killings of apex predators, keystone megaherbivores	Ecosystem disruptions
(*H. erectus*)	2.0–0.5 M			
(*H. sapiens*)	0.5 M to present			
H. sapiens	100,000 YA to present	Predator; hunter/ gatherer	Trade, migration, war	Commensals move with humans, some become invasive; disease, vectors, reservoirs move with humans, zoonoses
H. sapiens	15,000 to present	Hunter/gatherer	Domestication of wildlife (plants, fish, animals)	Domesticated spp. travel with humans, some become invasive and/or feral; some zoonoses emerge
H. sapiens	10,000 to present	Agriculturist	Ecosystems simplified by slash and burn, plowing, modern methods; crop protection begins	Insect, rodent, bird, other crop pests emerge; pest irruptions; feral plant releases, some invasive
H. sapiens	10,000 to present	Pastoralist/ nomad	Herding and roaming, livestock protection such as shepherding, guard dogs	Livestock predation; zoonoses; flock/herd diseases emerge
H. sapiens	4,000 to present	Urbanite	Cities spotty for most of history; 3% of humans live in cities at start of Industrial Revolution, 50% now	Commensal wildlife problems, zoonoses; increased suburban–wildlife contact; human/pet food needs intensify agriculture and livestock production; increasing energy needs impact ecosystems; wildlife damage problems follow
H. sapiens	2,000 to present	Industrialist (Industrial Revolution)	Use of natural resources, including fossil fuels; beginnings of modern wildlife damage management	Environmental degradation/simplification; global warming; wildlife damage problems follow
H. sapiens	recent	Conservationist	Island, other wildlife, endangered species protected; game species restored, wildlife damage management is part of effort	Increased contact with wildlife transfers diseases, zoonoses, facilitates movement of invasive species

*M = millions of years ago.

Hunting skills began improving some 2.5 million YA as early humans, such as *Homo habilis*, learned to use stone tools. Fossil records indicate, however, that at about four feet tall, they remained a staple in the diet of predatory animals, including leopards, lions, spotted hyenas, *Dinofelis* (a jaguar-sized sabertoothlike cat), *Megantereon* (also a sabertooth-type cat), and extinct hyenas (*Chasmaporithetes nitidula*). *H. habilis*, at least at first, probably still used tools mostly for scavenging. Self-preservation must also have dominated the wildlife damage practices of these early hominids and may

even have been a driving selective force for larger size and brain capacity.

Over time, *H. habilis* adapted, learning to create flint blades, stepping from scavenger to hunter, wearing hides, and understanding climate well enough to move from Africa into Asia and the borders of Europe. As *H. habilis* was replaced by *H. ergaster* and *H. erectus*, early humans continued to move into a more predatory ecological niche. A dramatic shift in their activity (and impact on the planet) occurred by 50,000 or so YA, a shift called by some the Great Leap Forward (Diamond

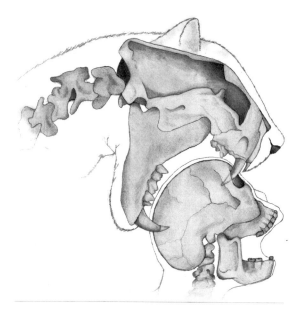

Figure 2.1 Early hominids were probably prey for predators such as sabertooth tigers. *Artwork by Ian Reid, based on earlier work by C. Rudloff, in Hart and Sussman 2005 and Cavallo 1991.*

1997). Some humans began burying their dead and making more sophisticated clothing, indicators of spiritual and conceptual thinking. By this time also, hunting and protection from predatory wildlife included predator pits as well as organizational skills with sufficient hunting and communication techniques to allow mass killing of large herds of herbivores, such as mastodons, by forcing them off cliffs. *H. ergaster* and *erectus* also controlled fire.

By the time of early modern humans, *H. sapiens*, both they and their precedent species had made forays from Africa into Asia and Europe, possibly following large herbivores such as woolly mammoths into the plains of Asia and subsequently into North and South America as times and climate changes either demanded or allowed. A relatively short time after leaving Africa, some 60,000 or 40,000 YA, humans had migrated all the way to Australia. By this time also, modern humans began having major impacts on ecosystems across continents, starting in Australia, then in North and South America, then on the islands of Oceania, then Madagascar, and finally New Zealand less than 500 years ago (Donlan et al. 2007). On each continent, major ecological impacts followed within 400 years of the advent of humans.

Based on information available at the time, Paul Martin proposed humans as the cause of the massive extinctions of Pleistocene megafauna, including its apex predators and larger herbivores, but he envisioned a direct predatory impact (Martin 1971). Today's information suggests that there were too few humans to directly cause such massive die-offs. For example, Clovis people probably numbered only about 50,000 in Pleistocene North America. Rather, today's thinking is that the impact was indirect, a form of additive mortality. Larger herbivores were already under heavy predatory pressure and perhaps were stressed by climate changes as well, when humans first arrived as a different (i.e., a weapon-toting, reasoning, socially organized) kind of predator. Humans, with their unusual knowledge, skills, and abilities, likely depleted the supply of larger herbivores living in localized areas and then moved to easier hunting elsewhere as part of a nomadic lifestyle. The absence of these keystone herbivores resulted in the subsequent demise of predators and abrupt and long-term changes in food chains and food webs. These changes set the ecological bases for some of today's problems with wildlife. In fact, some have suggested that "rewilding"—i.e., reintroducing large megafauna such as elephants, camels, and lions—might be the most effective way to restore the complexity of ecosystems, bringing back more interspecies interactions, species diversity, and stability along with increased resistance of ecosystems to wildlife damage (e.g., Alroy 2001; Donlan et al. 2005; Martin 2005).

As humans settled into a more pastoral existence, they domesticated wildlife that would serve them as food, pets, and beasts of burden. The dog was probably first domesticated about 12,000 YA (D. Morey 1994). Dogs and cats (domesticated about 4,000 YA; table 2.1), both domestic and later feral, aided by humans providing food and transportation, no doubt expanded their geographic ranges. Along with pastoral existence came shepherding, an occupation clearly part of wildlife damage management, as was breeding specialized varieties of dogs for guarding and working. Livestock-guarding dogs may have been used as many as 6,000 years ago, probably in upland regions of Turkey, Iraq, and Syria, where sheep and goats were first domesticated 9,000 to 10,000 years ago. Although the first dogs probably matched the black, gray, or brown colors of sheep, Romans preferred white, and thus breeds such as the Kuvasz or Pyrenean Mountain were selected for their white color. Early Roman authors suggested that white allowed distinction of the guarding dogs from wolves and other predators (Rigg 2001). By 10,000 YA, then, domestication of wildlife for both livestock and pets was well under way. But along with this pastoral existence came the release of domesticated animals back into the wild and the birth of wildlife damage due to feral animals.

Human domestication of wildlife also included plants and a more sedentary agricultural lifestyle. While an agrarian existence offered advantages to humans, freeing time for activities other than subsistence, it also brought problems. Growing crops and keeping livestock in set locations probably increased problems with insects, diseases, and other pests, including commensal rodents and other animals such as today's raccoons, foxes, and granivorous birds. Agricultural production provided the ecological **energy subsidies** that allowed the establishment of towns and cities, the beginnings of urbanization. Along with benefits offered to people living in towns and cities came urban wildlife problems. Diseases spread more easily among areas densely populated with people and pets. Diseases associated with wildlife, some transmitted to humans as zoonoses, constituted another form of wildlife damage.

Humans continued to disperse, establishing both land and sea routes that varied with time, changing patterns of land masses, and global climate changes (e.g., National Geographic 2009). Wherever they went, humans took with them livestock, pets, and agricultural crops as plants or seeds. Other organisms—unwanted but traveling nonetheless along with these domesticated plants and animals—included diseases and their vectors, insects, commensal rodents and weeds, and other wildlife that could damage crops and livestock. Sometimes humans purposely moved wildlife or domesticated plants and animals into new areas and released them to survive on their own, perhaps in hopes of harvesting later generations of survivors or simply out of a nostalgic desire to enjoy again the plants or animals of their homelands. For example, seafarers sometimes released goats or foxes on islands along their sea routes in hopes of having meat resources available in the future (e.g., Campbell and Donlan 2005). Other organisms simply migrated onto the islands during landings or shipwrecks (e.g., Sherley 2000). In these ways, exotic species were introduced into new areas, sometimes raising havoc in the ecosystems that they joined.

Trade was apparently under way as long as 100,000 YA (National Geographic 2009). Items traded might also have served as vectors for some wildlife problems, such as the transfer of weed seeds among those sold for crops or the transfer of commensal rodents and their associated zoonotic diseases among perishable goods. As with other early wildlife damage actions of humans, direct evidence for such transfers is wanting, but indirect evidence (in the form of modern-day examples) such as movement of weed seeds or zoonoses along trading routes is plentiful.

As humans dispersed into new areas, they slashed and burned or otherwise cleared land for agriculture, often simplifying complex checks and balances of natural ecosystems, turning them into simple **monocultural agroecosystems**. With species unleashed from their natural checks, sometimes the simplified ecosystems responded with their own perturbations, such as "outbreaks" of wildlife that destroyed crops, brought pestilence, or caused other wildlife-related problems for the settlers.

At some time within this conjectural narrative, humans began keeping written records more detailed than records found in parables, stories, and songs. Written records give us more detailed looks at human events and hone our understanding of the nature of offending wildlife as well as human responses. For example, the writings of Thucydides describe a plague that embraced the city-state of Athens in 430 BC and again in 429 BC and 427–426 BC. The human activity that caused the wildlife event was the Peloponnesian War. Athenians from surrounding areas were ordered to retreat behind the city walls. Crowded and unsanitary human conditions bred diseases, possibly plague, anthrax, or typhoid. Commensal rodents were likely carriers. Fear of getting the disease facilitated a retreat by the attacking Spartans but also encouraged the spread of the disease by Athenians who ignored extant laws because they believed they were doomed to die anyway.

While the time and order vary somewhat from country to country on a global scale, written records show a common pattern with regard to humans and wildlife. The pattern began with a period in which humans lived in relative ecological harmony with wildlife and other natural resources. Then followed a period in which the human population expanded, urbanized, and sometimes industrialized, with exploitation of natural resources, including wildlife. Finally, we have entered a period in which the limits of natural resources are recognized and efforts are made to conserve and sustain them. **Conservation** has therefore been one of the more recent efforts of humans, as has the understanding that conservation can itself exacerbate human-wildlife conflicts. Examples in North America include restoration, with concomitant increases in damage, of game species such as deer, bear, or river otter. Global examples include protected areas for wildlife and consequent damage by protected species to surrounding local communities (e.g., Distefano 2005; Treves et al. 2006).

Interspersed among these major human activities and events were wars, ranging in scale from tribal to global. These wars had enormous and long-term impacts on human populations and the ecosystems in which warriors traveled and in which wars occurred, whether due to the unintentional transport of wildlife; or the subsequent release of beasts of war such as horses, elephants, or camels into the wild; or direct impacts such as denuding of land with bombs, pyrotechnics, or herbicides. Sometimes ecosystems responded with irruptions of wildlife that damaged crops or spread diseases to animals or people. Nelson (2009) argued that because early human populations were relatively small and widely scattered in Europe, non-endemic diseases tended not to spread rapidly or easily. Justinian's invading army brought a pandemic to Europe in 540–565. No further pandemics occurred until the Black Death of 1347–1351. According to this account, the Black Death came to Europe after the disease broke out in the town of Kaffa, under siege by Mongols. The Mongols retreated, but only after the Mongol commander "loaded a few of the plague victims onto his catapults and hurled them into the town." Merchants, along with fleas and rats, left Kaffa shortly after this, carrying the plague with them. The course of the plague can be followed along established trade routes.

Wars generally denuded landscapes and reduced species diversity. Sometimes, however, warring actually protected wildlife from people. For example, Kay (2007) analyzed relationships between wildlife and contact with native North Americans described in the written records of the Lewis and Clark Expedition from 1804 to 1806. He found that abundance of wildlife was inversely related to contact with Native Americans. For example, wildlife was abundant in buffer zones between warring tribes, areas avoided by tribal people. In fact, he argued that, had it not been for warring tribes, European settlers would have seen hardly any wildlife during the period of western expansion.

The global industrial revolution, which brought broad-scale environmental impacts, was tied to urbanization, providing products at greatly reduced costs. Products included those associated with agricultural production and management of wildlife damage. The revolution also facilitated transport of people by air, land, and sea, just as it facilitated transport of wildlife,

intentional or not. An unprecedented use of energy, mostly from fossil fuels, drove the revolution. This part of the historical account continues today throughout the globe. It may culminate in the most profound of all environmental impacts caused by people: human-facilitated global warming (e.g., IPCC 2007). Wildlife and its damage, of course, respond to major climatic changes. With human-facilitated global warming, a greatly reduced overall biological diversity is predicted (Thomas et al. 2004). Predictions for wildlife damage are correspondingly profound and include range extensions of some offending wildlife, greatly increased numbers of invasive species around the world, extinctions of wildlife by other wildlife, and greatly simplified ecosystems subject to more frequent wildlife irruptions.

Underlying the latter period of written human history has also been an impressive geometric expansion of the human population. One consequence has been the settlement of land formerly occupied mostly by wildlife. This land is often marginal in its ability to support agriculture or stable human housing. Conflicts with wildlife often ensue. A burgeoning human population places further demands on natural resources, and their exploitation disrupts, simplifies, or destroys major natural ecosystems. These impacts appear particularly severe near the equator, in ecosystems such as tropical rainforests. Again, reduced biodiversity leads to oscillations of wildlife that damage crops, livestock, and property; facilitate disease outbreaks; and harm desirable wildlife and their habitats.

Beginning long before our relatively short existence (table 2.1) in the Pleistocene as hunters and hardwired into our brains during the long eons when our early hominid forebearers were mostly prey, we inherited instinctive fears of creatures such as snakes and wolves. We can suffer occasional nightmares wherein we are chased by a tiger, a bear, or an eerie creature of the dark such as the chupacabra of Puerto Rico, the orangpendek of Sumatra, the shunka warakin of the Great Plains of the United States, or the waheela of the Northwest Territories of Canada. These, too, are part of our wildlife damage heritage, because they influence our perceptions of damage and how we respond.

Examples

ANCIENT CIVILIZATIONS. Ancient Egyptians had a broad range of wildlife problems, from fleas and commensal rodents in their homes to crocodiles in their streams, birds in their fields, and lions in their pastures. The snake may have been introduced into homes and granaries to control rats (Talbot 1912). Dollinger

(2000) summarized many ancient problems with wildlife, paraphrased as follows: in the Bible's book of Exodus, reference is made to irruptions of frogs, locusts, lice, and flies in Egypt. An inscription in Menna's tomb (1400 BC) provided the following description: "The snake has seized half the grain, and the hippopotami have eaten the rest. Mice abound in the fields, the locusts descend and the herds devour; the sparrows steal—woe to the farmers! The remains on the threshing floor are for the thieves" (Brock 2005, p. 8 in Dollinger 2000). Herodotus (Histories II in Dollinger 2000) referred to children (and possibly priests) having their heads shaved to prevent lice and fishermen's use of casting nets as bed nets to exclude gnats. **Rinderpest**, a cattle plague that moves between livestock and wildlife, may have been present in Egypt since the third millennium BC (Spinage 2003).

Fleas (*Xenopyslla cheopis*) carrying bacilli that cause plague were found in the trash of an ancient workman's village at Amarna, Egypt. According to Panagiotakopulu (2004), the fleas may have co-evolved with the Nile rat (*Arvicanthis niloticus*) and then passed onto black rats (*Rattus rattus*) in cities during Nile floods. If so, plague originated in ancient Egypt and spread from there throughout the ancient world. As additional evidence, Panagiotakopulu (2004) cited squalid working conditions at the workmen's site and descriptions of epidemics from Amarna letters, Hittitic archives, and Ebers papyrus (including references to swelling buboes).

According to Dollinger (2000), sparrows and other songbirds attacked ripening grain of Egyptian farmers. The hieroglyph for sparrow meant "common" and "small" but also "bad." Nets were sometimes thrown over trees or held up by poles to capture and destroy the birds. Boys threw stones or used slingshots to protect crops from birds. Rats and mice got into homes, often made of soft (unfired) clay that was easily gnawed. Remains sometimes show rocks or stones placed in walls to block burrows. Petrie, a British Egyptologist working in Kahun between 1888 and 1890, found what may have been a pottery rat trap (Reeves 1992).

Rabies is known from ancient times, with written records of preventive measures dated as early as 1930 BC. In the Code of Eshnunna, owners were advised to confine dogs suspected of having rabies and to anticipate fines at different levels if aristocrats or slaves were bitten and infected. Although the viral nature of the disease was not understood in a modern-day sense, the saliva of the dog was thought to contain semen that transferred an organism carrying the disease (Tierkel 1975).

AUSTRALIA. The dingo (*Canis lupus dingo*), likely originating as a feral dog, was brought from southern Asia some 5,000 years ago and was followed by extinction from Australia's mainland of the Tasmanian hen (*Gallinula mortierii*) and apex predators including the thylacine (*Thylacinus cynocephalus*) and the Tasmanian Devil (*Sarcophilus harrisii*). Dingoes, increased hunting by humans, or both have been implicated in the extinctions (e.g., Johnson and Wroe 2003).

The list of invasive species released into Australian ecosystems is impressive and includes domestic cats (*Felis catus*), first released from Dutch shipwrecks in the 1600s, with feral populations established by the 1850s; domestic dogs (*Canis lupus familiaris*), donkeys, goats (*Capra aegagrus hircus*), horses (*Equus ferus caballus*), and pigs (*Sus scrofa*) that came with the First Fleet and subsequent landings; the European red fox (*Vulpes vulpes*) in 1855, for hunting; the European wild rabbit (*Oryctolagus cuniculus*), also for hunting; the camel (*Camelus* spp.) between 1840 and 1907, for riding and as a draft and pack animal for rail and telegraph construction, part of the effort to open up arid areas in central and western Australia.

Those species are just a start. Others include carp (*Cyprinus carpio*), first released in the 1850s, though the "Boolara" strain released in the 1960s has had the greatest ecological impacts (Koehn et al. 2000); the common myna (*Acridotheres tristis*), a member of the starling family, released in cane fields of Queensland in 1883 as a biological control for plague locusts, cane beetles, and other insects; the European starling during the late 1800s, with populations taking root in Victoria (1856–1871), New South Wales (1880), and South Australia (1881) so that most of Victoria was colonized by the 1950s; cane toads (*Bufo marinus*), native to South and Central America, in 1935 to control beetles that were infesting sugarcane crops; and over 220 plant species, over half of which were introduced as ornamentals.

Many of these species quickly became feral, establishing viable populations throughout Australia. Many, such as domestic dogs, the dingo, their hybrid, and the fox, became serious predators affecting the sheep, goat, and cattle industries. Others have become suburban and urban pests or pests of cropland. Many have impacted native Australian fauna and flora, raising concern for the long-term conservation of these resources.

Responses of the Australian people and their governments to these situations have also been impressive. One was the construction of fences, including rabbit fences beginning in the 1890s, to block the movement of wild rabbits from western Australia, and a dingo fence, starting in the 1880s, that extends about 5,320 km (3,306 mi) from Jimbour, Darling Downs, to Eyre Peninsula, Great Australian Bight. The fences, which have been maintained, have affected both the intended and some unintended species.

EURASIA. European fairy tales and other folklore abound with concerns about wildlife problems (e.g., the Big Bad Wolf), as do similar tales from other parts of the globe. Medieval Europeans developed poisons for rats (e.g., ratsbane) and wolves (wolfsbane) and built an array of wooden traps, sometimes elaborately designed. Rat traps were mentioned in the medieval romance *Yvain, the Knight of the Lion*, by Chrétien de Troyes, written in the 1170s. Birds were hunted during medieval times in Europe using box and access traps, snares, nets, lures, techniques such as "dogging," and poisons such as birdlime. Cornell University (undated) provided the following insights: dogging involved luring large numbers of birds onto water and into nets. Box and access traps usually involved propping a box on a stick with a pile of grain under it. A string attached to the stick was pulled to dislodge the stick and catch a bird. Traps having doors that opened in only one direction were also used. Netting was used with great variation in application; a common technique was to lure birds onto the ground, where nets were set to be drawn over them as they fed (fig. 2.2). Birdlime, glue made from boiled mistletoe berries and holly bark, was smeared on rods, in cones, or on layers of linen or grass, alternating with grain or other baits, and birds would stick to the birdlime. Sometimes bread soaked in beer or wine was used to stupefy birds so they could be captured. Alternately, poisonous mixes including henbane and foxglove were used to stupefy birds (Porta 1658). Although putting food on the table was probably the main focus of these hunting efforts, the same methods were undoubtedly applied to reduce damage to crops or property.

NORTH AMERICA. The Inuit made "spring-baits" of coiled baleen held by sinew and covered with blubber. Upon ingestion and digestion, the baleen would spring straight, cutting the stomach of a scavenger that would then die of internal bleeding. Reference to spring-bait traps is made in an account of English explorers visiting a central Eskimo village who were troubled by "13 ravenous wolves that proved troublesome to the English and Eskimos alike. The wolves did not attack people but were adept at seizing anything edible, including unguarded dogs" (Oswalt 1999, p. 173).

Two species, the mountain lion (*Puma concolor*) and the timber wolf (*Canis lupus*), were a ubiquitous

Figure 2.2 Bird netting technique illustrated by a fourteenth-century medieval artist. *From* Taccuino Sanitatis, *fourteenth century. http://commons.wikimedia.org/wiki. File: 34-caccia tortore, Taccuino Sanitatis, Casanatense 4182.*

part of colonial life in the United States from its beginnings. These species, accustomed to feeding on deer and other wildlife, found the fat and instinct-challenged livestock of the colonists to be easy prey and took them readily. Many colonists feared the predators would also attack them or their children. Many of them saw it as a God-given responsibility to tame the wildness of colonial America. William Wood (1634, p. 27) said of timber wolves "they be the greatest inconveniency the Countrey hath, both for matter of dammage to private men in particular, and the whole Countrey in generall." And nineteen years after the landing of the pilgrims in 1620, William Bradford (1856) still saw the timber wolf as one of the greatest threats to successful colonization, stating the hope that "pyson, traps and other such means will help to destroy them." (p. 163). In this context, therefore, the mountain lion and the wolf were killed both to eliminate their predation on livestock and for public safety.

Wolf pits surrounded many colonial towns. The pits had steep sides and often sharp stakes on the bottom. The pits were covered with light poles and leaves that gave way under the weight of a wolf or occasionally under that of an unwary traveler. Trefethen (1975) recounted the plight of a traveler in 1630 from Lynn, Massachusetts, who spent an uncomfortable night in one such pit—along with a fallen timber wolf—before she was rescued the next morning. Josselyn (1674) described how four hooks were tied with thread, wrapped in wool, and dipped in melted fat to form a ball. The balls were placed near recent wolf kills. Sometimes a gun or pistol was tied to a tree to be fired by the pull of a trip cord or a baited line leading to the trigger. In addition, wolves were caught with ropes from horseback and brought to town tied to the backs of horses. After wolves were relegated to swamplands or rocky areas as denning sites, wolf drives were used to further reduce their attacks on livestock. Townsmen carrying firearms or pitchforks would surround the denning area and advance toward it, killing the wolves that failed to escape.

Summary

- Human activities moved forward along with human evolution and include predatory tool-making, hunting, migrating, warring, trading, transporting, domesticating, herding, farming, urbanizing, industrializing, facilitating global warming, and conserving wildlife and other natural resources.
- Each activity affected human relationships with wildlife and wildlife damage. For example, trading and migrating led to exotic invasive species; domestication led to wildlife diseases and zoonoses as well as feral wildlife; a pastoral lifestyle led to zoonoses, shepherding, and guard animals; agriculture led to insect, disease, and wildlife damage, plus pest problems and irruptions from simplified agroecosystems; and establishment of cities and towns led to disease transfer within human populations, pressure on natural resources, reduced biotic diversity, and pest and damage irruptions and diseases.
- Responses to damage also evolved with humans: e.g., from self-defense as prey to humans as intelligent, socially organized, and weapon-bearing predators; to shepherding and breeding and use of guard animals; to sophisticated systems for management of pests and disease in managing crops, forests, livestock, and human diseases; and eventually to the conservation of wildlife species.
- Having begun long before humans were hunters and hardwired into our brains, our heritage of wildlife damage manifests itself today as instinctive fears of creatures such as snakes and wolves or occasional nightmares in which we are chased by a tiger, a bear, or eerie creatures of the dark.

Review and Discussion Questions

1. What do you think were hominids' first efforts at managing wildlife damage? Saving life and limb? Protecting crops, caves, or other wildlife? Defend your view by including appropriate references.
2. Relate the notions of the Pleistocene extinction of mammals, apex predators, keystone species, and foundational species to today's problems with wildlife damage. Relate "rewilding," proposed by some conservationists, to these notions.
3. It is said that "history teaches prudence." How might history teach us prudence in the management of wildlife damage? Give a specific example.
4. Wildlife damage management is said to be anthropocentric. How have past activities of humans influenced the development and expression of wildlife damage? Give some specific examples.
5. Explain how the historical activities of hominids, including humans, have contributed to today's global mosaic of wildlife damage problems.

3

Resources

We describe resources available to wildlife damage students and practitioners.

Statement

A vast literature, scientific and secondary, and direct expertise are now available globally to students and practitioners of wildlife damage management. This literature is accessible through private and public experts, libraries, databases, and the Internet.

Explanation

Students and practitioners of wildlife damage management include individuals with modest to great interest in the subject. Some are students enrolled in undergraduate or advanced natural resources programs, perhaps forestry, wildlife management, or wildlife damage management specifically, at colleges and universities around the world. Some are professional natural resource managers and members of public and private natural resources organizations. Others are farmers, ranchers, landowners, or homeowners trying to resolve wildlife damage problems.

Resources available to students and practitioners include both primary scientific literature and secondary literature that includes all forms of media and sometimes the direct assistance of experts. **Primary scientific literature** refers to original reports of studies following the **scientific method** and related observational or experimental designs (fig. 3.1). By using the scientific method, researchers follow a procedure that attempts to minimize opportunities for personal or other biases that might lead to misinterpretation of results and to maximize the odds that the results observed are actually due to the factors being controlled and studied. The scientific method is a series of steps followed sequentially (in theory, though sometimes not in practice): gather background information and current literature; form a hypothesis and its negative form, the "null hypothesis"; develop an experimental or observational design to test the hypothesis; conduct experiments or observations; gather data; analyze the data following appropriate statistical methods; interpret the results; and communicate the findings

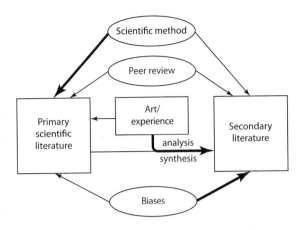

Figure 3.1 Relationships between primary and secondary literature in wildlife damage management. *Illustrations by Lamar Henderson, Wildhaven Creative LLC.*

to the rest of the world. While the scientific method does not eliminate bias, misanalysis of data, or misinterpretation of results, the method is one of humankind's better attempts at objectivity. Primary scientific literature therefore tends to be a useful initial source of information for both the student and the practitioner.

An important aspect of the scientific method, not revealed by simply listing and describing the procedural steps, is critical review by other scientists who have knowledge and experience of the subject. For example, findings may be presented at scientific meetings where methods or results are almost certainly challenged by peers. If the findings are to be published in a scientific journal, the journal editor sends the manuscript to at least one and often two or more knowledgeable peers for critical review. The peers are invited to challenge every aspect of the study, from inclusivity of appropriate background information and current literature, to formation of appropriate hypotheses, to correctness of experimental design, statistical analyses, and interpretation of results. Language is criticized, especially the use of technical words, "scientific jargon," and lack of clarity. The reviewing peers recommend to the editor acceptance, rejection, or revision of the manuscript. If revision is required, the editor may subject the revised manuscript to a second **peer review**. The final decision on publication or rejection lies with the editor, who is also trained and experienced in scientific methodology.

Original scientific reports may be published elsewhere—in proceedings of professional meetings, in books based on topical symposia, or in government technical reports. The results might be translated into an extension educational publication or put on a website for access by the public. Because there are few scientific journals devoted specifically to wildlife damage management and because much research on this topic has been done throughout the world, primary literature for wildlife damage management tends to be scattered globally.

Accessibility to scientific literature is greatly facilitated today through databases on wildlife management and related subjects, including wildlife damage management specifically, and even on specialized areas of wildlife damage management. Such databases may be provided by private vendors, professional associations, educational institutions, and governmental agencies. The databases often provide sufficient information for the student or practitioner to obtain the article from its published source, from the author, or from an agency or institutional website. Increasingly, the databases are electronic, connected through university or college information networks, and sometimes they even include electronic copies of articles.

A word of caution is needed here. Even though bibliographies and databases sometimes include as many as tens of thousands of references to primary literature and may offer a worldwide scope, the nature of wildlife damage management is such that many original research studies, often of high scientific quality, are in relatively obscure locations. For example, reports written by scientists at local, state, or federal agencies are archived in remote locations around the world. Creativity, diligence, and the assistance of information services professionals can help uncover this information. Assistance is available from the National Wildlife Research Center (NWRC), the National Agricultural Library, and universities. The University of Nebraska Extension Wildlife Program established and maintains the Internet Center for Wildlife Damage Management (ICWDM), an open database devoted exclusively to that subject. The ICWDM receives worldwide use, and its content is continually revised and updated. The University of Oxford Wildlife Conservation Research Unit also established and maintains a website devoted exclusively to wildlife damage management. The site serves as a link to the unit's research and management projects, which extend to much of the globe. Other websites are referenced in other chapters of this book. Some databases include secondary literature; when using secondary literature, the reader is cautioned to uncover biases, as described below.

Scientific literature often lacks information and findings related to the "art" or practical aspects of wildlife damage management. Feelings, intuitions, or "revelations" may be based on years of experience and firsthand knowledge of a management problem. Such feelings may lead to, indeed may be requisite to, particularly effective management actions. Given most scientists' focus on objective explanations, feelings are often systematically excluded from scientific literature through the scientific method or peer review (see fig. 3.1). Further, scientific literature may not contain specific information needed for a management situation; in such a situation, the practitioner is left to meld his or her personal experience-based judgment with available scientific and secondary information to formulate the most effective management action, a process best described as an "art."

Fortunately, wildlife damage management has abundant **secondary literature**, accumulations and summaries of information that often provide unique insights into wildlife damage problems and their management. Secondary literature is where one can find experience and knowledge relating to the art of wildlife damage management (see fig. 3.1). The literature may be general or focused on a specific area of wildlife damage management. It may be packaged as professional books, textbooks, control manuals or pamphlets, videos or documentaries, databases (described above), popular writings in newspapers or magazines, classroom lectures and other materials, and research-based extension educational activities that might include experiential/hands-on demonstrations, personal instruction, publications, and websites. Professional books and textbooks offer an amalgam of primary and secondary literature (e.g., Conover 2002).

Secondary literature tends to be focused on the practical aspects of management. Management manuals are widely available through governmental agencies or international organizations. For example, a partnership between the Wildlife Conservation Research Unit at the University of Oxford and the Born Free Foundation maintains a website titled "People and Wildlife." Accessed at the time of writing, the website offered over 50 manuals focused on human-wildlife conflict resolution. Included are manuals on resolving damage by crop-raiding primates and elephants in Africa and predation by lions and wild canids globally. Manuals offering humane approaches are emphasized. Governmental agencies, such as Wildlife Services in the United States, also offer manuals, as do state or provincial wildlife agencies and federal and state extension programs. Again, the Internet serves as a conduit for much of this information.

Secondary literature is usually subject to careful review and editing, often by peers experienced with the specific topic. However, these reviews can lack vigor regarding application of the scientific method. In fact, review and editing may include systematic inclusion or exclusion of information based on a particular view held by the author or the publisher, whether a private or a governmental organization. With secondary literature, the reader should look particularly for information based only on anecdote, or that follows exclusively the view or policy of an organization, or that is provided to support a particular cultural, social, religious, economic, or political perspective. The student and practitioner alike are cautioned to uncover such biases and consider them carefully before taking a damage management action that could be misguided.

Operational programs, in which experts are contracted to do the management, offer the advantages of both scientific base and artful experience, in that the management is practiced directly by professionals on behalf of the landowner or natural resources agency. Professional managers include those employed by for-profit entities as well as those working for governmental or private agencies or nonprofit associations. Within the constraints of the policies, politics, religion, and economic situation of the individual or organization, the manager provides the most appropriate wildlife damage management that his/her experience and knowledge allow.

How does one filter out the incorrect information while keeping the useful? How does one avoid the waste of time, energy, and expense associated with ineffective methods of control? Peer-reviewed scientific information available through primary literature seems an essential starting point. Reliance on experienced and knowledgeable expertise available through the secondary literature is also important in ensuring that the art of wildlife damage management is incorporated. Beyond this, the student and the practitioner alike are left to their own prowess.

Examples

PRIMARY SOURCES. Scientific articles on wildlife damage management appear in international journals such as *Crop Protection* and *International Pest Control*. Findings are also scattered in many countries among peer-reviewed journals, books, and proceedings of scientific meetings, sometimes in English, sometimes not. *Human-Wildlife Interactions Journal*, published by the Jack H. Berryman Institute, is an example of a U.S. publication. Three peer-reviewed scientific journals of The Wildlife Society, *Journal of Wildlife Management*,

Wildlife Society Bulletin, and *Wildlife Monographs*, serve as primary outlets for scientists publishing articles on wildlife damage management. Less frequently, articles can be found in journals specific to fauna or flora, such as *Journal of Mammalogy, Auk,* and *Ibis,* or those crossing faunal or floral boundaries, such as *Chemical Senses* and *Physiology and Behavior*. Articles on wildlife damage management also appear in applied scientific journals such as *Journal of Applied Ecology* and *Journal of Range Management*. Scientific publications representing regions or states, such as *American Midland Naturalist*, occasionally carry articles on wildlife damage management.

Peer-reviewed scientific articles on wildlife damage may be collected and published as proceedings from meetings or symposia. Notable among these are *Proceedings of the Vertebrate Pest Conference,* from an international conference held every other year in different U.S. West Coast venues, and *Proceedings from the Wildlife Damage Management Conference,* from a meeting conducted by the Wildlife Damage Management Working Group of The Wildlife Society, held in alternate years in various locations. The NWRC conducts symposia and publishes proceedings on selected topics in wildlife damage management. An example is the article "Repellents in Wildlife Management" (Mason 1997). Original research articles may also be published in books, particularly those having a specialized theme. Thus, the American Society for Testing and Materials has published books of original research to help standardize methods used in damage management. An example is *Test Methods for Vertebrate Pest Control and Management Materials* (Jackson and Marsh 1977).

Given the broad array of journals and other modes used for publishing scientific articles on wildlife damage and the scattered locations of these articles across the globe, a time-saving approach to access the information is using electronic databases. Globally, Biblioline Internet Services provides a database called Wildlife and Ecology Studies Worldwide. Included in the database are many scientific articles on damage management from around the world. SORA (Searchable Ornithological Research Archive) is a database containing articles from major ornithological journals, including authors from many countries, with full-text articles available. Within the United States, the Agricultural Network Information Center, a project of the National Agricultural Library of the USDA, land-grant universities, and selected associations, provides access to scientific literature on agriculture broadly, including a database on wildlife damage problems. The NWRC offers a series of databases on damage management organized around birds, rodents, aircraft hazards, and predators. The NWRC rodent database alone includes over 20,000 references.

The Internet Center for Wildlife Damage Management, described above, offers access to proceedings, including many full-text articles, on damage management. The Berryman Institute for Wildlife Damage Management in Utah provides access to proceedings and full texts of some articles published by the institute. As with all databases, however, be sure to check the range of journals searched by the database and the period of time covered, to see whether the research is broad and current; the NWRC bird reprint database described above has been inactive since 1985. The rodent bibliography of the NWRC, updated periodically, is available in printed form on request.

SECONDARY SOURCES. *Resolving Human-Wildlife Conflicts: The Science of Wildlife Damage Management* (Conover 2002) is a book that includes a mix of primary and secondary literature. In the manual *Prevention and Control of Wildlife Damage* (Hygnstrom et al. 1994), there are extensive practical references, including many to primary research publications, on managing wildlife damage. The species-specific information was written by over 50 wildlife damage experts. The handbook is organized around individual wildlife species; each species has a chapter, and each chapter has a selected reference list at its end.

Pocket Guide to the Humane Control of Wildlife in Cities & Towns (Hodge 1991) is an example of a manual offering potentially effective control methods, using choices constrained by a particular social view. Published by Falcon Press for the Humane Society of the United States, it emphasizes humane and particularly nonlethal approaches to damage management.

Articles and videos on wildlife damage management abound in newspapers, magazines, television shows, and other popular media. Again, the quality of information varies greatly, often depending on the intent of the individual or organization responsible for its production. "Living with Wildlife" (USDA 1994) is an example of a video that gives an overview of wildlife damage and its management from the perspective of an agency wanting to emphasize the importance of its role in the management of wildlife. Another video, *A Matter of Perspective* (Texas Agricultural Extension Service 1991), provides differing views on coyotes and their management in Texas.

COMMERCIAL COMPANIES AND GOVERNMENT ORGANIZATIONS. Offering both services and equipment, commercial vendors and public servants serve as additional resources for wildlife damage management.

Critter Control is an example of a franchised chain of wildlife practitioners that offer for-profit services to the public in mostly urban areas throughout the United States. Heart of England Raptors offers rabbit control, bird control, bird-proofing services, and other pest control services as one of many such private enterprises in England. Premier Environmental Ltd. is another such company. Leo Technova India, just one of many such services in India, offers bird and other control services. The major federal agency of the United States offering such services is Wildlife Services, although many natural resources management agencies, such as the U.S. Fish and Wildlife Service (USFWS) and the U.S. Natural Resources Conservation Service, have internal expertise as well. Wildlife damage management is often included as part of the programs offered by state agricultural cooperative extension services directly linked to land-grant universities. Most state fish and game agencies offer wildlife damage management services, either as technical programs or as direct assistance programs. Other federal agencies, such as the Federal Aviation Administration and the National Aeronautics and Space Administration, contract for damage management. Other countries have analogous agencies, as appropriate for their own political structures.

Products are also available to the public, including traps, pesticides, chemical and electronic repellents, scare devices, pyrotechnics, and exclosure nettings for birds and mammals. Again, the student and the practitioner need to be wary of questionable claims. For example, some ultrasonic or electromagnetic devices are sold to the public with claims that they repel only unwanted insects or rodents, not desirable animals or pets. Such claims are mostly devoid of any scientific basis and are suspect.

Summary

- Many resources are available globally to students and practitioners of wildlife damage management, including peer-reviewed scientific publications as primary literature and secondary literature in the form of professional books, textbooks, control manuals and pamphlets, videos and documentaries, databases, popular writings in newspapers or magazines, classroom lectures and other materials, operational programs, and extension educational programs that might include demonstrations, personal instruction, and research-based publications and websites.
- Students of wildlife damage management include those in colleges and universities who study wildlife damage in designated courses or as part of related courses.
- Practitioners include professional natural resource managers, individuals associated with public and private organizations with an interest in natural resources management and wildlife damage management specifically, and individual farmers, ranchers, landowners, and homeowners with a damage problem.
- Primary literature includes mostly publications that result directly from peer-reviewed studies that follow the scientific method.
- Secondary literature often includes a synthesis of scientific information with information that is based on experience and the insights it can provide. This literature might also include information based only on anecdote or follow exclusively the view or policies of an organization or a view that is provided to support a particular cultural, social, religious, economic, or political perspective.
- Access to resources is facilitated through local, regional, and international organizations, public or private, for-profit or nonprofit. Such entities include colleges and universities, governmental service and extension education programs, public associations, and private-sector businesses.
- Access to information is facilitated through various media, including print, video, and extension education or service programs, but particularly through electronic databases, often available on the Internet.
- Use of scientifically based information is an essential starting point. Reliance on experience and expertise found mostly in secondary literature is also important.
- Cleverness and scrutiny in separating useful information from biases based on culture, society, economy, religion, and political views are essential. To this end, the student and the practitioner are left largely to their own judgment.

Review and Discussion Questions

1. Explain why you are unlikely to find anecdotal recommendations for solving wildlife damage problems in primary literature.
2. Explain why you are unlikely to find tips, based on years of personal experience, on setting traps for Eurasian lynx in primary literature.
3. You have been told that Wrigley's spearmint gum is an effective agent for control of the eastern

common mole (*Scalopus aquaticus*). Is it? Base your opinion on both primary and secondary literature.

4. Find a website that contains useful information on the management of damage by larger mammalian predators and that carries biases. What is the citation? What are the biases? How might you use this information?

5. Using the Internet, locate literature that you would consider somewhat remote—e.g., an American student might check publications of the Bangladesh Agricultural Research Council or Institute. Does the source offer publications on wildlife damage or management?

PART II • **BIOLOGICAL AND ECOLOGICAL CONCEPTS**

In this section we probe concepts and principles of individual, population, and community ecology that underpin wildlife damage, its assessment, and its management.

4

Organismic and Species Systems

We describe habitat and ecological niche, sympatry and ecological equivalents, domestication and feral wildlife, biological clocks, and behavior as attributes of organismic and species systems that apply to wildlife damage management.

HABITAT AND ECOLOGICAL NICHE

Statement
Habitat and ecological niche describe places where organisms can be found and the roles of organisms in the environment and help practitioners understand, anticipate, and manage wildlife damage.

Explanation
Habitat is where an organism, species, or population lives. The broadest ecological term it encompasses is a **biome**, an assemblage of organisms distributed over an extensive area, such as a tropical monsoon forest or temperate grassland. One can also describe the spring habitat of the American robin (*Turdus migratorius*) as open meadows, often adjacent to deciduous forests. Depending on context, habitat here refers to the robin as a species, a population, or an individual organism. Wildlife damage practitioners use habitat in all of these senses.

Ecological niche is a larger concept than habitat. It is described as the multidimensional array of environmental factors (i.e., the address) within which an organism must live and how it interacts with its environment (i.e., its role or function). For example, some plants live in a constructed wetland and act to remove minerals and excessive nutrients as part of a water treatment program; the address and role combined describe the ecological niche. Muskrat (*Ondatra zibethicus*) may also live in the wetland and remove plants. Here, the address and role of the muskrat constitute its ecological niche, although the role may upset people and be viewed as damage.

The breeding habitats of weaver finches (*Lonchura* spp.) and redwinged blackbirds (*Agelaius phoeniceus*) are marshes and swamps near ponds, lakes, estuaries, or river deltas. However, the ecological niche of weaver finches includes their role as grain eaters, foraging extensively on rice during its milky-dough stage of development in southeast Asia and Africa. The niche of the blackbird includes its roles as a grain eater, feeding on rice, corn, or sunflower broadly in North and South America during emergent and maturing stages, and as an insectivore that sometimes benefits humans.

An organism's **fundamental** or **theoretical ecological niche** includes all possible roles in the broadest possible habitats. While most organisms never achieve the full potential of their theoretical ecological niche (i.e., some interactions between the individual organism and other biota reduce the size of the fundamental niche, resulting in the organism's **realized ecological niche**), the concept is useful. For example, theoretical ecological niche may describe the broad damage caused by newly invasive species, unrestrained by interactions of other organisms in a new ecosystem.

Organisms interact with both the **abiotic** (non-living) and **biotic** (living) components of their environment. Interactions with abiota are described by Liebig's law of the minimum and Shelford's law of tolerance. **Liebig's law of the minimum** states that the material available in lowest quantity in relation to need will limit the growth of an organism. Liebig was a botanist concerned mostly with plants and nutrients, but the concept applies as well to other organisms. **Shelford's law of tolerance**, often termed **range of tolerance**, states that excessive amounts as well as deficiencies can limit the habitat within which an organism can live. Again, the concept works well with plants but can be applied to other organisms. An application to wildlife damage management is use of climatic niche models to assess the potential range expansion of an exotic species invading new habitats. Another is exclusion of wildlife from habitats such as airports and landfills by reducing one or more factors needed for their presence (such as food or water).

An organism may be an ecological generalist and live within a niche having broad ranges, or it may be an ecological specialist and live within a niche that is defined by a narrow range of tolerance for one or more factors. The prefix "**eury**" describes a broad range or function, "**steno**" a narrow one. Thus, **eurythermal** organisms live within a broad range of temperatures, whereas **stenothermal** organisms are restricted to a narrow range of temperatures. Likewise, **euryhalic** organisms tolerate a broad range of salt concentrations, while **stenohalic** organisms live within a narrow range, perhaps in brackish marshes. Range of diet is described by adding "**phagic**," meaning "to eat." **Eu-**

ryphagic organisms have broad diets, whereas **stenophagic** organisms utilize specialized diets.

Some stenophagic wildlife affront people by, for example, the great economic and health costs due to the **sanguinivorous** habits and transmission of rabies by vampire bats in Central America. As a rule, however, "eury" wildlife species are the greatest offenders. Commensal rodents have some of the broadest ecological niches among mammals. Coupled with a high reproductive capacity, commensal rodents are among the most damaging of wildlife species today, both historically and globally. Other generalist mammals that affront people include bears (*Ursidae*), coyotes, and jackals. Although classified as carnivores, these mammals have broadly defined niches, are widely distributed, and are euryphagic, eating a broad range of plants and animals as opportunities allow. The feral pig, another euryphagic mammal, is now viewed almost universally as an offending species.

Among birds, blackbirds, starlings, and weaver finches tend to feed preferentially on grains but take insects and other plant materials given the opportunity. These adaptable birds live within broad niches and are distributed globally. One weaver finch, the red-billed quelea (*Quelea quelea*), is one of the world's most costly vertebrates, because of the damage it causes.

Fishes such as carp and suckers, which can tolerate extreme ranges of chemicals and nutrients, particularly low levels of dissolved oxygen, tend to be more broadly distributed in warm, freshwater ponds, lakes, and streams, sometimes replacing more desirable game or ecologically important species.

Niches differ not only among individuals and species but also with factors such as stage of metamorphosis, sex, stage of reproductive cycle, season, and geographic location. The ecological niche of a mosquito (*Culicidae*) larva is in stagnant water, feeding on nutrients and microorganisms within its size range, a euryphagic phase lasting from four days to a month or more for some species. During the pupal stage, mosquitoes cease feeding. The emerged female is stenophagic, seeking blood that provides protein for her first eggs, whereas the emergent adult male is short lived and **nectivorous** (eating nectar).

A practical constraint on wildlife damage management is the unlimited number of physical and biotic factors that describe niches. It is impossible to know if the most important factors have been discovered for a given organism. In practice, one focuses on factors that seem limiting, recognizing that a critical aspect may have been overlooked.

Example

Kikillus et al. (2009) used data on climate from the present distribution and known breeding sites (163 locations) of the red-eared slider turtle (*Trachemys scripta elegans*), plus 12 alternative models, to assess its "climatic envelope." The turtle is invasive, distributed globally as a pet. From the analyses, the researchers deduced that Southeast Asia still had large areas of unoccupied habitat suitable for the turtles. Sliders are farmed in China and commonly found in Chinese markets, potential sources of feral sliders. Further, the authors noted that religious ceremonies in Singapore resulted in vast numbers of sliders being released into the wild. The authors suggested other "hotspots" where particular effort might be made to detect the presence of sliders, thereby helping practitioners make strategic use of limited management resources.

SYMPATRY AND ECOLOGICAL EQUIVALENTS

Statement

Concepts of sympatry and ecological equivalents help to predict possible outcomes of efforts to suppress damaging populations, efforts to introduce species for biological control, and unintended introductions of exotic species.

Explanation

When two species occupy the same niche at the same time and place, they are called **sympatric**. When they occupy the same niche but are located in different places, they are termed **ecological equivalents**.

For sympatric species, **Gause's law** applies: if two organisms (or species) occupy the same niche at the same time, one of the organisms (or species) will move, change its role, or die. If the organism (or species) changes its role, morphological or behavioral changes may also occur over time within the population, known as **character displacement**. Because direct **competition** is not occurring with geographically separated ecological equivalents, selective forces may actually move the species closer together in morphological, physiological, and behavioral characteristics, a process known as **convergent evolution**.

The mammalian family Canidae includes wolves, coyotes, foxes, dingoes, dholes (*Cuon alpinus*), various jackals, and the dog. Canids have ecological niches that overlap with each other and serve as ecological equivalents in geographic regions where they occur separately (fig. 4.1). In locations where members of

Figure 4.1 The coyote (*Canis latrans, above*) and black-backed jackal (*C. mesomelasas, below*) as ecological equivalents. *Photos by USFWS and © Hans Hillewaert / CC-BY-SA-3.0, respectively.*

this family are sympatric, the ecological functions of one or both of the species may change.

For example, wolves probably consumed half of the ducks produced in the prairie pothole region of North America prior to western expansion and settlement in the 1800s. After settlement, however, most wolves were removed, and one might therefore predict an increased duck population for the region. The population actually remained steady, however, because sympatric red foxes, coyotes, and other, smaller predators assumed the ecological functions of the wolves. As surrounding habitat became increasingly agricultural, access to nesting waterfowl became even easier for smaller predators. With effective predator management, nesting success has reached 70–90%; without such management, nesting success remains about 50 percent (Sargeant et al. 1993).

Ecological equivalence can help to predict whether an introduced species will become invasive. The absence of an equivalent species and the presence of an unoccupied niche would favor success of the introduced species, for example, mammalian predators on islands previously without predators. The presence of an equivalent species would not necessarily ensure failure of the exotic; instead, the outcome would follow Gause's law, as with the introduction of dingoes onto Australia's mainland (see chapter 2).

Example

Gosselink et al. (2003) radiomarked 28 coyotes, 16 rural foxes, and 19 urban foxes in east-central Illinois. They utilized over 10,500 locations to gather information on habitat use, analyzing **home range** (the area within which the animal conducts its daily activities) from animal locations and resting sites within the study area. The researchers found that red foxes avoided coyotes by moving closer to humans and using less favorable habitats like tilled farmland, active farmsteads, and culverts. The authors saw urban areas as refugia where foxes could avoid coyotes, and urban landscapes provided more stable habitats for the foxes than did the rural landscapes. This situation is an illustration of Gause's law.

DOMESTICATION AND FERAL WILDLIFE

Statement

Domestication provided both guard dogs that help protect livestock from predators and also feral wildlife that can injure people and other wildlife, transmit diseases, and damage property.

Explanation

Domesticated plants and animals are often bred for characteristics other than ability to protect themselves. Domesticated plants tend to be susceptible to herbivory, and domesticated animals can be "instinctively challenged," prone to lose battles with wild predators. Domesticated species therefore exacerbate wildlife damage problems and make management more difficult. Even guard dogs, bred to protect themselves and livestock, are challenged by the capabilities of wild predators.

Feral wildlife result from release or escape of organisms into the wild after domestication and subsequent successful breeding. They often change to forms resembling their wild ancestors. Referred to by Garrett Hardin (1968) as "the tragedy of the commons," feral wildlife, including feral pigs, goats, horses, camels, cats, and dogs, cause some of the more serious wildlife damage problems worldwide today.

Examples

Examples of wildlife damage by feral animals appear throughout this book, but particularly in chapters 8 and 9. See chapter 17 for methods of eradication of feral wildlife on islands and in some mainland situations.

BIOLOGICAL CLOCKS

Statement

Understanding the biological clocks and rhythms of offending species can help practitioners time management actions; disruption of such rhythms can be a management tool.

Explanation

Those of us who dislike the sound of an alarm clock often find that we can wake up just before it triggers and turn it off. Possessing such internal **biological clocks** is common among organisms, serving as the bases for circadian, lunar, and circannual rhythms.

Circadian (*circa* meaning "about," *diem* or *dies* meaning "day") **rhythms** are internal biological clocks based on 24-hour cycles, in synchrony with the earth's daily rotation. These may be the most fundamental of rhythms. **Lunar rhythms** are based on 28-day cycles, tied to the rotation of the moon around the earth. **Circannual rhythms** arise from yearly cycles tied to the movement of the earth around the sun, including both seasonal and annual changes.

Although internal, rhythms are adjusted by external cues called **zeitgebers** (German for "time giver"). **Photoperiod**, i.e., daily changes in day length, is detected by organisms and is the primary cue for daily to seasonal changes. Organisms are classified by their responses to photoperiod. Those active by day are called **diurnal**; by night, **nocturnal**; and by dusk or dawn, **crepuscular**. Because day length varies less near the equator, tropical organisms are particularly sensitive to small changes in day length.

Whereas photoperiod might stimulate physiological and behavioral preparations for circannual events such as hibernation, reproduction, molting, or migration, their actual onset may be initiated by secondary cues such as rainfall or plant growth. Ali et al. (2008), for example, related outbreaks of rodents in parts of Bangladesh to flowering of bamboo, an event that may occur in cycles that span decades.

Cues and rhythms have been intensely studied (**chronobiology** or **ecophysiology**) over the years, but details of their relationships remain unclear. In experiments where animals were isolated from daylight, circadian rhythms continue but drift a bit longer or shorter each day. In such studies, subjects may still sense external cues that are difficult or impossible to eliminate, such as oscillations in magnetic fields or electromagnetic forces. Regardless of the source of cues, rhythms remain critical for the proper timing of many organismic activities, such as timing birth to availability of food for young or timing daily activity to availability of prey or lack of predators.

It would seem rudimentary to investigate circadian rhythms before introducing a new predator to manage a damaging prey animal. Yet wildlife damage literature attests that this fundamental consideration has often been overlooked or ignored. On the Marshall Islands, monitor lizards (*Varanus indicus*) were introduced before World War II, probably for their skins and food but also for rat control. Because the lizards were diurnal and the rats mostly crepuscular or nocturnal, intended predators rarely met intended prey. After lizards began raiding chicken houses, giant toads (*Bufo marinus*) were introduced to control the lizards. Rat populations continued to rise. Both the monitor lizards and the giant toads became pests (Bennett 1995). While photoperiod (diurnal lizards and toads, crepuscular and nocturnal rats) was not the only issue facilitating these consequences, it is one that could have been assessed prior to the introductions, and it would have weighed heavily against such introductions. The introduction of the small Asian mongoose (*Herpestes javanicus*) to control rats in Hawaii and other tropical islands in the Pacific had the same flaws and similar consequences; the mongoose turned to now-endangered birds and their eggs (Stone and Anderson 1988).

Although disruption of biological rhythms offers potential as a means of managing damaging species and some laboratory studies have supported the concept, we know of few practical uses. Haim et al. (2007) found that altering day length during winter impeded thermoregulation of the social vole (*Microtus socialis*) in confined populations under natural conditions, a lethal effect for the voles. The researchers suggested that precision agriculture (e.g., use of global positioning satellites to precisely locate burrows and assess microenvironmental conditions) might be used to pinpoint photoperiod alterations at active burrows. The social vole can cause extensive damage to alfalfa in Israel.

Example

Sicard et al. (1999) studied biological rhythms of four rodent pests in African habitats ranging from humid to arid. For each species and habitat, the researchers looked particularly for nonphotic zeitgebers, such as temperature, relative humidity, water, and chemical signals (e.g., signals that emanate from germinating plants), that stimulated endogenous circadian clocks. Methods included both long-term field monitoring and laboratory studies. From these studies, Sicard et al. were able to predict when critical biological events would occur, in turn suggesting times when specific rodent management actions might be most effective. For example, the researchers were able to help the Sahelian-Sudanese better time the use of sound and physical or chemical barriers to coincide with phases when rodents were dispersing or regrouping in new areas, i.e., when rodents were most mobile.

BEHAVIOR

Statement

Modification of undesirable behaviors, both wildlife and human, underlies all successful management of wildlife damage.

Explanation

Behaviors have been classified by relative complexity and include the following: **tropism**, general attraction to or avoidance of an environmental stimulus such as temperature or light intensity; **taxis**, attraction to or avoidance of environmental stimuli, but in a more directed manner; **reflex**, response of an organism or part of an organism to an environmental stimulus, similar to but more sophisticated than a taxis, and both modifiable with learning; **instinct**, unlearned response to an environmental stimulus consisting of a sequence of encoded, stereotyped behaviors, often observed with insects, amphibians, reptiles, and birds; **learning**, complex behavioral responses that are modified by experience; and **reasoning**, in which behavioral responses are based on rational thought and strategy (e.g., Dethier and Stellar, 1964).

Attenuation, reduced response to a stimulus after repetition (some believe because of pairing with safety rather than danger), is an attribute of certain behaviors important for wildlife damage management. For example, if an explosive sound is used continuously at a bird roost, birds will soon return and ignore the sound. Attenuation is a problem with most behaviorally based devices and approaches (e.g., VerCauteren, Lavelle, et al. 2003). Their effectiveness may often be measured in days. While benefiting from initial effectiveness, the practitioner should plan the next management actions. Some believe that the useful life of behaviorally based methods can be **enhanced**—that is, increased in intensity or extended in time—by occasionally pairing the stimulus with a stressful event, such as shooting or distressing an animal.

One way to alter the offending behaviors of a plant or animal is to kill it. This approach is straightforward and direct and, for the offending individual, permanent. The approach becomes complicated, however, because it involves potential social, cultural, and legal concerns that are best considered a priori. Complex ecological responses may also occur at the population and the community levels, such as increased fecundity or undesired adjustments in the food web, also best considered beforehand. These factors being stated, lethal control remains an important management option, often the best option, practically available for the practitioner. Further, the practitioner of lethal wildlife damage management often has a profound understanding of the behaviors of the offending plant or animal, an understanding that underlies success in conducting such activities as tracking, trapping, baiting, and shooting.

Some methods, such as exclusion, physically prevent an animal's undesired behavior. Fencing out coyotes can stop predation on livestock, just as netting can prevent bird damage to structures or crops. Other methods may be based on tropism, taxis, or reflex. Use of lights at night, for example, might discourage birds from roosting in a grove of trees or coyotes from entering a temporary corral. The effectiveness of **Mylar tape**, a red and silver plastic tape used to repel birds, is thought to possibly be related to its appearance as fire from the air (Bruggers et al. 1986). Sound systems designed to broadcast alarm or distress calls of birds have at least short-term repellent effects on **conspecifics** (other individuals of the same species) and closely related species.

Waterfowl hunters have long relied on decoys, using positions of their wings and heads to signal safety for landing, loafing, and feeding (e.g., Wright 1982). A repellent technique is to put decoys or dead animals in postures that signal danger to overhead conspecifics. Tillman et al. (2002), for example, found that hanging dead vultures or their effigies at roosts or sites of damage was effective in repelling black (*Coragyps atratus*) and turkey (*Cathartes aura*) vultures, protecting both property and agriculture. In one situation, a roost of about 800 vultures was dispersed using effigies.

The Electronic Guard, a combined strobe light and siren used to repel predators from sheep, has an effective life of 8 to 103 days (Linhart et al. 1992). The device, hung near a herd, takes advantage of randomness and multiple stimuli to slow attenuation (fig. 4.2). However, it may also serve as a "dinner bell" for some predators. In recent studies, devices placed directly on sheep or lambs can turn on during an attack, emitting a sound that startles the predator.

Methyl anthranilate and dimethyl anthranilate stimulate pain receptors in the trigeminal systems of many birds and therefore serve as repellents (Mason et al. 1992). The effect is unlearned and does not readily attenuate, so the birds continue to avoid treated areas or items. These chemical compounds are active ingredients in commercial repellents such as Bird-X GC-PT, Goose Chase Goose Repellent, and Liquid Fence Goose Repellent.

Figure 4.2 The Electronic Guard emits random light and high-decibel sounds. Placed with free-ranging livestock, it can repel predators for up to 80 days. *Photo by USDA.*

Odors can attract or repel offending animals. Odors may mimic those of a predator (e.g., a **semiochemical**), prey, or a conspecific (see information on pheromones below) or simply be attractive to the animal—e.g., the rotten odor of synthetic fermented egg is attractive to scavengers. Semiochemicals can repel herbivores directly or attract additional predators, further discouraging the presence of herbivores. Lindgren et al. (1995) reported that odors from feces, urine, and anal glands, particularly those from the weasel family, repelled voles, pocket gophers, and snowshoe hares in both laboratory and limited field trials. Damage to apple trees was reduced in the trials.

Management may also be based on reproductive, foraging, feeding, or social behaviors, such as grooming. Systems of hormones, nerves, and behaviors regulate reproduction in higher animals. These systems also regulate secondary sexual characteristics, including how animals look and behave. Some of these reproductive behaviors can harm people or their interests. For example, deer and elk bucks rub trees to remove velvet from their antlers, causing damage reported by forestry and Christmas tree industries. People have been injured when stepping between parenting wildlife and their young. Deer, elk, and moose collide with vehicles more frequently during **rut**, the fall mating season.

In birds, reproductive behavior is often strongly **stereotyped**, following a carefully orchestrated sequence of behavioral and physiological events that leads from mating to fledging. As Lehrman (1964) showed with ringneck doves (*Streptopelia risoria*), one specific behavioral event, courtship, produces physiological changes that induce the next behavioral event, nest building. Nest building then engenders physiological changes that induce egg laying, etc. Doves are hardwired in this sense and cannot avoid any steps. For example, doves could not accept a premade nest; instead, the pair would break down the nest and rebuild it, thereby stimulating the physiological changes needed for egg laying. Disruption of any event or alteration of a hormonal response or level could therefore disrupt the entire cycle, a weakness sometimes exploited by wildlife damage specialists.

Although less stereotypic than birds, mammals also have sequential physiological and behavioral events that lead them through the reproductive process. The behaviors can be affected by environmental factors. For example, Christian and Davis (1964) and Davis (1966) showed that Norway rats build strongly concave nests, effective in retaining pups, when population densities are low. With high densities, the nests are shallow and flatter, allowing pups to escape; in these situations, rats are more likely to share nests and less likely to retrieve wandering young that are emitting distress ultrasounds. Under high densities, mothers tend to snip into the gut of a pup when cutting the umbilical cord. The survival of young pups under high densities is greatly reduced when coupled with the mothers' reduced ability to lactate. Such adjustments to crowding have received only limited applications in wildlife damage management, but they seem to have great potential.

Synthetic hormones, such as artificial steroids, can directly induce sterility and reduce growth of overabundant bird or mammal populations. For example, baits containing the hormone-like compound **diethylstilbestrol** have been used to block egg production in birds (see chapter 11).

Feeding young can strongly motivate food gathering and predation (e.g., Blejwas et al. 2006; fig. 4.3). Bromley and Gese (2001a) found that surgical sterilization of coyote packs would reduce predation on sheep, a cost-effective method for small sheep operations.

Pheromones are substances, usually highly volatile, that are secreted into the environment and have behavioral or physiological effects on conspecifics. Broadly used for insect management, pheromones occur in other animals as well, including mammals. In the 1950s, pheromones in rodents were described that blocked pregnancy (Bruce 1959), synchronized estrus among females (Whitten 1956), and induced pseudopregnancy (Van der Lee and Boot 1955). Scientists have explored uses of pheromones as a wildlife damage prevention method—e.g., to keep beavers from building dams in New York State (Welsh and Muller-Schwarze 1989)—but the approach remains underexploited. Pheromones do influence management methods such as trapping. Stoddart and Smith (1986), for example, found that woodmice (*Apodemus sylvaticus*) were more likely to enter traps treated with conspecific odors than untreated traps set beside them.

Social behavioral concepts include dominance, subordination, home range, and territories. These behaviors are sometimes part of sexuality and reproduction. **Dominance** is a social "pecking order" wherein one or more members of a population claim more resources than are made available to others. The resources might be food, water, nesting or denning sites, or a mate. The arrangement can be linear, in which A is dominant over B, B is dominant over C, etc. At bait stations, for example, rats often assume linear dominance, allowing only the most dominant to feed at one time, often limiting feeding efficiency at bait stations.

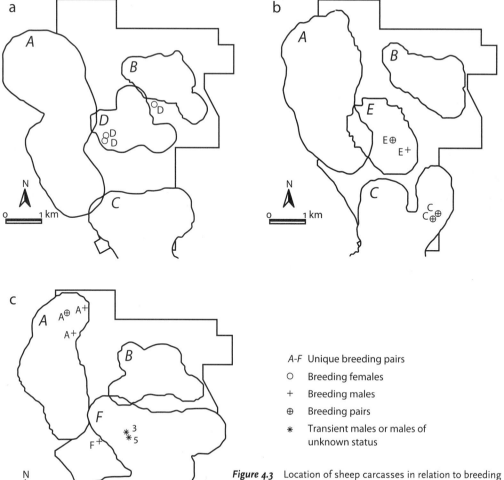

A-F   Unique breeding pairs
O   Breeding females
+   Breeding males
⊕   Breeding pairs
*   Transient males or males of
       unknown status

Figure 4.3   Location of sheep carcasses in relation to breeding coyotes at Hopland Research and Extension Center, California, during (*a*) Sep–Nov 1997, (*b*) Jan–Mar 1998, and (*c*) Apr–Jul 1999. Symbols denote status of coyotes responsible for kills. *Redrawn from Blejwas et al. 2006, with permission of John Wiley and Sons.*

This was remedied in the Philippine Masagana-99 national rat control program by adjusting the available feeding surface to feeding pressure at a particular site, effectively negating the limits imposed by social hierarchy (fig. 4.4).

Other hierarchies include despot, in which A dominates equally subordinates B, C, and D, etc.; circular, in which A dominates B, B dominates C, and C dominates A; and coalitions, in which B and C join to subordinate former despot A (Feldhamer et al. 2007).

Territorial behavior occurs when one member, usually a male, establishes a geographical area and defends it against other male conspecifics. Territories may help regulate overall population size by spreading members out according to available resources. Birds often use calls and visual displays to establish and defend territorial boundaries. Mammals, too, may use calls, but more commonly, they mark edges with scents such as pheromones found in scat or urine, using specialized scent glands and sometimes rubbing points such as branches or rocks.

Territories are defended, whereas home ranges (where an organism carries out its normal daily activities) are not. Both have core areas that provide food, water, and shelter. Blejwas et al. (2006), for example, found that most sheep kills occurred within home ranges of coyote pairs (see fig. 4.3).

Some management applications using territory and home range have been straightforward. For instance,

$A \rightarrow B \rightarrow C \rightarrow D \rightarrow$ etc.

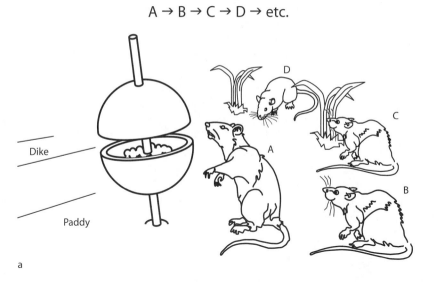

a

$A \approx B \approx C \approx D \approx$ etc.

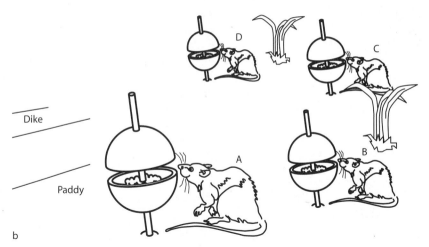

b

Figure 4.4   In *a*, the most dominant rodent is the only one feeding. In *b*, less-dominant rodents are able to feed at the same time as the more dominant ones by increasing available feeding surface at a baiting point. *Illustrations by Lamar Henderson, Wildhaven Creative LLC.*

practitioners sometimes choose trapping locations according to territorial boundaries and markings of wildlife. Other applications are more subtle. Conner et al. (2008; see also chapter 6) found that including territorial behavior and social structure as part of models of coyote populations led to predictions that sterilization can be a more appropriate long-term strategy than several lethal methods of management.

Foraging (searching for food) and **feeding** (eating) include herbivory and predation. Strategies for herbivory can be important to wildlife damage management (box 4.1), as can strategies for foraging. One foraging theory is called **optimal foraging**. Models based on this theory predict how animals maximize energy intake in relation to foraging behavior. Models usually include three factors: type of food or prey; currency, such as energy or time; and constraints or limits, including risk of predation. Another theory type is **marginal value**. These models are based on energy intake versus energy used. Marginal value models are designed to predict, for example, when an animal decides to abandon a patch where food resources are depleted and invest time and energy in locating a new patch with more food. **Food hoarding** (storing food) is another strategy used by some birds and mammals, including some rodents, shrews, and carnivores.

Foraging theories and strategies can predict movement and distribution of damaging species. For example, Feldhamer et al. (2007) suggested that optimal foraging theory might predict when feral pigs give

BOX 4.1 Herbivory and Wildlife Damage Management

Two basic strategies underlie herbivory. One uses a **digastric** (two-stomach) digestive system and is called **rumination** or **foregut fermentation**. This is used by artiodactyls (even-toed mammals) including camels, giraffes (*Giraffa camelopardalis*), hippopotamuses (*Hippopotamus amphibius*), antelopes, cervids, and bovids such as cattle and bison (*Bison bison*). Kangaroos (some *Macropus* spp.), sloths (*Folivora*), and colobus monkeys (*Colobus* spp.) also use foregut fermentation. Digestion of food through foregut fermentation is elaborate and time consuming, but plants are efficiently ingested by grazing or cropping. Relevant to wildlife damage management, microorganisms in the rumen detoxify poisons, including alkaloidal compounds used in plants that protect them from herbivory and active ingredients in some pesticides.

The other strategy is **hindgut fermentation**; it is **monogastric,** using one stomach. Hindgut fermentation is used by perissodactylian (odd-toed) mammals, including horses, zebras (*Equus zebra*, *E. quagga*, *E. grevyi*), asses, tapirs (*Tapirus* spp.), and rhinoceroses (*Rhinocerotidae*). Elephants (*Elephantidae*), lagomorphs (rabbits and pikas, *Lagomorpha*), and rodents also use this system. Almost all herbivorous birds use hindgut fermentation. Hindgut fermentation is faster and less elaborate than foregut fermentation but also less efficient. For some hindgut fermenters, the inefficiency is compensated for by **coprophagy**—ingesting partially digested feces for redigestion. Hindgut fermenters absorb toxins into the bloodstream and count on the liver to detoxify them.

Foregut fermentation allows the animal to forage and feed quickly, then find a place safe from predators to digest the food. Hindgut fermenters need to forage and feed more slowly, at greater risk of predation, carefully selecting foods that allow a more efficient digestion. Wildlife damage practitioners sometimes take advantage of this fact by visually "opening" areas of terrain, making foraging wildlife such as Canada geese feel unsafe.

up a depleted patch of acorns and move to the next. Amano et al. (2004) used optimal foraging theory to predict damage to rice grains or wheat by white-fronted geese (*Anser albifrons*) around Lake Miyajimanuma in Japan, and they used the predictions to reduce damage to wheat. Sterner (1994) used foraging theory to adjust baiting rates of zinc phosphide for vole control. Cached rice is removed from burrow systems in dikes in countries such as Bangladesh to reduce survivability of pest rodents during catastrophic events such as floods. Alternatively, hoarded foods are sometimes treated with toxicants to manage rodents.

Birds are baited or trapped at cattle feedlots, around feeding areas in farms, and near crops. Knowledge of feeding behavior helps the practitioner place baits where they are likely to be used selectively by targeted species. Treated grains may be placed at elevated bait stations in cornfields. Effectiveness of traps can sometimes be enhanced by placing live birds or silhouettes of feeding birds in the cages.

Feeding behaviors include preferences and avoidance and can be learned or unlearned. Feeding preferences underlie many wildlife damage management recommendations, from developing varieties of corn and sunflower seeds that resist bird damage to lists of "deer-resistant" ornamentals for landscapes. As a first rule, when wildlife species are given a feeding choice, they will select one food; use of "decoy" crops, where a crop of lesser value is planted to attract pest species and protect a crop of greater value, is an application of this rule. Refuge managers might plant crops preferred by wildlife to reduce damage to crops of neighboring farms. The approach is effective because it offers offending wildlife a better choice. There is an important second rule, however: given no choice, wildlife will take whatever is available. Hence, tops of young trees or bark and cambium of trees may be undesirable to deer until alternative foods are covered with snow. This situation is well known to horticulturists, foresters, and Christmas tree growers.

Feeding preferences may be expressed as **specific hungers** or drives, including for wildlife **geophagia**, the tendency to eat soil, and sodium or **salt drive**, the urge to eat salt. Salty baits containing a toxicant are sometimes used to control North American porcupines (*Erethizon dorsatum*; Anthony et al. 1986), a species exhibiting a strong salt drive. Deer, moose (*Alces alces*), and other animals lick de-icing salts off roadways, increasing chances for collisions with vehicles (e.g., Fraser and Thomas 1982). Nonsodium-based de-icers are sometimes used in areas with high deer densities to reduce potential for collisions (e.g., Bruinderink and Hazebroek 1996).

Feeding preferences and avoidances can be subject to **social facilitation** (i.e., when the presence of one animal improves the performance of another). Some young learn food preferences by associating flavors in mothers' milk with a positive feeding experience; weaned young may seek and eat foods having similar flavors (Galef and Henderson 1972). With some animals, young also learn to avoid food avoided by adults. For example, Avery (1996) showed that seeds avoided by adult house finches (*Carpodocus mexicanus*) were also avoided by their offspring. Social facilitation might be used in wildlife damage management by, for example, preparing baits having flavors preferred by adults of offending wildlife.

Behavioral defenses against dietary poisoning must be overcome if offending wildlife species are to ingest toxic or pharmacologically active amounts of pesticides. The defenses can also be used to induce repellency or avoidance of foods or crops. Garcia et al. (1974) summarized such defenses, subsequently describing them in the form of a prototypic gastronome. With this gastronome, Garcia argued that most organisms were consummate dietary gourmets within their own needs and environments. When encountering a new food, the prototypic gastronome first exhibits **neophobia**—fear of something novel—and avoids it. After neophobia is extinguished, the gastronome samples food for taste (**gustation**) or flavor (gustatory and olfactory components) but does not necessarily swallow the food. If the flavor is one instinctively associated with illness, usually bitter, or sufficiently stimulates pain receptors in the trigeminal system (cranial nerve V, the **common chemical sense**), the food is rejected without swallowing, called **primary flavor aversion**. If the food contains a poison and is swallowed, a post-ingestional gastric illness (stomach illness causing nausea or vomiting) will be associated with the food and, if the gastronome survives, result in subsequent avoidance of that flavor, called **flavor aversion learning** (fig. 4.5).

Flavor aversion learning has attributes important to application, including one-trial learning, attenuation, and association of the flavor with illness up to six hours after ingestion. Further, animals learn to avoid the sight (with visual learners such as birds), taste, or flavor but do not necessarily learn to identify the agent causing illness. Thus, a novel flavor followed by radiation (causing nausea) results in subsequent avoidance of the flavor, not radiation. Some flavors are more **salient**—i.e, more likely to be associated with the illness—than others. When several flavors are presented, the most salient **overshadows** the others, and the gastronome learns to avoid only the salient one.

The behaviors described by the prototypic gastronome apply directly to wildlife damage management.

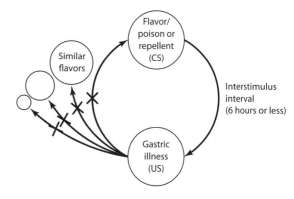

Figure 4.5 If an animal develops a gastric illness within six hours of ingesting a substance, the animal will learn to avoid the flavor of the substance as well as similar flavors, particularly if the flavor is novel. *Adapted from Reidinger 1997, with permission. Illustration by Lamar Henderson, Wildhaven Creative LLC.*

Food aversion learning, for example, describes **bait shyness**, avoidance of a bait following consumption of a sublethal amount. Neophobia explains the tendency to avoid novel baits, and primary flavor aversion involves avoidance of bitter-tasting baits and compounds. Prebaiting (providing bait without poison, allowing a positive post-ingestional consequence) is sometimes used to overcome these defenses. Flavor aversion learning has been explored extensively for applications to wildlife damage management (e.g., Reidinger 1997). Gustavson (1976) explored its applications to predation on sheep, while others used flavor aversion learning to induce avoidance of bird or turtle eggs by raptors. The concept has been used to help determine flavors of baits and to assist in the formulation of prebaits. While application of flavor aversion learning is an appealing concept, success of the approach has varied.

For example, at Camp Pendleton, a U.S. Marine base in coastal California, ravens take endangered least tern eggs. Surrogate eggs have been injected with methylcarbamate, a compound that induces nausea and learned aversions in ravens. The aversions generalize from surrogate eggs to other, nontoxic eggs including those of the least tern, thereby affording them protection (Avery et al. 1995). Similar approaches have been used previously by other scientists to protect eggs from predation (e.g., Nicolaus et al. 1983). Predation is a centrally biased stereotypic behavior that is difficult to stop. Even if consumption of a prey is prevented through aversion learning methods, predatory behavior may continue. Under those circumstances, a predator will attack and kill a prey, then become nauseated and avoid eating the quarry.

Grooming is another behavior that can be strongly stereotypic and sufficiently strong to overcome learned flavor aversions—during grooming, animals will ingest flavors that they have learned to avoid by flavor aversion learning. Grooming is underexploited in wildlife damage management (Reidinger et al. 1982).

Example

Bear damage was observed on forestry plantings in the western United States in the 1940s and became a concern as early as the 1950s. Bears peel the bark of Douglas fir, western hemlock, and western red cedar, causing economic losses valued at millions of dollars annually (Nolte and Dykzeul 2002). Douglas fir develop buds first and are usually the first damaged. The bears prefer berries when they become available during the early summer, and then they switch to cedars as berry abundance declines (Ziegltrum and Nolte 1995).

A supplemental feeding program was initiated as an alternative to lethal control in 1985. Bears were provided with a pelleted food supplement designed to be more attractive than the phloem of trees but less attractive than berries, and damage ceased (Flowers 1986). The program was expanded to other sites until it protected over 400,000 hectares of Douglas fir (Ziegltrum 2004). Supplemental feeding continues, with ongoing in-depth studies indicating no serious detrimental impacts on bears, the forest industry, or the safety of personnel. Supplemental feeding is also being used in Croatia, Gunma Prefecture in Japan, and elsewhere in Asia (Ziegltrum 2008).

Summary

- Ecological niche is used to assess potential range expansion of an exotic invasive animal and underlies exclusion of wildlife from airports and landfills.
- Ecological equivalents, sympatry, and Gause's law are used to predict outcomes of removals or introductions of wildlife, such as predators, in ecosystems.
- Some organisms with narrowly defined niches, such as the sanguinivorous vampire bat, can harm humans or their interests, but most offending wildlife, such as rodents, coyotes, feral pigs, and weaver finches, are ecological generalists.
- Domestication makes plants and animals susceptible to wildlife diseases and predation, exacerbating wildlife damage. Feral wildlife, including feral pigs and dogs, are among the most damaging species found worldwide today.

- Biological rhythms—circadian, lunar, and circannual—of prey should be compared with predators' behavior rhythms before the predators are introduced.
- Altering biological rhythms can disrupt behavioral events of offending wildlife. Understanding rhythms can help time management activities, such as baiting rodents during their dispersal, for optimal effect.
- Modifying offending behaviors, wildlife or human, underpins all successful management of wildlife damage.
- Physical exclusion, such as fencing, can prevent unwanted behaviors such as herbivory or predation.
- Some lights, sounds, odors, and flavors offer temporary repellency by taking advantage of the taxic or reflexive responses of wildlife; some, such as effigies and semiochemicals, are more biologically based.
- Reproduction underlies some wildlife damage, such as deer colliding with vehicles during rut. Reproductive behaviors can be disrupted by altering cues that stimulate reproduction or by using chemicals that interrupt reproduction, such as sterilants or pheromones.
- Foraging and feeding behaviors and theories help explain or predict some wildlife damage, such as excessive predation on livestock during the reproductive season and movement of feral pigs from one mast patch to another.
- Flavor preferences and aversions are used to manage damage, as when decoy crops attract damaging wildlife away from high–cash value crops; when preparing baits, repellents, or lists of ornamental or crop varieties avoided by damaging wildlife; and when understanding and managing bait shyness or inducing repellency by flavor aversions.

Review and Discussion Questions

1. What might be the ecological equivalent of the passenger pigeon (*Ectopistes migratorius*) in North America today? Explain, with references.
2. How does the domestication of wildlife species relate to their causing damage around the globe?
3. It is commonly believed that attenuation to auditory or visual repellents, such as "Scary-Man" and propane cannons to scare birds, can be reduced by occasional reinforcement with a negative experience, usually shooting some birds. Is there scientific evidence supporting this belief? Provide specific references and supporting evidence.

4. Has flavor aversion learning been used to induce coyote aversions to livestock? To reduce livestock depredation? Summarize scientific studies supporting and not supporting the application.

5. How might you use an animal's behavioral defenses against dietary poisoning to advantage in the management of wildlife damage? Give a specific example.

5

Populations

In this chapter, we look at population size and its critical minimum, population density, and crowding. We review the fundamental equation for population growth, types of growth curves, life tables and survivorship curves, age pyramids, and population models and their underlying equations. We show how these aspects of populations are used in assessing or managing wildlife damage. We examine relationships between population size and extent of damage.

POPULATION SIZE, DENSITY, OVERABUNDANCE, CROWDING, AND THE FUNDAMENTAL EQUATION FOR GROWTH

Statement
Attributes of populations, including size, density, growth, overabundance, and crowding, sometime underlie causes of wildlife damage and are the bases for assessing and managing damaging populations.

Explanation
Populations are actually or potentially interbreeding groups of organisms—i.e., species—living in the same place at the same time. The **population size** at any moment in time is usually designated as N. For example, if a total population of geese in a city park is counted at 145, then $N = 145$. If the intent is to reduce wildlife damage by reducing the size of the goose population, then N must be brought below 145. This might be done by rounding up the geese during the summer molt and relocating or harvesting them.

Sometimes populations are assessed by density. **Population density** is the number of individuals per unit area. **Ecological density**, the number of individuals per unit habitat, is sometimes more useful. For example, density of the geese in the above park is 145 total per 10 total acres of park, or 14.5 geese per acre. However, if half of the park is wooded and another quarter of it is used for parking lots and buildings, the actual habitat available to the geese may be only 2.5 acres, making an ecological density of 145 geese per 2.5 acres, or 58 geese per acre of habitat. Such a high density raises questions of overabundance or crowding.

Caughley (1981) said populations are **overabundant** when they threaten human life or livelihood, impact densities of (other) favored wildlife species, are too numerous for their own good, or impact ecosystems and cause their dysfunction. **Crowding**, related to overabundance, occurs when population density exceeds a level at which the habitat can maintain healthy population members, and they become stressed. Selye (1936) postulated that animals placed in an untenable environment, with no hope of escape, would become stressed, going through stages including alarm, resistance, and exhaustion. Selye later suggested that low levels of stress might be beneficial, expressed as **eustress**, but if continued without hope of relief would lead to **distress**. Selye tied stress to enlarged adrenal cortices and the hypothalamic-pituitary-adrenal axis, a concept picked up and expanded by Christian and Davis (1964). They found symptoms of crowding among many mammals, including rodents and red deer (*Cervus elaphus*; see chapter 4).

Factors other than density affect crowding, so it cannot be assigned an exact threshold density, above which crowding occurs. For example, animals may tolerate higher ecological densities in environments with high plant densities. One of the authors (RFR) observed densities of over 10,000 rats per hectare in marshes in the Philippines, and the rats seemed unstressed. Possibly the thick, tall vegetation in the marshes minimized contact among conspecifics and reduced stress. Regardless, crowding has been observed in situations where populations of mammals, e.g., deer or elephants, grow unchecked, often leading to sick animals, damage to the environment, and problems for wildlife damage managers.

Whereas N provides population size at a moment in time, it may be important to know whether a population is growing, stable, or diminishing over time. Such information can predict whether damage will likely abate or worsen and whether actions taken to reduce a population have worked. Growth rate measures such trends.

Growth rate (*r*) is the change in numbers of population members over time. It is also the sum of four other rates: **natality** (birth, *b*), **mortality** (death, *d*), **immigration** (*i*, movement into), and **emigration** (*e*, movement out of), described by the following **fundamental equation for growth**:

$$r = (b + i) - (d + e),$$

where (*b* + *i*) is sometimes called **recruitment**.

We illustrate this concept by hypothetically gathering more information on the geese in the city park.

We now decide to legally place numbered neck rings around the geese. Further, we mark each new member and record the loss of each old member and, having unlimited funds (this is *truly* hypothetical), we do this every day for a year. We find, summed for exactly one year, that 180 goslings were hatched and survived; 25 geese died; 20 new adult geese moved into the park; and 15 marked geese left the area.

There were 180 geese hatched and 20 immigrants, or 200 new members. Added to the original 145 members, this makes 345. But 40 members died or emigrated, so the actual size is reduced to 305. That is 160 new members per 145 total original members in one year, or a growth rate of 110.3% per annum! It looks like a burgeoning population of geese at the park. Maybe neighbors are feeding them, despite posted signs asking the public not to feed wildlife. Perhaps there are few local predators such as cats, foxes, dogs, raccoons, and urban coyotes.

As with natural changes in population size, those induced by management actions must also follow the fundamental equation for growth. Therefore, actions of wildlife damage practitioners must impact natality, immigration, mortality, or emigration rates or some combination of these. If the intent is to eradicate, the action must push population size below its **critical minimum** (the size below which the population fails). If the intent is to reduce damage, and population size is directly related to damage, the overall impact must reduce growth rate (*r*) even if one or more individual factors, such as natality or immigration rate, actually increase. For example, natality rate may increase as a consequence of reducing coyote populations; the consequence is nevertheless reduced damage if net growth is reduced and if population size relates to damage.

Each of the rates can underlie damage or be used to help manage it. Immigration is often important with exotic and invasive species (and may be seen as part of propagule pressure; see chapter 7). It can also underlie solutions, such as public policies that restrict movement of undesired wildlife between governmental units (e.g., states or provinces) or countries. Emigration and mortality rates contribute to reduced population size and can sometimes help to reduce damage. Increasing mortality is a common way to reduce populations. Wildlife damage managers sometimes use relocation as a form of emigration (see Fischer and Lindenmayer 2000 for a general review of animal relocations). Natality rates contribute to the growth of a population and can exacerbate damage. Reducing natality rate can sometimes reduce damage, as with addling eggs, sterilizing wildlife, or denning (see the section below, Measuring Population Size and Aging).

Examples

OVERABUNDANCE. In a ten-year study of impacts of deer density on forests of northwestern Pennsylvania, Tilghman (1989) and DeCalesta (1992) found that deer densities exceeding 7.9/km^2 resulted in significant decline in species richness, abundance, and diameter of saplings at breast height. Six woody tree species were missing among the saplings (DeCalesta 1992). Thus, deer at the higher densities negatively impacted the ability of the forests to regenerate. Cascading effects were found for songbirds (e.g., reduced woody vegetation meant reduced foraging, escape from predators, and nesting) and richness of other herbs and shrubs. Further, DeCalesta (1997) suggested that higher densities of deer might completely eliminate eastern hemlock regeneration and prevent American beech and sugar maple regeneration by furthering an insect and disease complex.

Deer health can also be impacted by overabundance. Davidson and Doster (1997) reviewed the two major health problems of deer in the southeastern United States, hemorrhagic disease and a syndrome that incorporates both malnutrition and parasitism. They concluded that hemorrhagic disease was not necessarily related to population density and, because of its complexity, was expressed as distinctive geographic variants. However, they also determined that the syndrome of malnutrition and parasitism was fundamentally density dependent. The syndrome met criteria established in a model for density dependency presented by Eve (1981). An example of a different form of impact, Nussey et al. (2007) found high rates of senescence (premature aging) in female red deer that experienced high levels of competition for resources early in life. In contrast, Kilpatrick et al. (2001) reported improved health of white-tailed deer herds and reduced deer browsing following a two-year program of deer reduction by shotgun hunting at Bluff Point State Park, Connecticut.

CULLING. Wildlife damage practitioners sometimes cull (kill individuals) overabundant populations, usually by hunting or shooting. For example, to minimize environmental damage and stress among densely populated herds of elephants in a 7,332-square-mile national park in South Africa, culling was begun in 1967. Based on the collective opinions of elephant experts from different parts of Africa, at the time culling was started, a limit of about 7,000 elephants was established for the park, roughly a density of 1 elephant

per square mile. Most experts felt that the park environment could be sustained with that density, as could the health of the herd. Culling continued until a moratorium was put on it in 1994, based on concerns from the international community on humaneness of culling and political pressure applied to the South African government. Alternative control approaches such as elephant contraception were considered (Whyte 2004).

Today's view (Whyte 2004) is that expanding and diminishing elephant populations add an important dynamic component to the park. Hence, Kruger Park was home to about 12,000 elephants in 2009, about twice the number that was the park's estimated carrying capacity in 1967. The dynamic nature of elephant populations is now seen to add ecological value in diversity and richness, provided that the populations are not extremely low or high for too long a period of time. The park continues to monitor and manage populations and to reassess their critical minimum and maximum densities as park scientists gather more information on herd conditions and impacts on habitat.

POPULATION GROWTH CURVES

Statement

Growth curves provide theoretical and practical information on patterns and limits of growth of offending populations; the curves are used by wildlife damage practitioners to predict when action is needed, to time management actions, and to monitor their effectiveness.

Explanation

Growth of wildlife populations is often a logarithmic (geometric) function of time rather than a linear one. Suppose, nonetheless, that a linear relationship exists between time and growth of the hypothetical goose population in the park. A linear relationship follows the equation for a straight line, as in fig. 5.1a, and is more readily understood than a logarithmic one. Here, the ordinate is N and the abscissa is t, and the relationship between the two is described as

$$N = mt + N_0,$$

where m is the slope, or change in N per change in unit t, and N_0 is the t-intercept, or the value of N when t is zero.

Reviewing our hypothetical goose population, there is an increase of 160 population members when one unit of time (a year) passes, or a slope of 160/1, or 160. At time zero (i.e., time N_0, the t-intercept), the pop-

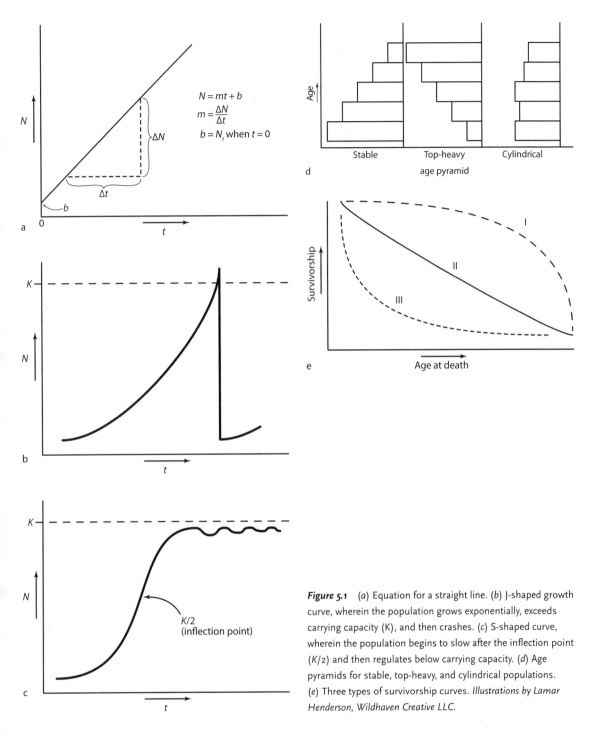

Figure 5.1 (a) Equation for a straight line. (b) J-shaped growth curve, wherein the population grows exponentially, exceeds carrying capacity (K), and then crashes. (c) S-shaped curve, wherein the population begins to slow after the inflection point (K/2) and then regulates below carrying capacity. (d) Age pyramids for stable, top-heavy, and cylindrical populations. (e) Three types of survivorship curves. *Illustrations by Lamar Henderson, Wildhaven Creative LLC.*

ulation had 145 members. Using the equation for a straight line, then, we would predict that in one year the population would be 305, as follows:

$$N = (\Delta N/\Delta t)t + N_0$$

or

$$N = (160/1)1 + 145 = 160 + 145 = 305$$

members (as we deduced in the first section of this chapter).

Further, if population size continued in the same linear fashion (although this is unlikely in reality)—that is, if the population increased at a steady rate of 160 geese per year—we would predict that in 6 years the park would have 1,105 geese, as follows:

$$N = (160/1)\, 6 + 145 = 960 + 145 = 1105.$$

Now, suppose we made the interval of time smaller and smaller for our measurements of population size, from once a year to every half-year, every three months, every month, every week, every day, every minute, every second, every millisecond, etc.. If this trend continued, the intervals would eventually become so tiny that they would be dimensionless and instantaneous (represented by d). We would then be measuring **instantaneous growth rate**, which gets us (and hopefully you) to dN/dt, the instantaneous change in size of the population for an instantaneous unit of time. (You could use calculus to calculate the same rate.) An advantage of instantaneous growth rate is that it can be used to calculate growth and related population parameters for relationships between N and time that are nonlinear, such as dimensionless points along **J-shaped** or **sigmoidal curves**. In such cases, the instantaneous growth can be visualized as the slope of the tangent to that point on the curve.

Two equations that describe population growth are unrestricted (Malthusian) growth,

$$dN/dt = r_{max}\, N,$$

and logistic or sigmoidal growth,

$$dN/dt = r_{max}\, N(K-N)/K.$$

In both equations, dN refers to the change in population size at any moment in time, dt, and dN/dt to its instantaneous growth rate. K is **carrying capacity**, and r_{max} (see the first section in this chapter) is defined here as the maximum growth rate, i.e., growth rate under ideal conditions. Note that, unlike instantaneous growth, which varies with time and which has been the basis for our discussions so far, maximum growth rate is a constant that remains unchanged. Growth rate is usually given on a per-member basis (or some round multiple, such as per 100 or 1,000 members). For example, if r_{max} is 0.25, each member contributes on average one-quarter of a new member to the population under ideal conditions during the unit of time (e.g., a year or a reproductive season).

When unrestricted, population size (N) passes right through carrying capacity (K) and exceeds it (fig. 5.1b). Because the size of the population now exceeds the number that the habitat can sustain (K), the population crashes. This was the point of Malthus' original description of the human population in which he predicted "the power of population is indefinitely greater than the power of the earth to produce sustenance for man" (Malthus 1798, p. 13). The pattern resembles a "J," and hence this is called a J-shaped curve. Many wildlife populations—e.g., microtine rodents such as voles—follow a J-shaped growth curve, also known as an **exponential growth curve**. So do some populations of "released" invasive species that have moved into new habitats with essentially unlimited resources and without naturally occurring diseases, parasites, and predators.

With populations having logistic or sigmoidal growth, the population responds to carrying capacity in a density-dependent fashion—i.e., environmental conditions that reduce growth, such as disease and starvation, collectively called **environmental resistance**, apply more and more pressure against ideal growth (and achievement of the biotic potential for the population) as the population density increases and the population size approaches K (Verhulst 1838). At one specific point, the inflection point (fig. 5.1c), or $K/2$, density-dependent factors impact growth sufficiently that growth rate (r) slows. This point is where maximum growth occurs, sometimes considered by wildlife biologists to be the **Maximum Sustainable Yield**, the population size at which maximum harvest can be sustained for the longest period of time.

Maximum Sustainable Yield is also based on **compensatory mortality**—that is, many game animals removed by hunting or trapping do not reduce net population size because these animals, had they not been hunted or trapped, would have been removed anyway due to other factors, such as disease or starvation. To take full advantage of compensatory mortality, timing of hunting or trapping is important. In temperate countries, hunting and trapping are usually permitted in the fall, before winter stresses further affect populations through starvation or disease.

For wildlife damage managers who sometimes want to reduce a population as much and as inexpensively as possible, additive mortality may be more important than compensatory mortality. With **additive mortality**, hunting, trapping, and poisoning are conducted so as to add to, rather than compensate for, other mortality factors. To accomplish this, management actions might be taken in late winter after disease and starvation have already taken their toll and compensatory mortality is no longer possible. Or, if practicable, the damage management actions might be taken before the expanding population reaches its $K/2$ inflection point, again to maximize impacts on N.

Density dependence is accounted for in the equation for sigmoidal growth by the expression $(K-N)/K$. The expression decreases geometrically as N approaches the carrying capacity, as illustrated below. If the carrying capacity of a habitat is 500 members for a given population, then as N increases, $(K-N)/K$ decreases as follows:

N	K	$(K-N)/K$
10	500	$(500-10)/500=0.98$
50	500	$(500-50)/500=0.90$
100	500	$(500-100)/500=0.8$
200	500	$(500-200)/500=0.6$
300	500	$(500-300)/500=0.4$
400	500	$(500-400)/500=0.2$
450	500	$(500-450)/500=0.1$
490	500	$(500-490)/500=0.02$

In fact, as N approaches K, the growth rate approaches zero. Now, putting the expression back into the whole equation for sigmoidal population growth, and assuming a maximum growth potential of 0.5 (i.e., r, the number of new population members contributed per member per unit time), the impact on instantaneous growth (dN/dt) can be seen as the following:

$dN/dt =$	r	N	$(K-N)/K$
4.9	0.5	10	0.98
22.5	0.5	50	0.9
40	0.5	100	0.8
60	0.5	200	0.6
62.5	0.5	250	0.5
60	0.5	300	0.4
40	0.5	400	0.2
22.5	0.5	450	0.1
4.9	0.5	490	0.02

Note that, although r_{max} remains constant, instantaneous growth is slow in the beginning, when there are few population members to find each other and mate, and it slows again as N approaches K.

In another form (a form that can be used in Excel spreadsheets as needed for Discussion Questions) the equation can be used to calculate population size (N) for any time interval (t), if the population size of the previous interval (i.e., $t-1$) is known, as follows: for unlimited growth,

$$N_t = N_{t-1} + r_{max} N_{t-1}.$$

For sigmoidal growth,

$$N_t = N_{t-1} + r_{max} N_{t-1} (K-N)/K.$$

So, what can an understanding of population growth curves add to the understanding of wildlife damage or its management? First, growth curves serve as the bases for most computer models of population estimation. Such models are used extensively in wildlife damage management to assess the status of offending populations, e.g., before and after management actions have been taken. Second, knowing the type of growth curve a particular species follows—e.g., J-shaped versus S-shaped—can help predict the likelihood of the population becoming a problem and provide insights into how to manage the population. For example, populations such as microtine rodents and hares, which follow J-shaped curves, also tend to exhibit irruptive cycles, damaging their environments and crops when their population numbers exceed the carrying capacity. Such irruptive cycles have also been observed with some ungulates such as deer, as summarized by McCullough (1997). Third, both inflection point and biological carrying capacity are useful concepts that have special applications in wildlife damage management. Using carrying capacity, for example, one can envision how wildlife populations might be reduced indirectly, without directly managing natality, mortality, emigration, or immigration rates. Recall that carrying capacity is determined by a density-dependent limiting factor (the factor needed for existence that is least available) that might potentially be food, water, or shelter. By severely restricting its presence, the manager can potentially make that factor the limiting one. In doing this, the manager would have also indirectly forced the population to self-regulate to a lowered N that now corresponds to its reduced carrying capacity.

David Lack, a prominent biologist who particularly studied populations, argued that food was the single most important factor limiting growth of many bird populations (Lack 1954). By reducing the presence of food in areas where a damaging bird population is present, the population would either self-regulate to a lowered number (for populations with sigmoidal growth) or crash when the carrying capacity for the population was exceeded (for populations with J-shaped growth). In practice, food might be limited by different methods, such as use of netting to exclude orchard grapes or other fruit from the birds, use of chemicals that make the flavor of potential foods distasteful, or use of soil to cover landfills and physically separate birds from edible wastes.

Regarding rodents, successful control in large cities such as Baltimore, Philadelphia, and Chicago has often hinged on public campaigns to sanitize areas by containing trash and garbage, thereby reducing their availability as food for commensal rodents. Repair of houses to prevent access by rodents further reduces habitat. Airports are examples of places where surrounding habitats are made as undesirable as possible to reduce carrying capacity for wildlife.

Example

Conner et al. (2008) developed a simulation model to evaluate the effectiveness of differing coyote management strategies. They began with a population model already designed by Pitt et al. (2003) that was nonspatial and stochastic (random based). The model also incorporated some behavioral features, such as dominance and sociality. Subordinate male coyotes are often nonterritorial and nonbreeding. They live along the interstices of other male territories. The subordinate males, however, serve as a population reserve, in that they can replace dominant males that have been removed, becoming both territorial and reproductive. Thus the behavioral responses, a social feature incorporated into the model, add resilience to coyote populations. The model was designed to simulate management actions with a pack of 100 coyotes, and in the model, parameters were adjusted to match those known from already-existing field data.

Conner et al. (2008) added a spatial component, also important for coyotes and coyote management, and refined the social components to make them even more realistic. The researchers then used the model to assess different management actions. They found that spatially focused and intense lethal removal was more effective and lasted longer than less intense, random removal of the same numbers of coyotes. Sterilization appeared to offer the longest-lasting and largest impact on the coyote population. They suggested that the model serve as a tool for developing more effective and socially acceptable predator management strategies and recommended some changes to further improve the model.

LIFE TABLES, SURVIVORSHIP CURVES, AND AGE PYRAMIDS

Statement

To facilitate interpretation, age-dependent natality, immigration, mortality, and emigration rates are often organized and displayed as life tables, survivorship curves, or age pyramids. The displays are used to portray the dynamics of offending wildlife populations and assess management impacts on populations.

Explanation

Growth rate (r) and its four component rates are influenced by ages of population members. Consider, for example, a hypothetical population having three age groups—young, middle, and old. Young and old members are often unable to reproduce and have natality rates near zero, leaving higher natality rates to the age group in the middle. Young members take more risks and therefore tend to have higher mortality rates than more experienced population members. Members of older age groups also tend to die, and so the older group also has a high mortality rate. And emigration from a population is often a characteristic of its younger members.

Life tables are helpful in understanding impacts of age on populations (box 5.1). The overall shape of a life table, portrayed as a graph with age groups as the y coordinates and numbers of members for each group as the x coordinates, can indicate the health of a population. For example, a pyramid-shaped life table (fig. 5.1d) would indicate a stable or even expanding population. However, a top-heavy or inverted pyramid (fig. 5.1d) indicates a declining population. A cylindrical-shaped life table (fig. 5.1d) would indicate a stable population or one that is about to decline.

Data from life tables can also be used to construct survivorship curves and age pyramids (fig. 5.1e). The curves are derived by plotting survivorship (see l_x in table 5.1) against age at death. Wildlife populations are often described as fitting one of three types of survivorship curves. In Type I, often after an initial die-off of young, the odds of surviving are high until an individual becomes old. Localized human populations often fit a Type I curve. With Type III curves, mortality rates are highest in earlier age groups. Type III curves characterize many plant, invertebrate, and fish species, but a Type III curve is rarely seen with populations of mammals and birds. With Type II, survivorship rates remain fairly constant over the age groups within the population; many mammals and bird species have populations fitting Type II curves.

Age pyramids are derived by stacking age groups on top of one another, from youngest on the bottom to oldest on the top. Sometimes the pyramid is split down the middle, with age groups of one sex on the left, the other sex on the right. A triangular pyramid usually indicates a stable population, whereas a top-heavy one indicates a diminishing population.

In wildlife damage management, understanding the age-related dynamics of populations can be important. Life tables, survivorship curves, and age pyramids can help assess the likelihood that a species will be invasive, the most appropriate control strategy, and the costs for control in relation to benefits for the control

BOX 5.1 Life Tables

Life tables have been used to estimate human population demographics since Roman times and were further developed by insurance companies to account for age in the costs of life insurance for people. The approach has been incorporated into models for wildlife populations.

One of the first wildlife studies that used life tables was by Murie (1944). He collected horns of Dall mountain sheep (*Ovis dalli*) from Mount McKinley, Alaska, and used the horns to establish the age of the sheep at death. Over time, he collected horns from 608 sheep, a sufficient sample to establish age-dependent mortality rates for the sheep, as reported by Deevey (1947). The whole of Murie's life table was based on horns collected opportunistically over years—i.e., all individuals, regardless of age, found dead over one period of time, often called a **static** table or **vertical** table. Static tables require an assumption that the population is stationary, with constant birth and death rates; however, a stationary population is unlikely. Sometimes all members of a population born within a period of time, say a year, are followed until all have died. A different type of life table, called a **cohort** life table or **horizontal** table, can be constructed from these data. Cohort tables require an assumption that the age group selected accurately represents the whole population, which is also unlikely. Attempts to fix these and other problems result in more complex mathematical models than we consider here.

When interpreting life tables, be sure to know the bases on which the rates are calculated, since population factors are often expressed as numbers per 1,000 females or sometimes reproductively active females; for example, a growth rate of 3.2 would represent 3.2 net new population members per 1,000 females, or 3.2 per 1,000 reproductively active females.

TABLE 5.1   *Life table for the raccoon dog in a primal forest in Poland*

Age (years)	n_x	l_x	d_x	q_x	p_x	e_x
0	250*	1.000	205	0.820	0.180	0.796
1	45	0.180	26	0.578	0.422	1.144
2	19	0.076	13	0.684	0.316	1.026
3	6	0.024	4	0.667	0.333	1.167
4	2	0.008	1	0.500	0.500	1.500
5	1	0.004	0	0	1.000	1.500
6	1	0.004	1	1.000	0	0.500
7	0	0	—	—	—	—

Source: Redrawn from Kowalczyk et al. 2009, with permission.

n_x = birth rate, l_x = survival rate, d_x = death rate, q_x = mortality rate per 1000, p_x = survival per 1000, e_x = expectation of life (i.e., lifetime remaining), all age dependent.

*Zero frequency calculated from fecundity ratio.

longevity up to seven years, and among the highest reproductive rate seen with raccoon dogs in Europe (the authors felt it might have been compensatory). Despite a high mortality rate, Kowalczyk et al. postulated that overall population dynamics supported the raccoon dog as a species well suited for invading new areas, including western Europe.

Examples

Dolbeer (1998) used four population models (PM1 to PM4) to predict relative responses of vertebrate pest species to reproductive and lethal control methods. He contrasted fruit bat populations with rat populations, both damaging species in the Maldives (fig. 5.2). He showed that fruit bat populations, with low reproductive rates, could be reduced more efficiently with lethal compared to reproductive control. Rat populations, with high reproductive rates, could be controlled more efficiently with reproductive rather than lethal methods. Dolbeer (1998) also compared brown-headed cowbirds to laughing gulls (*Larus atricilla*), again demonstrating that the animals with relatively low reproductive rates (the laughing gulls) could be more effectively managed with lethal control and that the animals with high reproductive rates responded better to reproductive methods. He suggested that red-billed quelea would respond as did cowbirds.

This study (Dolbeer 1998), based partly on analyses of age-dependent population dynamics of offending populations, provides important insights into both theoretical and practical approaches and limitations of differing management actions. For example, Dolbeer (1998) proposed that the relative efficacies of

effort. See, for example, the studies done in the mid-1990s in Finland on the invasive raccoon dog (e.g., Helle and Kauhala 1993). Kowalczyk et al. (2009) constructed a life table for the raccoon dog in the Bialowieza Primeval Forest, Poland, based on age of death recorded between 1996 and 2006 (table 5.1). These animals showed a high mortality within the first year,

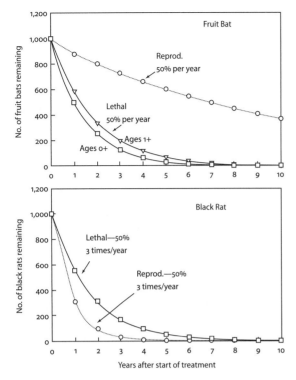

Figure 5.2 Relationships between reproductive potential and success of lethal or reproductive-based management strategies. Species with low reproductive potential, such as fruit bats, tend to respond to lethal methods, whereas species with high reproductive potential, such as black rats, tend to respond to reproductive methods. *Redrawn from Dolbeer 1998, with permission. Illustrations by Lamar Henderson, Wildhaven Creative LLC. With permission of the Vertebrate Pest Council.*

lethal and reproductive methods for vertebrate species could be generalized based on adult survival rate and age of onset of reproduction. Specifically, he pointed to the long period of time for effectiveness of contraceptive methods when used with species, such as deer, that reproduce relatively slowly.

MEASURING POPULATION SIZE AND AGING

Statement

Estimating actual population size depends on direct counting or indirect sampling methods, often expensive, whereas population indices can inexpensively provide information on changes in population sizes. Estimates of aging populations require some age-dependent factor such as reproductive status, tree rings, lens growth, tooth development or wear, or annular rings of scales or horns.

Explanation

Most ways to measure dynamics of wildlife populations are variations of a few basic methods. One approach is to use distance sampling methods such as line or point **transect** surveys. The technology can be as simple as a walk or drive along a transect during which one counts population members. The count might be of members themselves or of their signs, such as sightings of coyote tracks or scat or numbers of bird calls. Sometimes aerial photographs and digital analyses are used. The photography might be accomplished with fixed-wing aircraft or a helicopter or might come from satellite images. The photographs can be analyzed with methods used in geographic information systems to provide estimates of population density and size.

For example, the North American Breeding Bird Survey uses point counts to assess overall trends of bird populations in the continental United States, Alaska, and southern Canada (e.g., Peterjohn and Sauer 1993). The Audubon Christmas Bird Count also uses point counts, in this case to look at long-term trends in populations of birds in the United States and Canada and in parts of Mexico, Central America, and the Caribbean Islands (e.g., Dunn et al. 2005). The surveys provide trends useful for assessing wildlife damage. For example, long-term declines in songbird populations in North and South America have led to investigations of causes, including feral and domestic cats and brown-headed cowbird parasitism.

Gese (2004) summarized methods for estimating sizes of canid populations. Methods included counts of scat or tracks along transects, den and burrow surveys, response to vocalizations, frequency of complaints, harvest data, road mortality, drive counts (where the canids are driven past a point where an observer counts them), and spotlight surveys. Most of these methods could be used to estimate population size or to index populations (see below), but their precision and accuracy vary.

Blackwell et al. (2006) summarized applications of infrared technologies, including the use of forward-looking infrared (FLIR) devices for wildlife surveys in general and wildlife damage issues in particular. Infrared technologies sometimes allow better detection of targeted wildlife, improving the sensitivity of surveys (see, e.g., white-tailed deer surveys for management of a confined population at Plum Brook NASA Station, Sandusky, Ohio; Blackwell et al. 2006). FLIR devices have been successfully used to estimate seasonal abundance of raccoons (*Procyon lotor*) in north-central Ohio (Blackwell et al. 2006) and to detect polar bear (*Ursus maritimus*) dens in Alaska (Amstrup et al. 2004) and have been recommended for detecting feral goats and removing them from islands (Campbell and Donlan 2005).

As a second approach, animals can be temporarily removed, marked, released, and recaptured (box 5.2). Gese (2004) encouraged the use of mark and recapture methods when accurate estimates of population sizes are needed. Ear tags, radio collars, dyes, and physiological markers have been used, and alternatively, individuals might be recognized by facial or other features. Recapture might be physical or indirect with

---

BOX 5.2 Lincoln-Petersen Index

One capture-recapture method is called the **Lincoln-Petersen Index**. Assuming that the marked individuals stay within the population and mix randomly and that the marking does not differentially affect mortality of the members (e.g., by making them more visible to predators), the proportion of marked to unmarked members that are taken during subsequent collections can be used to estimate total population size, as below:

$$N = MC/R,$$

where

N = population estimate in numbers of individuals

M = total numbers of animals captured, marked, and released

C = total numbers of animals captured on the second visit

R = number of animals captured on the first visit that were marked and recaptured on the second visit.

For example, if 100 green sunfish (*Lepomis cyanellus*) were captured, tagged, and then released (M); and if 200 of the sunfish were captured on the second visit (C), of which 50 had tags (R), then

$$N = (100)(200)/50 = 400 \text{ sunfish}.$$

The population estimate is 400 sunfish. If the area of collection is known, an estimate of density might be given, such as 400 sunfish per surface acre of pond. Repeated samples might be used to give an estimate of variance. Although the concept is straightforward, models intended to accurately represent real populations can quickly become complex.

"**camera traps**" (i.e., individuals can be recognized from photographs taken at stations rather than actually captured). Karanth and Nichols (1998), for example, used camera traps to estimate populations of tigers (*Panthera tigris*) in India. Martorello et al. (2001) used bait-triggered stations and cameras to gather capture and recapture data on black bears (*Ursus americanus*) in coastal and mountain habitats. Goswami et al. (2007) used similar methods to assess population sizes and other demographic factors for the Asian elephant (*Elephas maximus*). Sequin et al. (2003) found that alpha males were wary of cameras and were underrepresented in their studies on coyotes, a factor that may require consideration when interpreting results for other territorial and social species.

With a third approach, sometimes called **depletion** or **removal trapping**, members of a population are intensely trapped and permanently removed. Trapping must occur over a brief interval, say three to five days, to minimize effects of immigration, emigration, or nontrapping mortality. Daily take is plotted against cumulative take totals, and regression analysis is performed to describe the relationship. Extrapolation of the resulting equation to the point where no more takes occur (i.e., one of the intercepts) will yield an estimate of population size (N; fig. 5.3). Again, repeated sampling gives an estimate of variance that can be used to calculate confidence levels around the curve. Related removal methods, including those based on change in ratio or catch per unit effort, have also been useful in assessing many wildlife populations, particularly those that are trapped or hunted (e.g., Wilson et al. 1996).

Depletion or removal trapping has limited use for some pest problems, such as those involving canids, because of social concerns (Gese 2004). Engeman and Linnell (1998) showed how trapping data for removal of brown tree snakes (*Boiga irregularis*) from Guam could

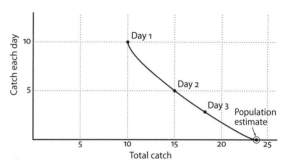

Figure 5.3 With depletion trapping, population size is estimated by plotting daily take versus total take to date and extrapolating to the point where daily catch would be zero. *Illustrations by Lamar Henderson, Wildhaven Creative LLC.*

be modeled as exponential decay and how the model could be used to plan control strategies. Specifically, the analyses showed that perimeter trapping around areas to be protected was relatively more effective in relation to trapping costs. And the analyses indicated that once snakes were removed from fragmented habitat, their recovery was slow. Hence, reduced trapping in these areas would be economical but still effective.

Population indices are a fourth, and often most practical, approach to assessing damaging wildlife populations. When it is too difficult or expensive to get estimates of absolute population size or densities, estimates of relative population size might still be obtained. A **population index** (Engeman and Witmer 2000) assesses a change in activity, such as movement to and from a station having an attractant, active burrows, nests, or feeding sites. One might measure the presence of signs or tracks around the stations, for example, or use tracking tiles to document footprints or tail movements. Although the approach does not yield absolute information, such as population size or ecological density, it allows indirect measurement of changes in population by measuring related changes in activity. For example, by counting the number of active burrows in an orchard before and after use of a rodenticide, one could assess the treatment's effectiveness. Population indices have been widely used to monitor damage and effectiveness of management actions (Engeman and Allen 2000; Engeman and Witmer 2000; Gese 2004, for carnivores).

Engeman and Campbell (1991) used reoccupation of burrows following application of rodent baits to assess the success of the treatment in reforestation areas of Oregon, where pocket gophers often damage seeds or seedlings. In North America scent stations, used over a 30-year period, have offered insight into the relative changes of coyote populations (Linhart and Knowlton 1975). Engeman et al. (2000) used a passive tracking index to monitor population changes due to trapping coyotes and bobcats on two ranches in Webb County, southern Texas. They established tracking plots along dirt roads at about 0.8 km intervals and examined them for spoor (a track, a trail, a scent, or droppings) for 2 to 4 consecutive days before and after periods of trapping. Allen et al. (1996) found that a passive activity index would reflect changes in dingo populations while it also monitored other species, including macropods, fat-tailed dunnarts (*Sminthopsis crassicaudata*), feral cats, brush-tailed possums, and rabbits.

Determining ages of wildlife is necessary when relating age to other population dynamics. To do this, one must find a factor that changes in a manner that can be related to age. Often, hard body parts such as bones, teeth, scales, or spines are used. Sometimes the relationship is developed by measuring the trait in captive wildlife of known age. For some wildlife, such as white-tailed deer, irruption, loss, and wear of teeth is commonly used. Bone growth is sometimes used to age rabbits and rodents, particularly the (epiphyseal/parepiphyseal) plates at the ends of long bones where growth occurs or the baculum (*os penis*) in some marine mammals. Weight of eye lenses increases over the lifetime of some animals and can be related to age. Sometimes annular rings can be used, as seen in trees, fish scales, and the cementum layers of teeth. Recession of gum lines from the incisors has been used to age mountain lions.

Example

Blackwell et al. (2006) used FLIR cameras to assess and manage white-tailed deer at the Plum Brook Station of the National Aeronautics and Space Administration in Erie County, Ohio. The station is 22 hectares in area and fully enclosed. Beginning in 1998, deer population estimates were standardized at the station using spotlight counts. Starting in 2005, Wildlife Services began comparing the spotlight method with one using a FLIR camera, finding 11% more animals with the camera. In the winter 2006 estimate, 271 deer were counted by spotlight, whereas 378 were counted with the FLIR cameras. The difference may be due to increasing vegetation within the station, which brought into question the reliability of the spotlight surveys. Public hunts are permitted to manage the herd, with harvest goals based on the most recent estimate of herd size.

POPULATION MODELS

Statement

Wildlife damage practitioners increasingly rely on population models to predict irruptions, to anticipate damage, to simulate impacts of alternative control methods, and to evaluate effectiveness of control.

Explanation

Populations can be modeled by modifying the fundamental equation for population growth to include an iterative component, as follows:

$$N_{t+1} = N_t + (b + i) - (d + e),$$

often called the BIDE equation.

While the BIDE equation is itself straightforward, bases for calculating component rates can be more difficult, complicating the model. For example, an estimate of N_t might be based on a ratio coming from one of the capture-recapture methods just described. Compensatory or additive mortality or breeding-season and age-group-related natality also add complexity, sometimes best simulated using matrix algebra.

Example

Chapron et al. (2003) used stochastic computer models of wolf population dynamics to examine two approaches to management of the wolves in Europe. One approach simulated zoning, wherein wolves were left untouched within predesignated areas called wolf zones. They were removed from areas near livestock, called nonwolf zones. The second approach simulated removal of a portion of wolves once a preset **population growth rate** was achieved. With this approach, there were no zones.

This research team used life information for wolves to create their computer simulations, refined with specific information on European wolves when it was available. They divided populations into four age groups, with pack leaders being at least 18 months old. In the model, winter mortality affected the whole population and accounted for annual mortality; subordinates dispersed only if the breeding pair survived; dispersing young sought vacant territory and partners; reproduction occurred in the spring, with first reproduction at 22 months; one litter was produced per year, with pup mortality occurring in the summer to simulate deadly disease; and a census of distribution and status of wolves was conducted each fall. Analyses and simulations were conducted using a computer program called Unified Life Models.

Chapron et al. (2003) drew two important conclusions: first, under zoning scenarios, long-term viability of the population of wolves is extremely sensitive to the number of packs; second, an adaptive strategy in which a moderate percentage of wolves is removed when the population reaches a threshold growth rate maximizes the effects of control to reduce livestock losses, while at the same time minimizing the risk of extinction of the population.

RELATIONSHIPS BETWEEN POPULATIONS AND DAMAGE

Statement

A clear relationship cannot be assumed between the size of a damaging population and the level of damage; such a relationship should be established before population reduction is undertaken.

Explanation

Management of offending populations—otherwise known as population reduction—is often viewed as the approach of choice for reducing damage. There are many examples of clear relationships between population increases and wildlife damage. Taylor and Dorr (2003) summarized general relationships between a burgeoning double-crested cormorant population and increased economic losses to catfish farmers and recreational fisheries in North America. Dolbeer et al. (1993) reported an increase in numbers of laughing gulls at Jamaica Bay Wildlife Refuge from 15 nesting pairs in 1979 to 7,629 in 1990, along with concomitant increases in birds striking airplanes at nearby John F. Kennedy Airport, New York City. These researchers also reported a decline in bird strikes that correlated directly ($r^2 = 0.97$) with a reduction of the birds by shooting ones attempting to enter airspace at the airport. Beasley and Rhodes (2008) assessed population densities of raccoons using mark-recapture methods and Program MARK at 14 locations in Indiana. They showed a clear correlation between population estimates and damage to field corn; in addition, these scientists demonstrated that the presence of edges between forests and crops was also important in predicting damage.

While such relationships between population size and damage often occur, they cannot be assumed. A direct relationship may not exist, or it may be sufficiently obtuse to prevent its effective use in limiting damage. Dolbeer et al. (1994a,b) warned that presence and relative abundance of a species does not necessarily prove guilt. They pointed to large flocks of grackles in sprouting winter wheat whose stomachs were filled with leftover corn residue, not wheat. Starlings, less abundant than grackles, were actually removing the germinating wheat seeds (Dolbeer et al. 1979).

Further, unless the population is considered in the broader context of the community (see chapter 6), reduction of one offending population might result in the "release" of another. Zavaleta et al. (2001) provided a useful summary of possible ecosystem responses to consider prior to eradication of an invasive animal, including food-chain and food-web effects, predator-prey interactions, and herbivore-plant interactions. We believe a wise manager considers such relationships before embarking on a major campaign of population reduction.

Example

On Stewart Island, New Zealand, exotic cats prey on the endangered flightless parrot, *Strigops habroptilus* (Zavaleta et al. 2001). The cats, however, prefer exotic rats. While it seems desirable to remove the cats, their absence might allow "release" of the rats and create the potential for an even greater impact on the parrot. That concern has prevented management action to eradicate the cats from the island. Zavaleta et al. argue that concerns for mesopredator release are not unique to Stewart Island; similar mixes of cats, rats, and mice occur on at least 22 other islands.

Summary

- Overabundance occurs when a population damages itself, humans or their interests, or the environment. Crowding is related and occurs when population density exceeds a level at which the habitat can maintain healthy population members and the population becomes distressed.
- Populations may be reduced when they become overabundant or crowded.
- If eradication is the goal, the population must be reduced below its critical minimum. More often, reducing damage is the goal, and there is an assumed relationship between extent of damage and size of the offending population.
- Natality and immigration add to populations, whereas mortality and emigration reduce populations. Each can contribute to damage, and any action to regulate populations must fall within the fundamental equation for growth and impact one or more of these rates.
- In wildlife damage management, when the intent is often to reduce an offending population to the lowest size practicable, additive mortality rather than compensatory mortality is often sought.
- Some wildlife, such as microtine rodents or some invasive species, follow a J-shaped growth curve wherein they exceed carrying capacity and crash, often causing damage. Other wildlife self-regulate just below the carrying capacity and can sometimes be managed by lowering the carrying capacity.
- Population growth and each of its component rates are affected by age of population members, often visualized using life tables, survivorship curves, or age pyramids; such displays help researchers to understand dynamics of offending populations and to assess effectiveness of management.
- Direct and sampling methods are used to measure population sizes and to determine whether management is needed or whether management was effective. Population indices, which provide information on change without measuring population sizes, are economical and often used in wildlife damage management.
- Aging a population requires knowledge of some age-dependent factor such as lens curvature or annual rings or scales.
- Computer-based population models can perform complex iterative calculations quickly and accommodate important ancillary factors such as compensatory or additive mortality; they are increasingly used in wildlife damage management.
- The relationship between abundance of a damaging population and level of damage should be determined before management actions are taken.

Review and Discussion Questions

1. If a population had a natality rate of 3.4 per thousand per year, an immigration rate of 1.2 per thousand per year, an emigration rate of 0.1 per thousand per year, and a mortality rate of 2.0 per thousand per year, what is its growth rate? Suppose a control action increased the mortality rate to 6.4 per thousand per year, but the population responded with an increased natality rate of 4.2 per thousand per year—what was the overall impact of the control action on growth rate? Does this necessarily have the same impact on the damage the population was causing?

2. Contrast the responses to carrying capacity of a population following a J-shaped curve with one following an S-shaped curve. Explain how this information might be helpful in applying a control method that is based on reducing carrying capacity of a population and indirectly controlling a damaging population.

3. If you were reducing damage by muskrats to a marsh used for treatment of wastewater, and you trapped the muskrats intensively for five consecutive days each month, might data on trap success be used to assess changes in size of the muskrat population? How? If you also gathered data on ages of the muskrats trapped, what shape of life table would you hope to see to confirm a diminishing muskrat population?

4. Using an Excel spreadsheet, show the growth of a population of sparrows that have been introduced onto an island that offers unlimited resources

and no limiting factors (see "A wee bit of math geekery—modeling populations with spreadsheets," by Brad Williamson, www.nabt.org/blog/2009/05/08/a-wee-bit-of-math-geekery/). Ten sparrows are introduced—five males and five females. Each pair produces 10 offspring per year, and all offspring survive to reproduce the next year. However, parents die before the next year (i.e., sparrows reproduce, then die). No new sparrows leave or enter the island. Plot the size (N) of the population over time (years), first using a graph where N is linear and then where it is logarithmic.

5. Use an Excel spreadsheet to show the growth of a population that follows a logistic or sigmoidal pattern and that has the following characteristics (see www.nabt.org/blog/2009/05/25/using-spreadsheets-to-introduce-the-logistic-population-growth-model/, "Using spreadsheets to introduce the logistic population growth model," by Brad Williamson): $r = 0.1$; $N = 10$ at time 0 and 11 at time 1; $K = 1000$. Graph this model. Now, using the model, show how a population might be reduced by lowering its carrying capacity.

6

Communities, Ecosystems, and Landscapes

More and more, wildlife damage practitioners focus on whole ecosystems or landscapes to both understand and manage wildlife damage. With this synecological approach, the practitioner effects change within the ecosystem or the landscape but often without regard to a specific damaging species. Checks and balances within the system then correct the problem.

Here, we look at attributes of communities, ecosystems, and landscapes that help us understand and manage wildlife damage. We consider community and landscape composition, diversity and stability, energy flow, food chains and food webs, and succession. We show how each characteristic is used in wildlife damage management.

COMMUNITIES, ECOSYSTEMS, AND LANDSCAPES AND THEIR APPLICATIONS TO WILDLIFE DAMAGE MANAGEMENT

Statement
Long-term solutions to wildlife damage often rely on understanding and manipulating ecosystems or landscapes, now viewed by some practitioners as functional units of wildlife damage management.

Explanation
Ecosystems are communities that interact with their abiotic components. A **community** is all organisms living in an area at a given time. An ecosystem is a concept that may represent a system that is huge or tiny, one that is either largely independent (i.e., gets most of its energy from the sun) or largely dependent on other ecosystems for its energy (e.g., a city). An example of a tiny ecosystem is a "balanced" aquarium (e.g., Townsend et al. 2003). The greatest ecological community on earth, constituting all living organisms on the planet, is the **biosphere**. The biosphere interacting with its abiotic components, powered by energy flowing from the sun, is called the **ecosphere**. Biomes, described in chapter 4 as types of habitat, are communities dominated by a particular type of vegetation and associated flora and fauna.

Ecotones are boundaries between ecosystems. The boundaries may be sharp or fuzzy, straight or irregular. Because they are edges of two or more ecosystems, ecotones usually have species from all contiguous systems. In addition, ecotones may provide habitat for unique species not found in any one of the contiguous ecosystems. Some species, such as white-tailed deer and coyotes, thrive in ecotones. The increasing presence of coyotes in the midwestern and eastern United States during the past few decades might be partly explained by ecotones created by humans. Other species, such as pileated woodpeckers (*Dryocopus pileatus*), martens, and flying squirrels, require space within the deeper recesses of an ecosystem, away from ecotones. The brown-headed cowbird (see chapter 1) is an ecotonal species whose presence in North America has increased along with the proliferation of fragmented landscapes.

Cities are examples of ecosystems that require enormous energy subsidies to maintain overall stability. For example, human and pet foods are energy subsidies moved from agroecosystems, sometimes located in distant parts of the globe, into the city. Transportation of the goods is also an energy subsidy, often a considerable one. City dwellers—humans and their pets—are therefore major economic forces underlying efforts to make modern agroecosystems more productive.

Within cities, the diversity (number of species) of wildlife species is often reduced. Remaining species are often eury-type generalists that cause damage problems (e.g., McKinney 2002). Thus starlings, the house sparrow (*Passer domesticus*), house finch, striped skunk, eastern gray squirrel, and commensal rodents—as well as feral animals such as cats and dogs—typify urban wildlife in many cities worldwide. Nonnative species are often more prevalent while species that do poorly with humans, such as the larger carnivores, tend to disappear (e.g., McKinney 2002). Overall, species diversity declines from suburbs to city centers. Urban wildlife can damage property, transmit diseases, cause fires (fig. 6.1), and sometimes attack people or take pets as prey. They also offer city dwellers glimpses of nature and remind them that humans are irrevocably tied to ecosystems. Cities can be viewed as novel ecosystems with wildlife adapting (e.g., gleaning food from bird feeders, streets, and exposed garbage; Clergeau et al. 1998) to their sometimes unique environmental circumstances. By understanding this, human residents can manage urban environments to increase the diversity of wildlife species as well as the number of positive experiences with them.

As cities sprawl into countrysides, suburbs overlap wildlife habitat. Thus the mountain lion, once

Figure 6.1   Some birds have learned to "ant" using smoke from a discarded cigarette. Bringing the lit cigarette into a nest on a building can cause a fire. *Artwork by Ian Reid.*

removed from urban areas, is again becoming familiar with *Homo sapiens* and their ways (and their pets). Add a protectionist inclination on the part of some city dwellers, as expressed in California's Proposition 117 (which protects mountain lions and was passed in 1990), and the stage is set for increased attacks on pets and people. It is unsurprising that mountain lion attacks on people are on the rise in California and along the eastern foothills of the Rocky Mountains.

Climate is a dynamic part of ecosystems. Planetary climate changes have occurred long before human presence and can force changes in ecosystems, from puddle-sized ones to biomes to the ecosphere. The changes can be dramatic, and they can result in perturbations and irruptions of populations as well as changes in communities that we view as wildlife damage. Global warming and changes in the earth's ozone layers are examples of ecospheric concerns that could dramatically impact wildlife damage management. For example, climatic shifts resulting from movement of the El Niño–Southern Oscillate increased rainfall and vegetation and subsequently caused a population irruption of the white-footed mouse (*Peromyscus leucopus*) in 1993 in the Four Corners region of the southwestern United States. The mouse carried hantavirus, which subsequently sickened and killed people in the area.

Experts for the United States Armed Forces Pest Management Board (AFPMB) use GIS and climatic models, including those of global warming at the ecosystem level, to predict future outbreaks of diseases in regions of the world that are of military interest. They also maintain an extensive database of scientific articles on diseases, including zoonoses (AFPMB undated). The information is used to anticipate diseases that might be encountered by troops, but it is available to the public, including students and practitioners of wildlife damage management.

Landscape ecology (Troll 1939) is a form of **biogeography** (the study of geographic distribution of organisms in space and time). **Landscape ecology** is the geographic description of ecosystems, their interrelations in time and space, and the interrelations of their functions. The term emphasizes spatial patterns in relation to process and function and often encompasses multiple ecosystems and large areas such as the Appalachian Mountains or Brazilian rainforests (Turner et al. 2001). Whereas ecosystems such as forests or grasslands are somewhat homogeneous, landscapes are heterogeneous (e.g., cropland including surrounding forests). Landscape ecology was first used to improve human-based landscapes in Europe, and humans often strongly influence one or more of the ecosystems within a landscape.

Landscapes do not always encompass large areas. Current usage focuses more on relationships of heterogeneity and geospatial distribution to process and function than on scale. More important than size is the notion of an ecosystem or habitat (see **patches**, below) submerged within a broader system or habitat—that is, a **matrix**. Like an ecosystem, a landscape has no clearly defined minimum size or acreage.

For example, DeVault et al. (2007) used landscape concepts to assess wildlife damage to corn and soybean fields in Indiana. Using global positioning satellite coordinates, the scientists related distances of damage to neighboring forests and human habitation, comparing actual distances with those randomly generated in a stratified sampling design. The researchers concluded that fields adjacent to forests would sustain the greatest damage and those near human habitations would sustain the least. Targeted removal of destructive species along crop/forest interfaces appeared to constitute cost-effective management. Ferraz et al. (2003) used a landscape approach to study damage by capybaras (*Hydrochoerus hydrochaeris*) to cornfields in Brazil. Fields closer to adjacent forests and water received the most damage. Tourenq et al. (2001) found that ricefields in France carried increased risk of damage by the greater flamingo (*Phoenicopterus ruber roseus*) when placed near wooded margins and natural marshes. As mitigation, the researchers suggested that scaring devices and hedgerows be strategically placed along the ecotones.

Landscape ecology has introduced other concepts and terminology, including patches, corridors, and habitat fragmentation. **Patches** are areas of relatively homogeneous habitat surrounded by a broader but different habitat, the matrix (fig. 6.2). For example, a patch of forest may be surrounded by broad expanses of cornfields, the matrix. The patches may or may not be interconnected by thinner runs of similar habitat, or **corridors**. The notion of patches stems originally from that of island biogeography (MacArthur and Wilson 1967) in that patches can be visualized as "islands." **Habitat fragmentation** occurs when a habitat needed for wildlife is continually made smaller or is isolated from previously interconnected habitats; this occurs slowly in nature but is accelerated by human activity. Habitat fragmentation may relegate a wildlife population to one or more patches. Often the patches themselves get smaller and smaller. As patches shrink, space along edges becomes relatively greater and interior space diminishes. If the patches are interconnected with corridors sufficient to allow movement and genetic exchange between the populations of wildlife, the interconnected individual populations can be thought of as a **metapopulation** (Levins 1969).

Landscape ecology has particular application in describing and managing wildlife problems associated with fragmented landscapes. For example, landscape fragmentation apparently increased contact between domestic and African wild dogs, allowing transfer of diseases and subsequent decline of the African wild dog (*Lycaon pictus*; Ward et al. 2009).

Sufficient deer metapopulations and corridors exist among fragmented forest habitat in some U.S. states, setting the ecological stage for reestablishment of mountain lions. Recent reintroductions or those underway—e.g., elk in at least five states—will provide additional resources. Free-roaming mountain lions have been confirmed in at least five midwestern states. Confirmed presence of reproducing mountain lion populations would represent a northward range extension from Louisiana and Arkansas, where established populations are already known or postulated (e.g., Knight 1994).

Root et al. (2009) used genetic information from raccoon metapopulations in Pennsylvania to assess effectiveness of an oral rabies vaccination program. The researchers found isolation, i.e., restricted gene flow in some metapopulations, inferring that changes in delivery of vaccines might be needed to reach these metapopulations in future programs. Losses of farm-raised channel catfish have been attributed to the

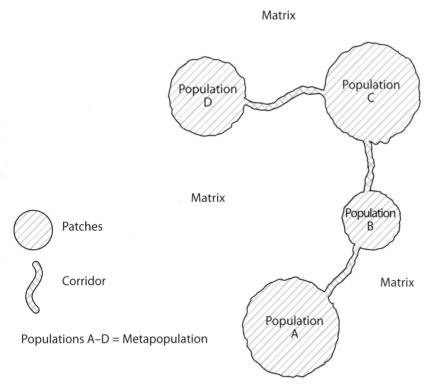

Matrix

Matrix

Patches

Corridor

Matrix

Populations A–D = Metapopulation

Figure 6.2 With metapopulations, each population resides in patches (such as forests) connected via corridors within a general matrix (such as cornfields). Genes can move between populations. *Illustrations by Lamar Henderson, Wildhaven Creative LLC.*

eastern metapopulation of the American white pelican (*Pelecanus erythrorhynchos*; King 2005; King and Anderson 2005) rather than the whole pelican population. Metapopulations have also served as the conceptual bases for megaparks in managing elephants and their damage in southern Africa (van Aarde and Jackson 2007). Such refinements improve efficiency and effectiveness of management while reducing unwanted environmental impacts.

Examples

Belant (1997) argued that gulls at landfills and airports should be managed at the landscape level. He reviewed integrated methods of control, including architectural design and construction methods to reduce roof substrate, manipulations of turf height in loafing areas, covering of refuse sites or compost facilities, temporary drainage of water, and erection of wire grid systems. He argued that effective management at one location, such as a landfill, would simply force birds to an unprotected site within the whole landscape. Belant (1997) suggested that management at city or county levels—that is, landscape scales—was needed.

Global climate change, as defined by the Intergovernmental Panel on Climate Change (IPCC 2007),

is any change in climate over time due to either natural variability or human activity. Measured as radiative forcing, changes in the abundance of greenhouse gases (carbon dioxide, methane, nitrous oxide) and aerosols (primarily sulphate, organic carbon, black carbon, nitrate, and dust) in both solar radiation and land surface properties drive warming or cooling of the global climate. Other radiative forces include tropospheric ozone and halocarbons. **Global warming** is the increase of the earth's air and oceanic temperatures. The IPCC summarized a broad base of evidence for a warming planet, probably facilitated by human activity.

Climate change can impact wildlife in at least five ways: (1) wildlife species change their geographic distributions, keeping within geographically shifting niches, as climatic factors change; (2) species alter timing of seasonal events, including migration, molting, and reproduction, in response to shifting climatic cues such as temperature and rainfall; (3) biotic relations within communities and ecosystems change in nature and complexity as new species, such as exotic invasive species and diseases, enter the mix; (4) more-frequent catastrophic weather events, such as droughts, floods, and precipitation, influence the structure and function (for example, the diversity) of populations and communities; (5) as humans respond to climatic changes with adjustments in land use, wildlife need to adjust as well.

We anticipate greatly enhanced wildlife damage if global warming proceeds progressively or for an extended period of time. For example, extinctions will reduce diversity of ecosystems, simplifying checks and balances, and facilitate surges in numbers of some species. Changing climate will favor introducing exotics, and some will be invasive. Climate change will also favor the spread of zoonotic diseases such as avian influenza, malaria, and dengue fever (e.g., Shope 1992). Finally, catastrophic climatic events also simplify ecosystems, making population irruptions and subsequent damage problems more likely.

COMMUNITY STRUCTURE AND DIVERSITY

Statement

Landscapes and ecosystems have structures and diversity that shape the nature, patterns, and extent of wildlife damage; understanding these factors can help predict damage and plan effective management strategies.

Explanation

Landscape or ecosystem structure is called its **physiognomy**. Structure exists in vertical and horizontal dimensions, in time, and in both aquatic and terrestrial ecosystems. Vertical physiognomy includes structure and patterns of physical attributes such as soil strata (e.g., horizons) as well as grosser structures such as canyons, mountains, and rock faces. Horizontal physiognomy includes structure and patterns of physical attributes such as soil types, rocks, ponds, and streams or rivers.

Physiognomy also includes how biotic communities within landscapes and ecosystems are structured. Vertical structure might mean separation of the biotic community into mature tree canopies, young tree subcanopies, and shrub layers. Vertical structure is the basis for classifying vegetation types into standard physiognomic layers, which can then be used for mapping. Horizontal structure includes patterns exhibited by biotic communities. Physiognomy of the biotic community may change with time, as with succession (see below) or with changes in community composition due to season, such as migration or aestivation (summer dormancy).

Abiotic physiognomy influences biotic physiognomy, as with altitudinal zonation (changes in struc-

ture and composition of living organisms as one advances up a mountain) or the distribution of trout in streams in relation to patterns of rocks used for hiding and predation. Biotic physiognomy can also influence the abiota, as, for instance, when invading species (i.e, the first species into a new area) break down rock with root hydraulics or alter the vertical structure of soil.

Structure often exists even though it may not be expected or obvious. For example, pelagic communities of seabirds display vertical structures while overflying oceans, with the smaller mollusc- and fish-catching birds such as shearwaters and petrels nearer the water and the larger, scavenging (and sometimes stealing) albatrosses above them.

Landscape and ecosystem ecologists look for patterns in such structures. Ecologists use tools such as satellite imagery to capture images and then conduct geospatial analyses using specialized software that reveals such patterns. The ecologists then relate patterns to processes (e.g., energy flow) and functions (e.g., production of timber) of landscapes and ecosystems.

Ward et al. (2009) discussed the importance of landscape structure in understanding and managing wildlife diseases. The researchers noted, for instance, that aerial delivery of rabies vaccines is less effective in hilly areas, because density of delivery is reduced on slopes (Vuillaume et al. 1997), and in urban and suburban areas where hand delivery or other approaches such as plastic bag baits may be needed (Boulanger et al. 2008). Features such as rivers and mountains can serve as natural barriers to movement of zoonoses or their vectors. Including such features in overall management strategies can improve effectiveness and efficiency of control. Ward et al. (2009) suggested that mathematical models, based on metapopulations in fragmented landscapes, can help plan details of zoonotic disease management, such as rabies vaccination strategies. Smith et al. (2005) used GIS mapping of rabies occurrence and mathematical models to design cordon sanitaires (barriers that stop spread of a disease) that minimize public exposure and serve as barriers to the spread of rabies. Smith et al. (2005) also suggested that landscape-based models improved the understanding of spatially transmitted foot and mouth disease on British farms, including aerial plumes from local farms and longer-distance movement by farmers, veterinarians, and contaminated vehicles.

Horizontal distribution of species can be **random, clumped, even (uniform)**, or some combination of these patterns (fig. 6.3). Distribution can be strongly influenced by available physical and biological resources. Information on distribution can be gathered from ground studies, by aircraft, or by using global positioning systems that allow precise location of objects, and they can be analyzed using geographic information systems. Knowing patterns of damage can help determine where to focus management actions, such as setting traps or baits. Often, information on species distribution is also needed to ensure statistical validity of sampling methods, e.g., in stratified sampling.

Biodiversity is related to stability of ecosystems and landscapes. Some biotic communities are dominated by a single or few species, whereas in other communities diversity is spread more evenly among species. If the ecosystem has a particularly abundant species, such as a prairie dog (*Cynomys* spp.), an elephant, or a wolf, that species may be a keystone species. **Keystone species** disproportionately dominate structure, diversity, and function of the entire ecosystem (Paine 1966). A predator at the top of a food chain is called an **apex predator** and is often also a keystone species. Actions by humans or nature that impact keystone species are of particular interest in wildlife damage management, because changes in the status of a keystone species can cause instability of the system and wildlife damage.

A keystone species differs from a **foundation species**, a dominant primary producer in an ecosystem. For example, kelp is a foundation species in kelp forests. Here the sea otter (*Enhydra lutris*) is the keystone species because it controls populations of herbivores feeding on the kelp and hence indirectly constrains the growth of kelp.

Donlan et al. (2007) argued that removal of larger predators, including such apex predators as the grizzly bear (*Ursus arctos horribilis*), the gray wolf, and the mountain lion (see chapter 2), caused a predator-prey disequilibrium in North America, resulting in some of today's damage problems. In northern Yellowstone Park, for example, overpopulation of elk from lack of wolf predation has led to heavy browsing of willows and loss of beaver wetlands. A similar disequilibrium involving overabundance of moose, willow loss, and reduced wildlife diversity has occurred in the southern Greater Yellowstone Ecosystem. Kauffman et al. (2010) and others, however, have presented evidence that contradicts aspects of the theory—e.g., fear as a mechanism for amplifying impacts of predators. More generally in North America, mesopredators, especially the coyote, have replaced the wolf as apex predator, leading to a broad range of wildlife damage problems such as those already described for duck production in the prairie pothole region and for the fox in Illinois (see chapter 4).

Johnson et al. (2007) found that survival of small to mid-sized marsupials was greater in areas of Australia where dingoes were present as apex predators. These researchers argued that high densities of dingoes protected small ground-dwelling marsupials from extinction by preventing release of foxes and cats, intense mesopredators of smaller marsupials. The authors argued conversely that removal of dingoes from wide areas of Australia since settlement indirectly facilitated extinction of marsupials by releasing mesopredators, a consideration for future conservation strategies worldwide. Impacts of **mesopredator release** have been documented broadly in oceanic, freshwater, and terrestrial ecosystems globally for species including lizards, rodents, birds, rabbits, fishes, and sea turtles (Prugh et al. 2009).

Species diversity is often measured with indices—e.g., the Simpson's or Shannon's Diversity Indices—or functions that measure diversity, such as dominance (the relative importance of one or more species), **richness** (number of species and distribution of individuals among species), and **evenness** (also the distribution of individuals among species). These indices are also important in identifying

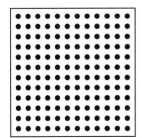

Uniform

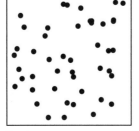

Random

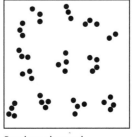

Random-clumped

Figure 6.3 Patterns in distribution of living organisms influence how one estimates and manages populations. *Illustrations by Lamar Henderson, Wildhaven Creative LLC.*

damage, its distribution, and its management (see chapter 10).

Example

In the Muddy River drainage of the Mojave Desert, the exotic invasive plant saltcedar (*Tamarix ramosissima*) has taken over riparian habitats. Aggressive efforts have been made to eradicate saltcedar; those efforts, in turn, have reduced both structural and compositional diversity of riparian ecosystems. Fleishman et al. (2003) explored diversity and vertical physiognomy of affected habitats. They used species richness indices to assess both avian and floral diversity and measures of total vegetation volume to assess vertical physiognomy. These scientists found that avian diversity related more closely to vertical physiognomy than to species richness of flora. Species contributing to richness sometimes had overlapping functions, serving mostly as "insurance"—i.e., as backups—rather than in daily ecosystem function. The researchers found that birds used exotic saltcedar for nesting and protection, especially if native flora were nearby. They concluded that bird diversity did not suffer from the presence of saltcedar. They suggested gradual removal of the saltcedar, allowing time for replacement with native vegetation, rather than aggressive removal and concomitant loss of the volume of vegetation needed by the birds.

ENERGY FLOW

Statement

Trophic-level cycling needs to be considered when using pesticides to prevent **ecological backlash** due to food-chain concentration; trophic interactions described by Lotka-Volterra equations can predict outcomes, useful in managing damage, of predation or competing species.

Explanation

Movement of energy through ecosystems begins with sunlight (most common), an energy subsidy from another ecosystem, or from chemically derived energy in a few benthic oceanic ecosystems. Energy is captured by producers using **photosynthesis** (or, rarely, chemosynthesis), which converts the sun's energy into a chemical form. Energy is then passed via food chains and webs to other consumers including herbivores, omnivores, carnivores, and detrivores. Herbivores eat plants, whereas **carnivores** eat animals. **Omnivores** consume both plants and

animals. And **detrivores**, scavengers, and decomposers eat **detritus**, the dead remains of formerly living organisms, usually a mix of materials from the body and fecal matter.

By the **first law of thermodynamics**, energy is neither created nor destroyed but can be converted from one form, such as sunlight, to another, such as the chemical energy bound in adenosine triphosphate (ATP). ATP serves as currency for chemical energy exchanges in cells of all known life. By the **second law of thermodynamics**, however, some energy is lost as heat during each transfer. Major paths of energy flow through ecosystems, including **grazing circuits** (food webs based directly on sunlight, such as prairies or savannas) and **detritus food circuits** (food webs based on consuming dead organisms).

Energy flows through organisms in communities or ecosystems from one trophic (feeding) level to the next; these collective interactions are described as food chains, food webs, and food pyramids. Constrained by the first two laws of thermodynamics, only about ten percent of energy is transferred from one **trophic level** to the next; the rest is lost as heat. **Food chain concentration** or **biomagnification** can occur for some compounds, such as organochlorine pesticides (e.g., those that are lipid soluble) or radioactive materials, usually concentrating by an order of magnitude at each level of transfer. Food-chain concentration should be considered whenever pesticides are used in wildlife damage management.

For example, evaluations for the use of Compound 1080 in the Livestock Protection Collar, used to control coyote attacks on sheep, included animals that might scavenge carcasses of sheep, such as skunks and eagles. Findings determine how a pesticide can be used. Strychnine baits may only be used to control pocket gophers in the United States. Baits must be placed in burrows to prevent hazards to nontarget animals in the food web that might access baits placed above ground. In California, for example, concerns have included the threatened California red-legged frog (*Rana aurora draytonii*), the California tiger salamander (*Ambystoma californiense*), and the federally endangered San Joaquin kit fox (*Vulpes macrotis mutica*).

Predation and **parasitism** (wherein one species benefits at the expense of another) are specific types of trophic interactions. These are described by Gause's law (see chapter 4) and Lotka-Volterra equations (box 6.1; fig. 6.4).

BOX 6.1 Lotka-Volterra Models

Lotka (1925) and Volterra (1926) independently derived useful models to describe relationships between two species. The models explain predator and prey relationships as well as competitive relationships between species and are relatively easy to understand. They can also be derived from equations that model logistic growth of populations, as below:

$$dN/dt = r_{max}N(K-N)/K \text{ (from chapter 5).}$$

The $(K-N)/K$ term is a mathematical way to model intraspecific competition, or competition within the population or species. As N approaches K, intraspecific competition leads to a geometric decline in rate of growth r. Lotka-Volterra replaced this term with one that models both intraspecific and **interspecific** (between-species) competition. In a **Lotka-Volterra model**, N_1 represents the population size of one (the first) species, and N_2 represents the population size of the other (the second) species. In analogous ways, carrying capacities are represented by K_1 and K_2 and growth rates by r_1 and r_2.

Although one might expect competition to increase as N_1 and N_2 increase, one cannot expect the competitive effects of each species to be the same. For example, species 1 might have twice the competitive effect on species 2 as species 2 has on species 1, the total competitive effect being $(N_1 + 0.5N_2)$. To account for this as a general factor, the Lotka-Volterra model adds a constant α_{12} (the value 0.5 in the preceding example) to represent the effect of species 1 on species 2 and a constant α_{21} to represent the effect of species 2 on species 1. Thus,

$$dN/dt = r_{max1}N_1(K_1 - (N_1 + \alpha_{12}N_2))/K_1$$

represents the total impact of competition on growth rate for the first species, and

$$dN/dt = r_{max2}N_2(K_2 - (N_2 + \alpha_{21}N_1))/K_2$$

represents the total impact of competition on growth rate for the second species. The two equations together make up the Lotka-Volterra model.

Now, using the model, consider the range of pressures that are exerted on the sloped line (i.e., the zero **isocline** where $dN_1/dt = 0$, fig. 6.4a). An analogous zero isocline can be calculated for species two—that is, $dN_2/dt = 0$ (fig. 6.4b). The equations can be mathematically manipulated and reduced, so that when $dN/dt = 0$,

$$N_1 = K_1 - \alpha_{12}N_2$$

and

$$N_2 = K_2 - \alpha_{21}N_1.$$

To understand what is happening conceptually, it might help to do iterative (mental or actual) calculations using the Lotka-Volterra models, substituting some hypothetical values for the constants and variables. The arrows in fig. 6.4a indicate the impact of size of one population on the other, sometimes increasing, sometimes decreasing, but always directed toward the sloping line.

If one were to combine the zero isoclines (box 6.1) of two interacting species on the same graph, one could envision four possible combinations (fig. 6.4b–e). In viewing the graphs, pay particular attention to the size of K relative to the size of K/α as they intercept each axis. In the first situation (fig. 6.4c), K_1 is greater than K_2/α_{21} along the N_1 axis, indicating that species 2 is a relatively weak competitor. K_1/α_{12}, however, is greater than K_2, indicating that species 1 is a relatively strong competitor. Following the directions of the arrows, interactions will move along the isoclines until species 1 achieves its carrying capacity, and in this case species 2 becomes extinct; species 1 wins (circled isocline intercept with N_1). In the second situation (fig. 6.4d), just the opposite occurs. Here, carrying capacity for

species 2 is greater than the competitive strength of species 1, indicating that species 1 is a relatively weak competitor. Following the arrows in this situation, species 2 achieves its carrying capacity and species 1 becomes extinct. In the situation in fig. 6.4e, both species have carrying capacities that exceed their relative competitive strengths. Here, an unstable equilibrium is achieved, with some arrows pointing toward the intercept of the isocline and some away from it. Eventually, one species will force the other to extinction, but it is not possible to predict the winner from this model. In the last situation (fig. 6.4f), the competitive strengths of both species exceed their respective carrying capacities and, as indicated by all the arrows moving toward the intercept of the isoclines, a stable equilibrium is

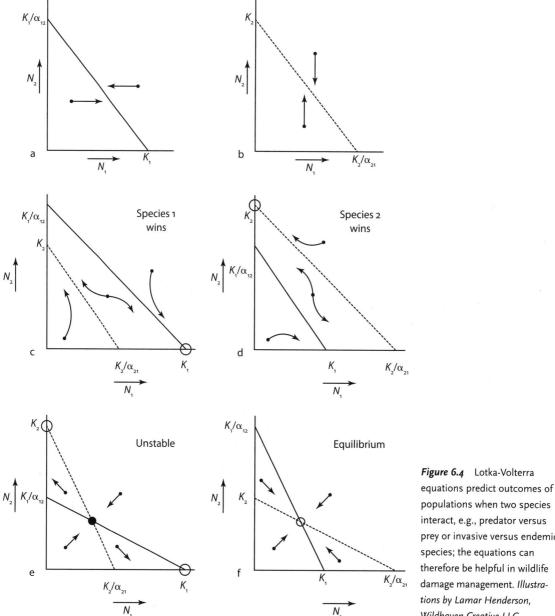

Figure 6.4 Lotka-Volterra equations predict outcomes of populations when two species interact, e.g., predator versus prey or invasive versus endemic species; the equations can therefore be helpful in wildlife damage management. *Illustrations by Lamar Henderson, Wildhaven Creative LLC.*

achieved. This is what might occur, for example, when some niche differentiation has occurred. Although the models presented here are relatively simple and only approximate actual interactions between species, more sophisticated versions are available, with better predictive capabilities.

Lotka-Volterra models can predict outcomes of the removal of predators for prey or for other predators (i.e., the mesopredator release effect). Examples are the outcomes of the removal of wolves for populations of other, lesser predators such as coyotes or skunks.

Similarly, the models can predict the outcomes of removing a species when it competes with another (i.e., the **competitor release effect**). The models have been used to help predict competition between kangaroos and livestock in Australia (Moloney and Hearne 2009), impacts of removal of snow leopards on mesopredator release in Sagarmatha (Mt. Everest) National Park in Nepal (Thakuri 2009), means of minimizing trapping strategies for management of beaver problems in North America (Bhat et al. 1993), impacts of white-tailed deer culling on management of Lyme disease in

North America (Li and DeMasi 2009), and impacts of wolf management on moose populations in Norway (Skonhoft undated).

Example

Witmer et al. (2007) reported the outcomes of a successful campaign to eradicate black rats from Buck Island Reef National Monument, U.S. Virgin Islands. The campaign occurred over two years. Rats were not found in six years of post-treatment trapping, but trappers noted a growing population of house mice (*Mus musculus*) after removal of the rats.

Caut et al. (2007) used population data and a generalized Lotka-Volterra population model to explain the irruption of mice as a competitor release effect. These scientists used one equation to model rat population dynamics and a second to model population dynamics of the mice. The researchers added a third equation to simulate prey species common to both the rats and the mice. From the equations, they found four points of equilibria: when both species, rats and mice, were eradicated; when the rat was eradicated, allowing the mouse to achieve its maximum population size as a point of equilibrium; when the rat was controlled but not eradicated, triggering competitor release; and intense control but not eradication of either, also triggering competitor release under some conditions.

The study suggests possible irruption of minor competitors during periods when both pests are being controlled. Such irruptions would occur when the control methods are beginning to impact the superior competitor but before they impact the inferior one. Consequences could be serious, in that the irruptions could lead to unintended effects on endangered flora and fauna. Prior knowledge of such releases might help to design management strategies that accommodate responses from both species.

SUCCESSION

Statement

Wildlife found in earlier successional stages tend to have characteristics of offending wildlife, whereas those at later stages are less likely to be irruptive; pest irruptions sometimes follow efforts to push back ecological succession.

Explanation

Invader species—the first species into an area—eventually alter their environment so they can no longer survive in it. These species are then replaced by others, replacement continuing in an orderly manner until a **climax** (energetically balanced) **community** emerges, a process called **succession**. Each successional community is called a seral stage, or **sere**. **Climatic climax communities** are mature seres for major biomes or regions. **Edaphic climaxes** are seres made stable for prolonged periods by recurrent factors such as fires or floods. **Primary succession** occurs when organisms enter an area previously devoid of life. **Secondary succession** occurs when an ecosystem is pushed back to an earlier sere by some factor, such as fire, herbicide use, or intensive grazing. Earlier seres are more productive than later seres but tend to have less diversity and stability.

Species characterizing early seres are sometimes called **R-growth species**. They tend to reach reproductive age early, produce large numbers of offspring but invest little in their care, and have short generation times (i.e., be short-lived). They are often opportunistic. Species characterizing mature seres are called **K-growth species**. They tend to produce fewer young but invest more in parental care, reproduce late in life, and have long generation times.

Wildlife damage problems are sometimes the consequence of pushing an ecosystem to an earlier sere, simpler and more productive but also characterized by R-growth species. People use various energy subsidies to do this, from slash-and-burn to herbicides, brush-hogs, and "chaining." People remove fencerow vegetation or consolidate smaller farms into larger ones, all cropped in a single high-yield variety to boost production. Although such systems are productive, they are also less stable than mature systems. Species may be released from the complex community interrelationships that had kept them in check, such as predation, parasitism, or abiotic limiting factors associated with nutrient cycling.

Ecological offenses of some wildlife can also push ecosystems to earlier seres. Rooting by feral pigs has this effect. Overabundant populations of herbivores such as deer, goats, or rabbits push ecosystems to earlier seres. Browsing by overabundant deer can eliminate the regenerating capability of a forest (Russell et al. 2001; Rooney and Waller 2003).

Example

In the late 1950s, a portion of the island of Cotabato, Philippines, was set aside for human settlement as part of a governmental transmigration program. The new settlers cleared large areas and began growing crops. An outbreak of ricefield rats (*Rattus argentiventer*) ensued in which the rodents destroyed much of the harvest; international assistance was required to provide food relief and prevent massive human starvation. Similar outbreaks continue to occur for analogous reasons throughout many parts of the globe.

Summary

- Some practitioners view ecosystems and landscapes as fundamental units of damage management.
- Ecotones provide habitat for offending species such as deer, coyotes, and parasitic cowbirds.
- Cities are ecosystems that support generalist wildlife such as starlings, house sparrows, commensal rodents, and feral cats and dogs. Cities require major energy subsidies that provide economic incentives to push agriculture into intense production, exacerbating wildlife damage.
- Climates impact wildlife damage. Global warming portends dramatic increases in wildlife damage.
- Landscape concepts—such as ecosystems within matrices, habitat fragmentation, and metapopulations—are useful in understanding wildlife damage and planning management strategies.
- **Community physiognomy** greatly influences patterns and distribution of damage in ecosystems.
- Species interact in ecosystems through food chains and webs; interactions of predators and competitors can be described by Lotka-Volterra models.
- Food-chain concentration should be considered before using any pesticides; Lotka-Volterra models can predict whether removal of predators or competitors will cause ecological problems such as the mesopredator release effect.
- Many wildlife damage problems are explained by efforts to push communities from mature successional stages to earlier stages, as is done with most agriculture; early communities tend to have more irruptive species and simplified interactions.

Review and Discussion Questions

1. It has been argued that diversity equals stability in communities and ecosystems. How might this notion apply to wildlife damage and its management?
2. Landscape ecology, habitat fragmentation, metapopulations, and corridors are concepts often used in conservation biology. How might they be applied in the understanding and management of wildlife damage? How might these concepts point to ties between conservation biology and wildlife damage management?
3. Choose three characteristics of seral stages, and contrast early versus late seres for each characteristic. Then, relate each characteristic and stage to likelihood of wildlife damage. Be specific.
4. Choose a specific study where Lotka-Volterra equations were applied to wildlife damage management. Assess the contribution of the models to the study. Were they helpful? Give some specific examples of other applications to wildlife damage management.
5. Contrast zinc phosphide and strychnine, both vertebrate pesticides, in terms of food-chain concentration. How do these compounds compare to organochlorine pesticides such as DDT?

PART III • SURVEYS OF DAMAGE AND DAMAGING SPECIES

In this section, we look at damage caused by wildlife both globally and regionally in North America. Included are wildlife diseases and zoonoses.

7

Exotic Invasive
Species Worldwide

We look at species that affront humans worldwide in this chapter. To achieve global notoriety, the species must have attained sufficient distribution to gain worldwide attention and then created an impact that seriously conflicts with people and their interests. The story of globally damaging wildlife is therefore largely a story of exotic invasive species.

STATEMENT

Worldwide, exotic invasive species are found among all taxa, are often generalists, and have characteristics of species at early successional seres; they are viewed by people as being among the most damaging of wildlife.

Explanation

Some invasive species, such as the European starling, have seen management actions for years. Only in recent decades, however, has the full extent of exotic invasive species come to the forefront and been added to the many responsibilities already within the purview of wildlife damage management. Successful invaders are found among all types of wildlife, not just vertebrate species. Invertebrates such as the zebra mussel (*Dreissena polymorpha*) in the United States and American crayfish (e.g., the signal crayfish *Pacifastacus leniusculus*) in Europe have come under management scrutiny, as have invasive plants such as the blackberry (*Rubus* spp.) and the mimosa (*Mimosa* spp.).

All species continually attempt to invade new areas. A species of wildlife leaving an area is an emigrant, but it is an **exotic** when it appears in the new habitat. Most invading species either fail to make it to new lands or, once there, are unable to become established. Failures occur for many reasons, but all lead to an inability to complete a necessary life cycle and spread in the new area. For example, the protist (protists are unicellular or multicellular organisms without specialized tissues) causing avian malaria was unable to colonize the Hawaiian Islands until an appropriate vector, the house mosquito (*Culex quinquefaciatus*), was introduced.

Likelihood of a successful invasion depends partly on the number of individuals that emigrate and partly on the persistence (frequency) of invasions, called **propagule pressure**. Propagule pressure, combined with factors such as pathway of invasion, the species' characteristics (for example, flight of birds facilitates invasions), and resource availability (e.g., Britton-Simmons and Abbott 2008), underwrite successful invasions (Fine 2002; Lockwood et al. 2005).

Successful invasions occur in at least three stages: introduction to the new area; establishment of one or more colonies via naturalization (adapting to or altering environments to reliably complete life cycles); and invasion of surrounding areas from the colonies (after Di Castri 1990). Invaders tend to be generalists that can survive for long periods of time with limited resources. They are often small, mobile, agile, and inconspicuous. Invaders tend to be R-selected species (see chapter 6) with high fecundity and growth rate, limited genetic variation, and short and simple life cycles. They are effective in dispersing seeds or young either themselves or with the help of environmental factors (**phoresis**) such as wind or other organisms. Though experience with a similar environment helps, the presence of fewer competitors, predators, or pathogens are probably more important to success (Di Castri 1990). Although these attributes characterize some successful invaders, there is sufficient case-by-case variation that the traits have little predictive value.

Environmental conditions can facilitate successful invasions. Open, simple, cleared areas seem particularly vulnerable, whereas more complex areas with a diverse array of organisms seem less vulnerable. For instance, areas settled by humans or having recent geological or evolutionary disturbance are more likely than undisturbed areas to allow successful invasions. Thus, a recent volcanic eruption or flood that clears land can make it accessible to invasive species. Areas where humans transport people and products for trade, colonization, or war are also vulnerable to invasion (Di Castri 1990). Rapid transit, such as use of aircraft, particularly facilitates transfer of exotic wildlife diseases because the speed of transit is sometimes faster than the incubation times of the diseases.

At least two notions have been proposed to explain vulnerability to invasive wildlife: the biotic resistance hypothesis and the biotic acceptance hypothesis. The biotic resistance hypothesis is based on the relationship between diversity and stability of an ecosystem; ecosystems with the greatest species diversity will also be the most resistant to invasions by exotic species. Conversely, ecosystems with less diversity will be more susceptible. If an ecosystem has many niches but they are already filled by diverse organisms, none are available for newcomers. This hypothesis has been used to explain isolation (i.e., some niches remain unoccupied) as a factor in vulnerability of islands, such as New Zealand, Australia, and the Hawaiian Islands, and some peninsulas to successful invasions (e.g., Fagerstone 2003).

The biotic acceptance hypothesis supposes that some ecosystems are sufficiently diverse to allow for new niches for newcomers. The new niche might be split from an existing one, or perhaps not all niches are fully used. Biotic resistance and biotic acceptance appear to be opposing hypotheses. We do not favor one over the other, but we present both because they help conceptualize how successful invasions may occur.

A third theory on successful invasions has been proposed—the human activity hypothesis. Human activity affects propagule pressure. For example, invasion of a new plant species on the Hawaiian Islands might occur naturally every 20,000 or 30,000 years. Assuming that wildlife can hitch a ride, human transport by boat and airplanes might allow repeated introductions over years, thereby increasing propagule pressure by a factor of thousands. Hence, human transport could greatly facilitate invasive species completing their first stage, moving to the new area.

Once there, human disturbance of habitat might help by simplifying ecosystems and making them susceptible to invasion. Economic indicators, such as real estate sales, and demographic factors, such as percentage of urban area or human population density (e.g., Taylor and Irwin 2004; Pyšek and Richardson 2006), have been used to index human activity. Studies (e.g., Leprieur et al. 2008) have shown that indicators of human activity relate directly to abundance of invasive species. We favor this third hypothesis.

Most successful invasions involve exotics of little interest to humans. Some contribute to overall diversity at local scales and may be otherwise beneficial. A relatively few are damaging and are termed exotic invasive species.

Table 7.1 lists 100 of the world's worst exotic invasive species, as proposed and presented by the Global Invasive Species Database (GISD 2010) of the International Union for the Conservation of Nature (IUCN). The listing does not rank species. Rather, it chooses species that represent a balance of taxa causing problems, limiting any genus to one example, even though there may be other species within the genus that also cause problems for humans. For instance, the western

TABLE 7.1   *One hundred of the world's worst invasive alien species*

MICROORGANISMS

avian malaria	*Plasmodium relictum*
banana bunchy top virus	*Banana bunchy top virus*
rinderpest virus	*Rinderpest virus*

MACRO-FUNGI

chestnut blight	*Cryphonectria parasitica*
crayfish plague	*Aphanomyces astaci*
Dutch elm disease	*Ophiostoma ulmi*
frog chytrid fungus	*Batrachochytrium dendrobatidis*
phytophthora root rot	*Phytophthora cinnamomi*

AQUATIC PLANTS

caulerpa seaweed	*Caulerpa taxifolia*
common cord-grass	*Spartina anglica*
wakame seaweed	*Undaria pinnatifida*
water hyacinth	*Eichhornia crassipes*

LAND PLANTS

African tulip tree	*Spathodea campanulata*
black wattle	*Acacia mearnsii*
Brazilian pepper tree	*Schinus terebinthifolius*
cogon grass	*Imperata cylindrica*
cluster pine	*Pinus pinaster*
erect pricklypear	*Opuntia stricta*
fire tree	*Myrica faya*
giant reed	*Arundo donax*
gorse	*Ulex europaeus*
hiptage	*Hiptage benghalensis*
Japanese knotweed	*Fallopia japonica*
Kahili ginger	*Hedychium gardnerianum*
Koster's curse	*Clidemia hirta*
kudzu	*Pueraria montana var. lobata*
lantana	*Lantana camara*
leafy spurge	*Euphorbia esula*
leucaena	*Leucaena leucocephala*
melaleuca	*Melaleuca quinquenervia*
mesquite	*Prosopis glandulosa*
miconia	*Miconia calvescens*
mile-a-minute weed	*Mikania micrantha*
mimosa	*Mimosa pigra*
privet	*Ligustrum robustum*
pumpwood	*Cecropia peltata*
purple loosestrife	*Lythrum salicaria*
quinine tree	*Cinchona pubescens*
shoebutton ardisia	*Ardisia elliptica*
Siam weed	*Chromolaena odorata*
strawberry guava	*Psidium cattleianum*
tamarisk	*Tamarix ramosissima*
wedelia	*Sphagneticola trilobata*
yellow Himalayan rasberry	*Rubus ellipticus*

AQUATIC INVERTEBRATES

Chinese mitten crab	*Eriocheir sinensis*
comb jelly	*Mnemiopsis leidyi*
fish hook flea	*Cercopagis pengoi*
golden apple snail	*Pomacea canaliculata*
green crab	*Carcinus maenas*
marine clam	*Potamocorbula amurensis*
Mediterranean mussel	*Mytilus galloprovincialis*
Northern Pacific seastar	*Asterias amurensis*
zebra mussel	*Dreissena polymorpha*

LAND INVERTEBRATES

Argentine ant	*Linepithema humile*
Asian longhorned beetle	*Anoplophora glabripennis*
Asian tiger mosquito	*Aedes albopictus*
big-headed ant	*Pheidole megacephala*
common malaria mosquito	*Anopheles quadrimaculatus*
common wasp	*Vespula vulgaris*
crazy ant	*Anoplolepis gracilipes*
cypress aphid	*Cinara cupressi*
flatworm	*Platydemus manokwari*
Formosan subterranean termite	*Coptotermes formosanus shiraki*
giant African snail	*Achatina fulica*
gypsy moth	*Lymantria dispar*
khapra beetle	*Trogoderma granarium*
little fire ant	*Wasmannia auropunctata*
red imported fire ant	*Solenopsis inicta*
rosy wolf snail	*Euglandina rosea*
sweet potato whitefly	*Bemisia tabaci*

AMPHIBIANS

bullfrog	*Rana catesbeiana*
cane toad	*Bufo marinus*
Caribbean tree frog	*Eleutherodactylus coqui*

FISHES

brown trout	*Salmo trutta*
carp	*Cyprinus carpio*
large-mouth bass	*Micropterus salmoides*
Mozambique tilapia	*Oreochromis mossambicus*
Nile perch	*Lates niloticus*
rainbow trout	*Oncorhynchus mykiss*
walking catfish	*Clarias batrachus*
western mosquito fish	*Gambusia affinis*

BIRDS

Indian myna bird	*Acridotheres tristis*
red-vented bulbul	*Pycnonotus cafer*
starling	*Sturnus vulgaris*

REPTILES

brown tree snake	*Boiga irregularis*
red-eared slider	*Trachemys scripta*

MAMMALS

brushtail possum	*Trichosurus vulpecula*
domestic cat	*Felis catus*
goat	*Capra hircus*
grey squirrel	*Sciurus carolinensis*
macaque monkey	*Macaca fascicularis*
mouse	*Mus musculus*
nutria	*Myocastor coypus*
pig	*Sus scrofa*
rabbit	*Oryctolagus cuniculus*
red deer	*Cerus elaphus*
red fox	*Vulpes vulpes*
ship rat	*Rattus rattus*
small Indian mongoose	*Herpestes javanicus*
stoat	*Mustela erminea*

Source: After Lowe et al. 2000, with permission.

mosquito fish (*Gambusia affinis*) is included on the list, while the eastern mosquito fish (*Gambusia holbrooki*), similar in population size and distribution and also a concern to people, is not. The list guided our choice of examples.

Examples

We start with plants, progress to the smallest animals, and move upward in size and complexity.

PLANTS. Many successful plant invasions are probably the product of human-facilitated transport (e.g., Pauchard and Shea 2006) and characteristics of the plant. Baker (1965) saw an "ideal" plant invader as a plastic perennial that germinates under most conditions, grows fast, flowers early, is self-pollinating, produces an abundance of seeds that disperse widely, reproduces vegetatively as well, and competes effectively. He thought that only some of these characteristics were needed to be successful. Plants invading natural areas tend to be aquatic or semiaquatic, nitrogen-fixing legumes, grasses, climbers, or clonal trees, mostly different (about 25% overlap) from species invading agroecosystems (Daehler 1998).

Water hyacinth (*Eichhornia crassipes*), originally from South America, is now found throughout the globe. It grows and spreads quickly, choking waterways and blocking the penetration of sunlight into water, thereby reducing the diversity of aquatic ecosystems (Lowe et al. 2000). Water hyacinth can achieve sufficient density and coverage to disrupt waterflow and production of hydroelectric power, as occurred at the Owen Falls Dam facility in Uganda in the late 1980s (fig. 7.1; Kateregga and Sterner 2007). Concerns now extend to China, where the plant is already in the watershed of the Three Gorges Dam (Ding et al. 2008). Commercial sale and trade (Padilla and Williams 2004) of water hyacinth and other aquatic invasive weeds may facilitate their spread. Kay and Hoyle (2001) listed 27 such species that are available commercially through the Internet.

Miconia, a South American tropical plant with huge red and purple leaves, is sold as an ornamental throughout the tropics. Fruit-eating birds helped it become established in the wild in Tahiti beginning around 1937. It has now invaded over half of the island and is a dominant canopy tree in many areas. Several endemic species are either threatened or extinct because of it. Miconia is now on islands in the Pacific and was introduced as an ornamental into Hawaii in the 1960s (Lowe et al. 2000).

INSECTS, ARTHROPODS, AMPHIBIANS, AND REPTILES. Exotic invasive arthropods, amphibians, and reptiles

Figure 7.1 A ferry wading through water hyacinth (*Eichhornia crassipes*) in the Alppuzha Canal, India. *Photo by P. K. Niyogi.*

tend to be omnivores, have high reproductive rates, be capable of high population numbers and densities, and be hard to detect because of their behavior or small size (Pitt et al. 2005). Some, such as the cane toad, are poisonous to people, domesticated animals, and other wildlife. Others, such as the Caribbean tree frog (*Eleutherodactylus coqui*), make loud, incessant noises, reducing the value of real estate. In sufficient numbers, released exotic arthropods, amphibians, and reptiles can outcompete their native counterparts, dramatically altering both aquatic and terrestrial ecosystems.

Humans often provide the initial transportation within or between land masses for invasive arthropods, amphibians, and reptiles, sometimes intentionally and sometimes not. Thus, the bullfrog (*Rana catesbeiana*) was transported throughout the globe for its value as a food and for fishing. Introductions of Burmese pythons (*Python molurus bivittatus*) and cobra species into the Florida Everglades probably followed their transport into the United States as pets. The release of the brown tree snake in Guam was accidental, but again, humans probably provided the transportation.

The crazy ant (*Anoplolepis gracilipes*) has caused extensive damage to ecosystems on islands (McGlynn 1999). The ants form colonies in the canopies of tropical forests, allowing multiple colonies with multiple queens. Supercolonies with 300 queens have been observed, e.g., on the Christmas Islands, with infestations covering about 28% of its 10,000 hectares of rainforest (Abbott 2005). Crazy ants are omnivores with both predatory and scavenging habits. They eat grains, seeds, and detritus. They "farm" scale insects and aphids by both overseeing crawlers and protecting them from predators. Scale and aphids can consequently also invade some island forests, resulting in mold growth, canopy dieback, and death of canopy trees.

Crazy ants have destroyed red land crab (*Geocarcoidea natalis*) populations on the Christmas Islands. The ants spray the crabs with lethal doses of formic acid as their paths cross. Crab carcasses then become a high-protein food for the ants (O'Dowd et al. 2003). Fifteen to twenty million crabs have been killed since 1989, eliminating the crab as a keystone species on parts of the islands. In those areas, litter cover has doubled, as has seedling species richness. Changes in seedling densities and canopy holes have cascaded through the food web, reducing the presence of some endemic species. Similar impacts have been observed on other islands. Presence of the ant also favors introduction of exotic rats and cats (Advice undated).

Cane toads were introduced from Central America into sugarcane-growing parts of the world to control beetles. The beetles avoided the toads by climbing to higher parts of the plants; the toads then preyed on other wildlife. From egg to adult, cane toads contain toxic bufadienolides, so predatory wildlife and pets, and occasionally people, are poisoned. Cane toads eat some threatened species, outcompete other native frogs for breeding sites, and transmit diseases such as *Salmonella* (Shanmuganathan et al. 2010).

About 100 cane toads were introduced into Australia in 1935 to control sugarcane beetles. Today the cane toads are found throughout Australia's tropics and subtropics and have reached western Australia. Models of global warming predict a further extension southward. Following the toad's recent arrival at the Kakadu National Park, native predators, including the quolls (*Dasyurus* spp.) and large goannas (monitor lizards), declined markedly.

Bullfrogs are endemic to the eastern United States but were introduced into other states and at least 40 countries over the last century (Lever 2003), including parts of Asia, Europe, and Central and South America. Populations occur in South America, including Argentina, Brazil, Costa Rica, Ecuador, Uruguay, and Venezuela. In Europe, populations exist in Belgium, France, Germany, Greece, and Italy (Ficetola et al. 2007). Tadpoles of bullfrogs are large and outcompete larvae of native species, and adults prey broadly on native species, including other amphibia (Kats and Ferrer 2003). Blaustein and Kiesecker (2002) found that the presence of the tadpoles can affect how native tadpoles use their environment, making them more vulnerable to predation by fish. Bullfrogs also carry *Batrachochytrium dendrobatidis* and frog chytrid fungus (see chapter 9) and are partly responsible for distributing this emerging infectious disease.

Established wild populations of the red-eared slider turtle (*Trachemys scripta elegans*), native to the Mississippi Valley region of the United States, are now found in Australia, France, Italy, Spain, Japan, and Taiwan (Ficetola et al. 2009). These turtles are shipped worldwide as popular reptilian pets (Connor 1992). Sales peaked in the late 1980s and early 1990s, along with the popularity of the "Teenage Mutant Ninja Turtle" television series and movies. Since 1975, the sale of red-eared slider turtle eggs and turtles smaller than four inches (small enough for a child to put in his/her mouth) as pets has been prohibited in the United States, because the turtles can transmit *Salmonella* bacteria (e.g., Nagano et al. 2006). Sliders are still raised for sale in other countries, both as pets and for other purposes (Williams 1999).

Docile hatchlings and young turtles become aggressive adults, growing up to 13 inches long and living for over 30 years. After becoming unmanageable, the pets are discarded in local ponds and waterways by former turtle enthusiasts. Discarded turtles can outcompete local wildlife for basking sites (fig. 7.2). For example, increased numbers of turtles coincided with decreased numbers of birds in ponds of several parks in London, UK, after the "Ninja Turtle" craze subsided. These turtles also consume local plants and algae, invertebrates, fish, frog eggs and tadpoles, and aquatic snakes. Cadi and Joly (2004) described impacts on European pond turtles (*Emys orbicularis galloitalica*). In addition, the turtles can hybridize with others such as the indigenous yellow-bellied slider (*T. s. scripta*) in Florida or the Big Bend slider (*T. gaigeae*) in New Mexico (Stuart 2000).

FISH. Most exotic fish are introduced as food sources, for sport angling or commercial fishing, or for aquaria or ornamental ponds. Aquaculture is a growing industry worldwide and also contributes to introductions of exotic fish and diseases into native ecosystems. Predatory fish such as bass and trout compete with native species. Carp, introduced globally for sport and food, muddy water and otherwise degrade aquatic habitat. Diseases carried by introduced fish can broadly affect native species.

Leprieur et al. (2008) found six major drainage basins worldwide where exotic fish represented more than a quarter of the total number of species per basin: the Pacific coast of North and Central America, southern South America, western and southern Europe, Central Eurasia, South Africa and Madagascar, and

Figure 7.2 Red-eared sliders (*Trachemys scripta*) competing with a mallard (*Anas platyrhynchos*) for a basking site. *Photo by Mbz1.*

southern Australia and New Zealand. They note that these areas, based on the World Conservation Union Red List, also have the highest proportions of fish species facing extinction. The Northern Hemisphere has the greatest number of exotic fish. Leprieur et al. found significant correlations between presence of exotic fish and human activity but not with either biotic acceptance or biotic resistance. Among human factors, gross domestic product of river basins best predicted the presence and impacts of exotics.

Common carp have achieved global distribution, partly as a source of protein, because they are inexpensive, are resistant to handling stress, and can survive in waters with low oxygen concentrations (Arlinghaus and Mehner 2003). They have also been distributed worldwide as ornamentals (e.g., as nishikigoi or koi), often being put in ponds or local waterways, and sometimes also used as baitfish (Nico 1999; Aguirre and Poss 2000). In addition, carp are introduced to clear waterways of aquatic weeds. Once in a waterway, common carp are capable swimmers that can negotiate turbid waters and leap over obstacles a yard high (Koehn 2004). The carp then move into unintended areas, becoming effective invasive species.

Carp destroy vegetation and increase water turbidity by rooting and dislodging plants. They may uproot or consume aquatic macrophytes or prevent light from reaching them by increasing turbidity. Carp prey on eggs of other fish and may have been responsible for the decline of the razorback sucker (*Xyrauchen texanus*) in the Colorado River basin (Taylor et al. 1984).

Brown trout (*Salmo trutta*) are native to Europe and Asia, but pure native populations probably now exist only in a few remote places, such as Corsica. They were introduced for angling worldwide, including North and South America, Africa, Asia, Australia,

and New Zealand. First introductions in the United States were into the Pere Marquette (now Baldwin) River, Michigan, in 1883 (Courtenay et al. 1984). Because of heavy angling and low or nonexistent natural reproduction, they are propagated in fisheries and restocked in streams and ponds annually in the United States.

These trout are aggressive predators that strongly impact other fish. For example, they displaced native adult brook trout (*Salvelinus fontinalis*) throughout the northeastern United States (Fausch and White 1981), replaced cutthroat trout (*Oncorhynchus clarki henshawi*) in some large rivers (Behnke 1992), and contributed to the decline of golden trout (*O. aguabonita*) in the Kern River and Lahontan cutthroat trout (*O. clarki henshawi*) in Lake Tahoe (McAfee 1966). In upland waters of Australia and New Zealand, brown trout prey on indigenous endangered or vulnerable fish (Wager and Jackson 1993) or exclude them competitively so they become fragmented populations. Brown trout may have been involved in extirpation of an endemic New Zealand grayling (*Prototroctes oxyrhynchus*; McDowall 1990) in New Zealand. In Japan, introduced brown trout have displaced native white-spotted char (*Salvelinus leucomaenis*) in a part of the Ishikari River in Hokkaido (Takami et al. 2002) and are aggressive piscivores in other lakes where they have been introduced.

Mozambique tilapia (*Oreochromis mossambicus*) is native to southern Africa. It has become popular for mosquito control and is relatively easy to raise; it grows uniformly, quickly, and survives under a broad range of environmental conditions. This tilapia has therefore been introduced broadly, on purpose or accidentally from aquaculture facilities, so that it is now found in many subtropical and tropical habitats throughout the world. Once released, it competes indirectly with native fish for food and nesting habitat and directly by preying on smaller fish. It reduces plant growth because of a rooting behavior that increases water turbidity. The striped mullet (*Mugil cephalus*) is threatened in Hawaii because of competition with this tilapia, as is the desert pupfish (*Cyprinodon macularius*) in California's Salton Sea. The fish has caused havoc in Australian streams since its introduction for mosquito and weed control in the 1970s. Tilapia were found in El Junco, a volcanic lagoon in a volcanic mountain in the Galápagos, in 2006. The population of tilapia was expanding and consuming a copepod important for eating algae and preventing blooms in the lagoon. Rotenone, a piscicide, was used to eradicate the fish and allow the lagoon to restore itself (Carrion 2009). Ironically, the Mozambique tilapia may itself become threatened in its native range, where it is being harmed by hybridization with an introduced fish, the Nile tilapia (*O. niloticus*).

The Nile perch (*Lates niloticus*) is native to the Nile River system, Lake Mariout, and some West African river systems and occurs in the Zaire (Congo) River system and Lakes Albert and Turkana. It was introduced into Lake Victoria in 1954 and, after several decades, exploded in biomass as an apex predator. In the process, about 200 species of smaller fish, including many endemic and unique haplochromine cichlids, were either greatly repressed in numbers or disappeared (Kaufman 1992; Witte et al. 1992). The perch became the basis of a major export industry to Europe, United States, Australia, and New Zealand as well as a local food source. Because its flesh is more oily than other local species, this fish requires a long smoking time, increasing demand for firewood and consequent effects on coastal forests.

An aggressive fish, the Nile perch can grow over six feet in length and weigh over 400 pounds. Its presence in Lake Victoria has supported not only a commercial but also a growing tourist sport fisheries industry, so ecological impacts are considered in relation to economic benefits by governments of countries that adjoin the lake (Ogutu-Ohwayo 2004). This perch has been introduced into lakes in countries around the world. Attempted introductions into reservoirs in Texas in the 1970s and 1980s are believed to have been unsuccessful (Howells and Garrett 1992).

The mosquito fish (*Gambusia affinis*) is a small fish native to the eastern and southern United States. It was introduced into many waterways during the early 1900s as a way to control mosquitoes and is now believed to be the most widely spread freshwater fish in the world. An opportunistic omnivore, it eats mosquito larva but also a broad range of algae, crustaceans, insects, and amphibians, including the larvae and eggs of indigenous species. The mosquito fish influences trophic structure, including rare fish and invertebrates. Impacts on invertebrates may be a particular concern when this fish is introduced for mosquito control into normally fishless waters. Mosquito fish are probably no more efficient at eating mosquito larva than are local predators (Haas et al. 2003).

BIRDS. Birds are among the most common natural invaders of new areas. They are often found on islands because their ability to fly facilitates moving within and between land masses. Even small birds such as the arctic tern (*Sterna paradisaea*) or hummingbirds (*Trochilidae*) routinely fly great distances as part of their migratory life cycles. Occasionally storms and winds move birds away from their accustomed migratory paths and into new territories where they might succeed as invasive species.

People have also transported birds. Some were taken to newly colonized lands as reminders of homelands. Others were imported to solve problems, e.g., to control insects that damage crops. Many were imported because they are exotic, aesthetically pleasing, and serve as pets. Once in the new land, flight can help birds find areas suitable for colonization and for further invasions.

As with other globally exotic invasives, successful bird species tend to be generalists with broad diets that are able to adapt to many types of environments, including those disturbed or inhabited by humans. Some are aggressive, outcompeting local species for food, nesting space, and cover; some can have several broods in good seasons and make large investments in parental care. Three bird species are listed among the world's 100 worst invasive pests: the European starling, the red-vented bulbul (*Pycnonotus cafer*), and the Indian myna bird (see table 7.1).

The European starling originated in Europe, Asia, and North Africa. Starlings, often introduced for aesthetic reasons, are now distributed globally except in the neotropics. These birds were introduced into New Zealand partly to control agricultural insects, and they subsequently became pests in both New Zealand and Australia. They were introduced in the United States as a bird referenced by Shakespeare and have since spread throughout the country and into Canada and Mexico.

Starlings congregate in large flocks and can cause economic damage to crops, including fruit and grain.

These birds, particularly females during the reproductive season, feed on a broad range of insects, including some that attack corn and other crops. Although beneficial in this context, starlings are generally seen as more damaging than helpful in most cropland settings. Starlings eat some ground invertebrates, and they have the evolutionary advantage over frugivorous birds of being able to probe with an open bill (GISD 2010). At livestock feeding lots, starlings compete for feed, selectively taking the higher-protein components and contaminating feed with droppings. Because they often travel between farms and mingle with livestock, starlings can transfer diseases between farms (Gough and Beyer 1981).

This species competes aggressively for nesting cavities, either natural or those constructed by other species. Ingold (1994, 1998) found that starlings usurped nest-site cavities to the detriment of northern flickers (*Colaptes auratus*) and red-bellied woodpeckers (*Melanerpes carolinus*). Koenig (2003), however, found no overall impact on native cavity nesters in North America (except possibly with some sapsuckers).

Starlings damage property with nest construction and are messy defecators, defacing properties where they nest. Their droppings can serve as a medium for the fungus histoplasmosis (*Histoplasma capsulatum*). Because starlings often forage in flocks near airports, they are a concern for aircraft (Dolbeer et al. 2000).

The red-vented bulbul originated on the South Indian subcontinent but has become established throughout the Pacific region, including Fiji, Samoa, Tonga, the Hawaiian Islands, and New Zealand, as well as in Dubai, United Arab Emirates. Bulbuls inhabit dry scrub, plains, and cultivated lands in their native habitat. These birds prefer dry lowlands when they are introduced, becoming an agricultural pest by damaging fruit, flowers, beans, tomatoes, peas, and ripening soft fruit, such as bananas. Bulbuls may facilitate dispersal of seeds of some invasive plant species; for example, in Tahiti, bulbuls disperse seeds of the invasive tree miconia (*Miconia calvescens*; Meyer and Florence 1996).

The common or Indian myna is native to India. It was introduced throughout the tropical world to control agricultural insect pests. The myna is now found in Southeast Asia, Africa, and many of the Pacific Islands, including Australia and New Zealand. As with starlings, mynas compete with native species for nest sites in cavities. The birds also destroy young chicks and eggs and evict small mammals (e.g., Tindall et al. 2007). They sometimes "mob" other birds or mammals, for example, native possums in Austra-

lia (Tidemann 2005). The mynas also damage grape and other fruit crops including apricots, apples, pears, strawberries, and gooseberries. Their droppings can carry diseases such as histoplasmosis.

MAMMALS. Globally exotic invasive mammals seem particularly tied to human activity. These mammals can be grouped into four categories. First are the commensal mammals, including the ubiquitous house mouse and Norway rat, but also four other rat species that have traveled with humans. Second are the feral mammals, released from being agricultural animals, livestock, or pets. Third, we have the game mammals, introduced into new lands because they offered sport. Finally, there are the species introduced for other reasons, such as pest control, research, medicine, pets, or aesthetics. For example, rhesus monkeys have been grown and used for research but have become a problem on some islands where they reside.

Once established, exotic invasive mammals can inflict serious damage on humans, their activities, their livestock, and their pets. They can damage both agroecosystems and natural ones, sometimes destroying ecosystem functions and endangering other wildlife. Possibly because of their long ties with humans, they can also be among the most difficult to manage.

The Norway rat probably originated in China but is now found worldwide. This rat is the largest of the three common commensals and is fossorial (burrowing). It lives with humans, thriving in urban areas. Strains have been domesticated for research and as pets. The Norway rat is euryphagic and can adapt to local resources. For example, it eats fish when living near fisheries and can dive for molluscs in the Po River of Italy. It eats seeds, seedlings, and agricultural crops and damages property. These rats can eat sufficient seeds and seedlings at times to reduce ranges of desirable plants. They consume and contaminate stored foods and spread diseases such as trichinosis, rat bite fever, viral hemorrhagic fever, and hantavirus. Norway rats caused or contributed to the extinction or reduced ranges of many birds, mammals, reptiles, and invertebrates through predation and competition, particularly on islands.

Next to humans, the house mouse is probably the world's most widely distributed mammal. Originally from Eurasia and northern Africa, this mouse has traveled with humans for at least 8,000 years (Cucci et al. 2005). It has been a pet as well as commensal for at least 3,000 years (Royer undated). This commensal mouse damages crops and consumes or contaminates stored foods. The house mouse has contributed to the decline of some species, including some albatrosses (*Diomedeidae*) and petrels (e.g., Cuthbert and Hilton 2004). This mouse carries parasites and diseases such as bubonic plague, *Salmonella,* and tularemia but often does not transmit the diseases to humans (see chapter 9).

Another invasive mammal, red deer are among the largest deer species. Native to southwestern Asia, Europe, and northern Africa, red deer were introduced into Argentina and Chile, Australia, and New Zealand for hunting and food. These deer impact native flora and fauna such as those found in the national parks in Argentina. Red deer compete with native ungulates such as guanaco (*Lama guanicoe*) and Patagonian huemul (*Hippocamelus bisulcus*) in Chile and Argentina (Flueck et al. 2003). This deer species also competes with livestock for forage (Lowe et al. 2000).

Pigs (*Sus scrofa*; fig. 7.3) are native to Europe and Asia as far south and east as the Malaysian Peninsula, Sumatra, and Java. They have traveled along with humans as a domesticated species and have

reached virtually all regions of the world, including many islands. In much of the world, pigs have also been released and have established themselves in feral populations. Feral pigs root, form wallows, compete with native wildlife for food, damage streams, prey on ground-nesting native wildlife, alter seed banks by seed predation, and change soil temperature and leaching characteristics; thus they have major impacts on ecosystems by reducing overall vegetation. Rooting by feral pigs, for example, has slowed oak regeneration in parts of eastern North America. Digging and rooting can sometimes be sufficiently intense to move an ecosystem to an earlier sere, sometimes exacerbating exotic plant invasions (Mungall 2001). Diminished numbers or even extinction of native species is sometimes a consequence.

In Hawaii the introduction of feral pigs remained unremarkable until earthworms (*Oligochaeta*) were also released and became feral. The earthworms became a staple food for the feral pigs, allowing an

Figure 7.3   Feral pigs in a park in Germany. *Photo by 4028mdk09.*

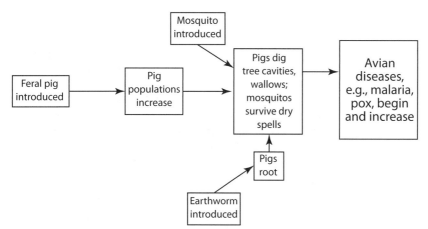

Figure 7.4 Introduction of feral pigs, mosquitoes, and earthworms worked together to help spread avian diseases among Hawaiian birds, severely impacting some, such as honeycreepers. *Illustrations by Lamar Henderson, Wildhaven Creative LLC.*

increase of populations and their ensuing damage. The pigs formed wallows that serve as microhabitat for insect vectors of diseases such as avian malaria and avian pox, indirectly contributing to the decline or extinction of native Hawaiian birds (fig. 7.4).

Feral pigs are aggressive euryphagic omnivores, eating young land tortoises, sea turtles, and sea birds, thereby contributing directly to the demise of these populations on both continents and islands (see chapter 4). For example, feral pigs were believed to play a major role in wildlife extinctions on the Galápagos Islands (Loope et al. 1988). Concerns were sufficient that intensive efforts were continued over a period of 30 years until feral pigs were eradicated from Santiago Island, one of the protected Galápagos islands (Cruz et al. 2005).

In addition to their impacts on wildlife and natural ecosystems, feral pigs can negatively affect agricultural crops, timber, and pastures. Crops damaged in the United States include hay, small grains, corn, and peanuts as well as some vegetable crops, watermelons, soybeans, cotton, tree fruits, and conifer seedlings (West et al. 2009). Feral pigs directly damage infrastructure, including fences, roads, dikes, and irrigation canals. Equipment can be damaged and operators injured from holes made by the feral pigs. These pigs prey on young livestock and can carry diseases such as brucellosis, pseudorabies, leptospirosis, and foot-and-mouth disease.

Another worldwide invasive mammal, the long-tailed macaque (*Macaca fascicularis*) comes from Southeast Asia but has been introduced into parts of Indonesia, Mauritius, Palau, and Hong Kong. Macaques have few enemies in the new areas and are considered invasive or potentially invasive. This mammal does well in disturbed habitats and competes with local avifauna

for fruit and seeds. Macaques disperse seeds of exotic plants and damage agricultural crops in areas where they have become established. This monkey may carry Ebola virus and monkeypox, and the plasmodium causing malaria in the macaque can also infect humans.

Summary

- All species continually attempt to invade new areas, but most attempts fail. The intensity of the effort at invasion can be measured as propagule pressure.
- Successful invasions occur in three stages: movement to the new area, establishment of colonies, and invasion of surrounding areas from the colonies.
- Likelihood of success may be facilitated by at least three factors: biotic resistance, biotic acceptance, and human activity. Human activity can be measured using economic indicators such as real-estate sales or demographic factors such as area of human habitation and human population density; it is probably a major contributor to success of invasions.
- Most successful invasions are of little concern to humans, and some benefit them.
- A relatively few species are both successful and damaging in the eyes of humans. These species are termed exotic invasive species.
- Exotic invasive species are found among all taxa and are often generalists; many have characteristics of species in early successional seres. They are viewed by humans as among the most damaging of wildlife.
- Examples of some of the most damaging exotic invasive species are provided in table 7.1.

Review and Discussion Questions

1. Review the list of 100 of the world's worst exotic invasive species. Can you find one that has only negative characteristics? If this is the list of "worst" invasive species, can we assume that all species have at least some positive characteristics in the eyes of humans?

2. Why do you think that, at the global scale, the story of damaging wildlife is really the story of exotic invasive species?

3. Why might an island or peninsula be more susceptible than a continent to successful invasion by an exotic species? Cite some literature to support your view.

4. From the list of 100 worst exotic invasive species, choose the 10 worst and rank them in severity as pests. What criteria did you use for ranking? How evenly were these distributed among taxa?

5. Choose a species not discussed in chapter 7 but nonetheless an exotic invasive species. Find the species in a database such as that provided by the Invasive Species Specialist Group. Summarize the information you find, including common and scientific name, origin, present distribution, and problems associated with it.

8

Damaging Species of North America

Here we highlight wildlife in North America (Central America, Mexico, Canada, and the United States) that persist in causing damage to humans and their interests. For variety, we avoid duplication of species listed in chapter 7, while recognizing that many of them occur in North America.

Statement

All categories of wildlife damage occur in North America, caused by endemic as well as locally and exotic invasive species. Damaging species include plants and invertebrate and vertebrate animals such as reptiles, birds, and mammals.

Explanation

The Center for Invasive Species and Ecosystem Health (CISEH undated) lists over 1,500 plants, 187 pathogens, 472 insects, 31 arachnids, 29 crustaceans, 67 molluscs, 94 fishes, 7 amphibians, 82 reptiles, 92 birds, and 32 mammals as exotic in North America. Many cause damage and are therefore exotic invasives. For example, the Russian olive (*Elaeagnus augustifolia*), originally from Asia, was introduced into North America in the late 1800s. It is now found in both the United States and Canada. This plant has fragrant flowers, edible fruit, and attractive foliage that made it a desirable ornamental. Unfortunately, it escaped cultivation and is now found in natural riparian habitats. Its seeds live a long time, and the plant crowds out native vegetation, making it undesirable in the wild.

Local species can also be damaging and invasive. The osage orange tree (*Maclura pomifera*) is an example. This tree had a historical range mostly within the Red River drainage of Arkansas, Texas, and Oklahoma. They were planted for windbreaks under the "Great Plains Shelterbelt" Works Project and, having sharp thorns, were also planted along hedgerows to deter cattle. The trees became widely distributed throughout the central plains states and have since moved beyond the plains into other states, state parks, and other protected areas. Osage orange trees shade areas beneath them, exacerbating erosion. The osage orange is now reported as invasive in at least 13 states, from Washington to North Carolina.

The coyote is also locally invasive. For example, this mammal is listed as invasive in Florida, where it has occurred by human introduction and range expansion since at least 1925. Native to western North America, the coyote replaced the wolf, gradually expanding its range to include most of Canada, the United States, and Mexico. The coyote has also moved from natural areas into suburbs and cities (Morey et al. 2007).

Endemic species can also be problematic. For example, Hygnstrom et al. (1994), in their manual for wildlife damage practitioners (see chapter 3), list many endemic species, such as crayfish, alligators, nonpoisonous snakes and rattlesnakes, eagles, gulls, waterfowl, woodpeckers, beaver, mountain beaver, muskrat, porcupines, prairie dogs, foxes, raccoons, skunk, bats, and deer.

One hundred of the most serious and persistent damaging species for North America are included in table 8.1. In creating this list, we aimed for a balanced approach that included representative taxa rather than a comprehensive listing or ranking.

Examples

We start with plants, continue to the smallest animals, and then move upward in size and complexity to mammals.

PLANTS. Most damaging plant species in North America are exotic invasives. Thousands of plants have been imported and sold as ornamentals, used in agriculture, or brought for purposes such as marsh draining. Many find their way from backyards or agricultural fields into natural ecosystems. Plants can cause ecological and economic problems. Ecological impacts include alteration of fire regimens (e.g., cheatgrass, *Bromus tectorum*; Young and Clements 2009), nutrient cycling (smooth cordgrass, *Spartina alterniflora*; e.g., Osgood and Zieman 1998), and water dynamics (tamarisk, saltcedar, *Tamarix*; Mack et al. 2000). **Genetic pollution** (mixing genes and thereby causing hybridization or introgression, affecting survivability of a species) can also impact plants, sometimes endangered ones. Economic costs can be astronomical. For example, invasive weeds in pastures alone cost almost $1 billion annually (Pimentel et al. 2005).

Purple loosestrife (*Lythrum salicaria*) is an herbaceous perennial that probably arrived in North America by the early 1800s in ballast water of ships from Europe, Asia, or Africa. By the late 1800s it had spread throughout the northeastern United States and south-

eastern Canada. It invaded estuaries throughout this region, becoming a prominent problem in the 1930s when it took over floodplain pastures along the St. Lawrence River. This plant crowds out native vegetation along riverbanks and waterways, and it may replace monotypic stands of cattails with an exotic monospecific community that has limited value to wildlife (Thompson et al. 1987).

Paperbark tree (*Melaleuca quiniquenerva*) originated in Australia, where aboriginal people used it for its antibacterial and antifungal properties. It was first offered for sale as an ornamental tree in Florida in 1887. This plant was also used to help dry up shallow water basins. It grows aggressively, crowds out other vegetation, and catches fire readily. This species now occupies over 202,000 hectares of flatwoods, marshes, and cypress swamps in southern Florida (Turner et al. 1998). Biological and physical control efforts are underway to contain the damage. Physical methods include herbicides, flooding, prescribed burning, and mechanical or hand removal of plants. Leaf weevils (*Oxyops vitiosa*), released for biological control in Florida during the spring of 1997, are now well established (Center et al. 2000). The melaleuca psyllid (*Boreioglycaspis melaleucae*) was released in the spring of 2002, but its effectiveness for biological control is still unknown (Serbesoff-King 2003).

Kudzu (*Pueraria montana*) is a pea-family vine that came to North America from Japan in 1876 for the Philadelphia Centennial Exposition. It was planted in the eastern United States for control of erosion during the depression years (Bailey 1939). Nicknamed "the vine that ate the South," it quickly spread over millions of acres of farmlands and forests, covering and smothering native vegetation wherever it gained roothold and leaving a "desert of vines." Its vines can extend over 100 feet and cover entire canopies, telephone poles, or abandoned houses (fig. 8.1). It grows from runners at nodes and crowns but rarely from seed. Kudzu carries soybean rust fungus, a fungus that may have arrived in the United States via hurricane in 2004. Used as forage for livestock, kudzu may have some medicinal uses, and it contains a starch used for food in parts of Asia. It has been utilized in the United States to make soaps, lotions, jelly, and compost (Everest et al. 1999). A small patch of kudzu was discovered in 2009 near Leamington, Ontario, Canada.

INSECTS AND OTHER INVERTEBRATES. Insects, many of them exotic invaders, affect agroforestry and natural forest ecosystems. Invaders include the emerald ash borer (*Agrilus planipennis*), gypsy moth (*Lymantria*

TABLE 8.1 *One hundred of North America's damaging species**

AQUATIC PLANTS

Giant salvia	*Salvinia molesta*
Hydrilla	*Hydrilla verticillata*

TERRRESTRIAL PLANTS

Brazilian pepper tree	*Schinus terebinthifolius*
Common broom	*Cutisus scoparius*
Garlic mustard	*Alliaria petiolata*
Japanese barberry	*Berberis thunbergii*
Kudzu	*Pueraria montana*
Leafy spurge	*Euphorbia esula*
Osage orange	*Maclura pomifera*
Paperbark tree	*Melaleuca quiniquenerva*
Purple loosestrife	*Lythrum salicaria*
Spotted knapweed	*Centaura stoebe*

ARACHNIDS

HONEY BEE MITE

	Acarapis woodi
Honey bee varroa mite	*Varroa destructor*
Reptilian mite	*Amblyomma sparsum*

INSECTS

Africanized honey bee	*Apis mellifera scutellata*
Emerald ash borer	*Agrilus planipennis*
Formosan subterranean termite	*Coptotermes formosanus*
Gypsy moth	*Lymantria dispar*
Large elm beetle	*Scotylus scotylus*
Pine shoot beetle	*Tomicus piniperda*
Red imported fire ant	*Solenopsis invicta*
Small poplar borer	*Saperda populnea*

CRUSTACEANS

Fishhook water flea	*Cercopagis pengoi*
Rusty crayfish	*Orconectes rusticus*
Spiny waterflea	*Bythotrephes cederstroemi*

MOLLUSCS

African giant snail	*Achatina fulica*
Green mussel	*Perna viridis*
Zebra mussel	*Dreissena polymorpha*

AMPHIBIANS

American bullfrog	*Lithobates catesbeianus*
Cane toad	*Bufo marinus*
Coqui	*Eleutherodactylus coqui*

REPTILES

American alligator	*Alligator mississippiensis*
Burmese python	*Python molurus*
Curious skink	*Carlia ailanpalai*

FISHES

Black carp	*Mylopharyngodon piceus*
Green sunfish	*Lepomis cyanellus*
Northern snakehead	*Channa argus*
Red shiner	*Cyprinella lutrensis*
Sea lamprey	*Petromyzon marinus*
Silver ("flying") carp	*Hypophthalmichthys molitrix*

BIRDS

Acorn woodpecker	*Melanerpes formicivorus*
American crow	*Corvus brachyrhynchos*
American white pelican	*Pelecanus erythrorhynchos*
Black-billed magpie	*Pica hudsonia*
Black vulture	*Coragyps atratus*
Brown-headed cowbird	*Molothrus ater*
Canada goose	*Branta canadensis*
Common grackle	*Quiscalus quiscula*
Common pigeon	*Columba livia*
Double-crested cormorant	*Phalacrocorax auritus*
European starling	*Sturnus vulgaris*
Golden eagle	*Aquila chrysaetos*
Great blue heron	*Ardea herodias*
Great horned owl	*Bubo virginianus*
House sparrow	*Passer domesticus*
Laughing gull	*Leucophaeus atricilla*
Mallard	*Anas platyrhynchos*
Northern flicker	*Colaptes auratus*
Red-tailed hawk	*Buteo jamaicensis*
Red-winged blackbird	*Agelaius phoeniceus*
Sandhill crane	*Grus canadensis*
Snow goose	*Chen caerulescens*
Turkey vulture	*Cathartes aura*
Yellow-bellied sapsucker	*Sphyrapicus varius*

MAMMALS

Badger	*Taxidea taxus*
Beaver	*Castor canadensis*
Black bear	*Ursus americanus*
Black-tailed prairie dog	*Cynomys ludovicianus*
Bobcat	*Lynx rufus*
California ground squirrel	*Otospermophilus beecheyi*
Cotton rat	*Sigmodon hispidus*
Coyote	*Canis latrans*
Eastern chipmunk	*Tamias striatus*
Eastern cottontail	*Sylvilagus floridanus*
Eastern gray squirrel	*Sciurus carolinensis*
Eastern mole	*Scalopus aquaticus*
Feral cat	*Felis catus*
Feral dog	*Canis familiaris*
Feral pig	*Sus scrofa*
Little brown bat	*Myotis lucifugus*
Long-tailed weasel	*Mustela frenata*
Meadow vole	*Microtus pennsylvanicus*
Moose	*Alces alces*
Mountain beaver	*Aplodontia rufa*
Mountain lion	*Puma concolor*
Muskrat	*Ondatra zibethicus*
Nine-banded armadillo	*Dasypus novemcinctus*
Northern pocket gopher	*Thomomys talpoides*
Norway rat	*Rattus norvegicus*
Nutria	*Myocastor coypus*
Opossum	*Didelphis virginiana*
Ord's kangaroo rat	*Dipodomys ordii*
Porcupine	*Erethizon dorsatum*
Red fox	*Vulpes vulpes*
Raccoon	*Procyon lotor*
Rocky mountain elk	*Cervus elaphus nelsoni*
Striped skunk	*Mephitis mephitis*
White-footed deer mouse	*Peromyscus leucopus*
White-tailed deer	*Odocoileus virginianus*
Woodchuck	*Marmota monax*

*Selected examples. Every genus is represented by only one species, although several species may cause damage.

Figure 8.1   Kudzu, "the vine that ate the South," suffocates native vegetation in a field near Port Gibson, Mississippi. *Photo by Gsmith.*

dispar), and pine shoot beetle (*Tomicus piniperda*; see table 8.1).

Most damaging aquatic invertebrates have moved into new areas where they outcompete local fauna, disrupting food production such as commercial fishing and aquaculture (CISEH undated). For example, the spiny water flea (*Bythotrephes cederstroemi*; see table 8.1) moved into Lake Huron from Great Britain or northern Europe. First found in 1984, its full impact on the lake's ecosystem is still unknown. The water flea reproduces rapidly and because of its spine can be consumed only by larger fish. It competes with young perch and other small fish.

Given that about 8.6 million reptilian and amphibian pets existed in the United States in the year 2000, many imported, it is unsurprising that exotic reptilian ticks and mites are arachnids of concern. We found that reptilian ticks constituted 14 of 31 records of arachnids listed as invasive and exotic in North America by CISEH (undated). Some reptilian ticks (e.g., *Amblyomma sparsum* and *A. marmoroeum*) are known vectors of bacteria that cause heartwater in both domestic and wild ruminants (Burridge 2005) and may occur in humans. Four are bee mites, including the honey bee mite (*Acarapis woodi*) and the honey bee varroa mite (*Varroa destructor*; see table 8.1), known problems for both wild and domesticated honey bees.

Many molluscs of concern are also exotic, transported to North America for use in aquaria, on plants, or in ship ballast. All breed prolifically, and some have already caused severe problems in natural and agricultural ecosystems. Each has its own story. For example, giant African land snails (*Achatina fulica, A. achatina,* and *Archachatina marginata*) have been spread widely in North America as pets and for study in schools. The green mussel (*Perna viridis*) was first reported in 1999 in U.S. coastal waters following an accidental release, probably from ballast water, with a subsequent occurrence near St. Augustine, Florida, in 2002. This mussel is believed to be an aggressive competitor with local species and, according to Power et al. (2004), may become the marine equivalent of the freshwater zebra mussel. The veined rapa whelk (*Rapana venosa*), a carnivorous gastropod that has spread rapidly into parts of Europe, was discovered in the Chesapeake Bay in 1998; this predatory whelk could seriously harm the shell and oyster industry there.

The gypsy moth, *Lymantria dispar*, is one of North America's most devastating forest pests. The moth was imported to Boston, Massachusetts, from Europe in 1868 or 1869 by a silk producer to improve the resistance of silkworms to disease. Feral populations emerged within 10 years. State and federal eradication efforts began in 1890. The moth had moved into most of the northeastern United States by 1994 and is expected to go as far west as Iowa and as far south as South Carolina by 2025. Within its range, the gypsy moth erupts in cycles that are difficult to predict. During a high cycle, the moth defoliates trees, especially oak trees (*Quercus* spp.). With repeated defoliation, up to 20% of the oaks die, with occasional very heavy die-offs. Long-term impacts on forest vegetation are unknown, but replacement of oaks by less susceptible species is likely. Several million acres of trees are sprayed with insecticide annually to control the moth. Aggressive efforts are made to monitor for the presence of gypsy moths in new areas and to eradicate the moths where they are found. Between 1985 and 2003 alone, over $194 million was spent on monitoring and managing this pest (Mayo et al. 2003).

The rusty crayfish (*Orconectes rusticus*) is native to the Ohio River drainage system of the United States, but it has moved beyond its historical range into Illinois, Wisconsin, Minnesota, parts of 11 other states, and central Canada (Olden et al. 2006; Phillips et al. 2009). Rusty crayfish have large chelae (pinchers) and use their size advantage to outcompete other species. They reduce vegetation so that habitats for other native species are lost. Predators tend to take less threatening crayfish, so the rusty crayfish prevails, having more and more dramatic impacts on the aquatic ecosystems that it invades.

Twenty years after the first zebra mussels were found, Strayer (2009) reviewed their status in North America. Zebra mussels probably first arrived in ballast water of ships from Russia and were first seen in the Great Lakes in 1988. Early predictions were correct that the zebra mussel would spread rapidly throughout major waterways of North America (Strayer 1991). Zebra mussels moved from the Great Lakes into many major waterways of the eastern half and midwestern part of the United States. The invasion is slowing but continues, with models now predicting its spread into waters with sufficient soluble calcium (i.e., not the Pacific Northwest or New England), except for waters that are very cold (e.g., northern Canada) or very warm (e.g., the southwestern United States and Mexico). The predicted complete collapse of the Great Lakes fishery has not occurred.

Zebra mussels absorb and retain nutrients such as phosphorus and calcium. They take the nutrients along as they recede to the lower benthos, in this manner clearing waterways such as the Hudson River. The net effect on the Hudson River was an enriched littoral or benthic zone with reduced productivity of the upper-level photic (planktonic) and pelagic systems and fisheries dependent on them. Pelagic fish in the Hudson River were reduced by 28%. Plankton-based food chains often disappeared, whereas systems based on rooted plants and attached algae flourished. In the Hudson River, littoral fish increased by 97%.

AMPHIBIA AND REPTILES. Pitt et al. (2005) suggested that while many amphibians and reptiles were globally declining, a few were expanding rapidly and could be problematic. Species of concern were euryphagic, had high reproductive rates, and could achieve large population sizes. Many could hide undetected in transit. A mix of exotic and endemic amphibians and reptiles are a concern in North America. For example, imported pet snakes such as boa constrictors and pythons cause problems when they get into natural ecosystems. But native snakes are also a concern. About 45,000 people are bitten by snakes each year in North America, about 8,000 to 10,000 by venomous snakes. Children are bitten by rattlesnakes, especially when playing in grass (Howard 1994). About 6,500 require medical treatment, and about 15 people die from the bites (Parrish 1966; Russell 1980; Juckett and Hancox 2002). Livestock and pets, including hunting dogs, may also be bitten by snakes, sometimes causing death.

Alligator (*Alligator mississippiensis*) attacks are increasing in the United States. Langley (2005, 2010) found that there were 567 injuries and 24 deaths between 1928 and January 1, 2009. In a 2005 report, Florida had over 334 documented attacks, with 14 fatalities (and over 17,000 nuisance complaints per year); Texas had 15 attacks, and Georgia had 9 attacks, with 1 fatality. Most people injured were either attempting to capture (or pick up or exhibit) an alligator (17.4%), swimming (16.7 %), fishing (9.9%), or retrieving golf balls (9.5%). Alligators usually seize an appendage and twist it off by spinning (Woodward and Davis 1994). Alligators also serve as vectors for at least one zoonotic disease. West Nile virus infection was confirmed in three infected American alligators following an epizootic event at an alligator farm in Florida in 2002 (Jacobson et al. 2005).

Curious skinks (*Carlia ailanpalai*) are terrestrial lizards that were introduced into Guam in the 1960s. In areas without snakes, this skink has achieved population densities approaching 10,000 per acre, both in forested areas and near human habitation (Pitt et al. 2005). By numbers alone, this skink is a force outcompeting native lizards. It has also become a food source for brown tree snakes, furthering their population growth. The curious skink has also served as prey for the yellow bittern (*Ixobrychus sinensis*), a bird whose numbers around airports now pose risks for collisions with aircraft (Pitt et al. 2005).

Burmese pythons are native to Southeast Asia. They were accidentally and intentionally released in Everglades National Park, Florida, and established breeding populations there by the 1990s. Often, pet snakes become too large for their tanks and are released in the wild by the owners. Other snakes, such as Siamese cobras (*Naja siamensis*), have also become established. The snakes compete with native species and prey on native birds, including the federally endangered wood stork (*Mycteria americana*; Dove et al. 2011) and mammals including Key Largo wood rats (*Neotoma floridana smalli*), round-tail muskrats (*Neofiber alleni*), and some reptiles, including alligators. Both alligators and pythons prey on the threatened and endangered Key deer (*Odocoileus virginianus clavium*). Because of their large size, pythons can frighten Everglades Park visitors (fig. 8.2). Based on increased numbers of road kills and increased numbers removed

(over 1,300 so far), the population may be expanding. Traps, hand capture, and detection with dogs are being utilized in attempts to control the snakes in the park (Pitt et al. 2005).

FISHES. Aquatic ecosystems are particularly susceptible to wildlife damage. As with terrestrial ecosystems, people have often facilitated movement of fishes between bodies of water. Some species, including largemouth bass, rainbow trout, and common carp, have been introduced by public agencies to improve sport fishing or for biological control of weeds. The butterfly peacock bass (*Cichla ocellaris*) was introduced into Florida for both sport and control of tilapia. Often placed there for aesthetic value, goldfish are widespread in water bodies of North America. Bighead (*Hypophthalmichthys nobilis*), black (*Mylopharyngodon piceus*), and silver carp (*H. molitrix*) were brought into aquaculture facilities because they might meet a niche food market or control aquatic weeds. Many, such as the red-bellied pacu (*Colossoma bidens*) and some tilapia, were aquarium pets that escaped or were released. Some introductions occurred with fish used as baitfish or contaminants of baitfish.

Exotic fish can carry diseases. For example, stocking rainbow trout has introduced whirling disease in about 20 states (Fuller 2011c). Introduced fish may outcompete native ones, such as rainbow trout forcing the decline of cutthroat trout. Exotic fish may be a source of genetic pollution. For example, blue catfish, by hybridizing, contributed to the decline of the

Figure 8.2 An invasive Burmese python (*Python molurus*) meets a native American alligator (*Alligator mississippiensis*). *Photo by Lori Oberhofer, National Park Service.*

threatened Yaqui catfish (*Ictalurus pricei*) in Mexico (Fuller 2011b). Rainbow trout likewise hybridized with native cutthroat trout, golden trout, and redband trout (*O. mykiss* subsp.; e.g., Behnke 1992 in Fuller 2011c). Exotic fish often alter habitat, as with the effects of carp or suckers on water turbidity, sometimes dramatically changing community composition and dynamics.

The sea lamprey (*Petromyzon marinus*; fig. 8.3) is a jawless fish native to the eastern seaboard of the United States. It is usually marine but moves into fresh water to spawn. It was probably introduced into the Great Lakes region from the Finger Lakes or Lake Champlain in New York and Vermont. By the 1930s or 1940s, the sea lamprey had spread to Lakes Michigan, Huron, and Superior, where it parasitized fish including lake trout (*Salvelinus namaycush*), lake whitefish (*Coregonus clupeaformis*), chub, and lake herring (*C. artedi*). The sea lamprey has had major ecological influences on the fisheries of the Great Lakes. Millions of dollars are spent each year to control the lamprey and restore fish to the lakes. Costs to the fishery industry are estimated in the billions of dollars (Hansen 2010).

Like the sea lamprey, the white bass (*Morone americana*) found access from drainage areas along the eastern United States into the Great Lakes and waterways of upper midwestern states, this time via the Erie Barge Canal in the 1930s. The white bass is now found in about 20 states outside of its native range. The bass eat fish eggs during spring, preying on eggs of the walleye (*Stizonstedion vitreum vitreum*), another white bass (*Morone chrysops*), and others (Schaeffer and Margraf 1987 in Fuller et al. 2011). They also feed on eastern shiners (*Notropis spp.*; Schaeffer and Margraf 1987 in Fuller et al. 2011) and may contribute to the decline of some of these species (Parrish and Margraf 1994 in Fuller et al. 2011). White bass are sympatric with other white bass (*M. chrysops*) species and hybridize with them in western Lake Erie (Todd 1986 in Fuller et al. 2011).

The brook stickleback (*Culaea inconstans*) and the red shiner (*Cyprinella lutrensis*) were moved into new areas as baitfish or as contaminants in the shipment of bait or stocking fishes. Its native range in the northern United States, the brook stickleback has been reported in at least 14 additional states. Zuckerman and Behnke (1986) felt that its presence in Colorado may have been due to accidental stocking by the Colorado Division of Wildlife, and Pigg et al. (1993), finding brook sticklebacks among fathead minnows in a shipment from Minnesota, suggested that accidental stock might also explain occurrence of the fish in Oklahoma. Bait-bucket releases have probably introduced red shiners into at least 15 states outside of the Mississippi River basin. The shiner is extremely aggressive and may have carried the Asian tapeworm into the Virgin River, Utah, where it infected the endangered woundfin (*Plagopterus argentissimus*; Nico and Fuller 2011).

At least four species of snakehead (*Channa* spp.) have been introduced into North American waters. Snakeheads originated in Asia, where they are important commercial fish, grown in aquaculture, for sport, and for aquaria. Introductions in the United States are sometimes accidental from aquaria and sometimes intentional, as potential food fishes. The blotched snakehead (*C. maculata*) has been established in Oahu, Hawaii, since before 1900 (Courtenay et al. 2004). The giant snakehead (*C. micropeltes*) and northern snakehead (*C. argus*) have been reported in at least 6 and 11 states, respectively. A reproducing population of northern snakehead was found in Crofton Pond, Maryland, and in the Potomac River (Courtenay and Williams 2004). The bullseye snakehead (*C. marulius*) has been illegally stocked in residential lakes and canals in Broward County, Florida. Snakeheads are predatory fish and may harm competing native fish and crustaceans (Fuller 2011a). Overall ecological impacts are unknown.

The silver carp was introduced in 1973 by a fish farmer in Arkansas who thought they could improve water quality at sewage treatment lagoons and aquaculture facilities and serve a niche market for food

Figure 8.3 Sea lampreys (*Petromyzon marinus*) have parasitized lake trout (*Salvelinus namaycush*) and other fish since they invaded the Great Lakes in the early 1900s. They have dramatically altered the Great Lakes ecosystems. *Photo courtesy of the U.S. Geological Survey.*

(Conover at al. 2007). The fish were evaluated by a state agency and a university in Arkansas and tested in at least four lagoons and at Mallard Lake. The present population in the Mississippi River Basin may have come from one of these locations. Feral populations have been confirmed in at least 10 states outside the Basin (Nico 2005), with a general movement toward the Great Lakes. Silver carp jump (fig. 8.4), are difficult to handle during production, are not cultured in the United States, and are listed as injurious under the Lacey Act (see chapter 15). The commercial harvest of silver carp has increased in the Mississippi and Illinois Rivers with efforts by the industry to establish more markets. Silver carp are plankton eaters, affecting mussels, larval fish, and some adult fish. Sampson (2005) found that silver carp compete with gizzard shad and bigmouth buffalo in these rivers. He believed they might reduce large zooplankton, with cascading effects on the river ecosystems. In addition, silver carp might be a source of Asian tapeworm.

BIRDS. Flight and gregarious or predatory behavior underlie much of the damage caused by birds in North America. Flight allows gregarious birds, such as many **granivorous** or omnivorous species, to move great distances daily and to congregate in large flocks for feeding and then assemble or reassemble for loafing or roosting. Predatory birds such as raptors tend toward individual rather than gregarious lifestyles, but with these birds, the ability to fly enhances the accessibility of livestock such as poultry and young calves. Individual birds also cause damage, such as woodpecker damage to house siding or utility poles. Birds can carry diseases such as West Nile virus or avian influenza. Invasive birds such as European starlings compete with native cavity-nesting birds and otherwise harm natural ecosystems.

Red-winged blackbirds are one of a group of about 10 blackbird species in North America, including the starling since its introduction from Europe in the late 1800s. Also included are common grackles (*Quiscalus quiscula*), great-tailed grackles (*Q. mexicanus*), yellow-headed blackbirds (*Xanthocephalus xanthocephalus*), brown-headed cowbirds, and Brewer's blackbirds (*Euphagus cyanocephalus*; Dolbeer 1994; Dolbeer et al. 1994a, b). Blackbirds are the most abundant birds in North America. They damage ripening corn, sunflower, and rice and can damage sprouting rice in localized areas. Starlings contaminate feed at cattle feedlots, selectively taking higher-protein portions of cattle feed. They also damage fruit, including cherries and grapes. Blackbirds congregate in large flocks and roosts, especially in the winter, making noise and

Figure 8.4 Invasive silver carp from Asia, now found in some North American waterways, often respond to speeding boats by jumping. *Photo courtesy of the U.S. Geological Survey.*

defecating (fig. 8.5). The resulting guano can host the fungus that causes histoplasmosis (Dolbeer et al. 1994a, b).

Blackbirds also collide with aircraft. Barras and Wright (2002) and Barras et al. (2003) reviewed records in the National Wildlife Strike Database of the U.S. Federal Aviation Agency. They found 1,704 strikes involving blackbirds and starlings between 1990 and 2001. Collisions occurred in 46 states and the District of Columbia. Damage was reported in only 5.9% of the incidents, but costs totaled $1,607,317.

Double-crested cormorants (*Phalacrocorax auritus*) are part of a group of herons, egrets, bitterns, pelicans, and other cormorants that feed on fish in aquaculture facilities, including publicly operated hatcheries and commercial fish farms, where they can cause serious economic losses (Glahn et al. 1999; Glahn and King 2004; King 2005). Cormorants recovered from a precipitous decline in the 1950s caused partly by use of organochlorine pesticides such as DDT. Since then cormorants have increased dramatically, adjusting both habit and breeding locations to take advantage of the growing aquaculture industry in states such as Mississippi and Arkansas (Taylor and Dorr 2003). For example, many previously migratory cormorants now remain near southern aquaculture facilities year-round. Cormorants roost in rookeries in numbers sometimes over 10,000, populations that often cause serious depredation to recreational fisheries, lakes, and reservoirs throughout North America. Guano

accumulation in permanent roosts causes secondary problems such as odor, destruction of vegetation including large mature trees, erosion, and sometimes loss of desirable areas for tourism and recreation, such as at Lake Guntersville in Alabama (Barras 2004–5).

Red-tailed hawks (*Buteo jamaicensis*) are raptors that, along with goshawks and great horned owls (*Bubo virginianus*), sometimes take poultry or other livestock (Dolbeer et al. 1994a, b). Raptors generally leave puncture wounds in the back and breast and often pluck the feathers, and owls often also decapitate the prey. Raptors usually take only one prey animal daily, unlike some mammalian predators. Golden eagles (*Aquila chrysaetos*) sometimes take lambs or kids on open ranges, leaving puncture wounds that are about one to two inches apart from three front talons and a wound four to six inches behind, made from the hallux (rear toe; Dolbeer et al. 1994a, b). Turkey and black vultures also take livestock, sometimes poking eyes out of calves before killing them (Lowney 1999). Hawks and vultures accounted for almost 30% of reported down time from bird-aircraft collisions (Blackwell and Wright 2006); their relatively large size is a concern for damage resulting from such collisions.

The acorn woodpecker (*Melanerpes formicivorus*) is one of several woodpeckers that peck holes in wooden siding and similar structures to find insects or store acorns. They sometimes annoy homeowners by pecking siding that amplifies the pecking sounds and

Figure 8.5 Blackbirds form large flocks, often damaging crops and becoming a nuisance and a health hazard in urban and rural settings. *Photo by Edibobb.*

aids the woodpecker in establishing a territory. A particularly notable example was holes pecked in foam insulation at Launch Pad 39B, Cape Canaveral, Florida, in June 1995. The woodpeckers, probably flickers, pecked about 200 holes in the insulation, some four inches deep. Launch of the Space Shuttle Discovery was delayed by five days until the insulation could be replaced and pyrotechnics deployed. Woodpecker spotters were utilized to watch the launch pads around the clock. Sapsuckers damage ornamental or commercial trees, such as Christmas trees, by feeding on sap, bark, or insects in the tree trunks (Dolbeer et al. 1994a, b).

MAMMALS. Mammals are often crepuscular or nocturnal and are less easily connected than birds to the damage they cause. Nonetheless, damage can be extensive.

White-tailed deer (*Odocoileus virginianus*) are mammals that, along with other ungulates including moose and elk, browse agricultural crops and trees such as ornamentals, fruit, Christmas trees, and forests. Ungulates can also cause damage by rubbing and trampling. Deer and elk damage agricultural crops such as alfalfa, oats, and winter wheat; elk raid haystacks; and deer may consume stored crops (Ver-Cauteren, Pipas, et al. 2003). In addition to cattle grazing, the ungulates can affect long-term availability of forage and reforestation efforts. Moser and Witmer (2000) found reduced shrub and mammal biodiversity in areas of the Blue Mountains of eastern Oregon where elk and cattle were present in high densities. Deer feed on buds during the winter, affecting the following year's production, and take the lead growth of smaller Christmas trees, often eliminating their commercial value. Deer rub trees with diameters of about one-half inch to four inches, affecting ornamental, forest, and Christmas tree industries.

When overabundant, deer and elk can affect natural ecosystems by selectively foraging, thereby modifying patterns and distribution of vegetation. Effects cascade to other species, including insects, birds, and other mammals (e.g., Côté et al. 2004). Deer, elk, and other ungulates collide with aircraft, automobiles, and other vehicles, causing economic losses that are estimated at over a billion dollars annually and injuring and killing both humans and themselves (fig. 8.6; Huijser et al. 2007). Deer are vectors of diseases such as Lyme disease, serving as hosts for the tick that transfers the disease to humans, and can transfer other diseases, such as tuberculosis, to livestock (Dolbeer et al. 1994a, b)

Nutria (*Myocastor coypus*), an exotic invader introduced in 1899 for fur farming, feed on aquatic plants and crops close to waterways, damage wooden structures and floating docks, and eat and damage young woody vegetation, creating major problems in bottomland forest regeneration efforts. They can be a particular problem in southern states with crops such as rice and sugarcane and in wildlife areas such as refuges, where eradication is often desired (Jojola et al. 2005). Muskrats (*Ondatra zibethicus*), too, damage wetlands and ponds, whether natural or manmade, throughout North America (Miller 1994). They burrow extensively into dams and levees, adjusting burrow design and height according to fluctuating water levels. They can have important impacts on the structure and species composition of wetland marshes, both natural ones and those constructed for water treatment, changing dense vegetation into patchworks of open areas (Higgins and Mitsch 2001; Kadlec et al. 2006).

Figure 8.6 Major efforts are made at military and commercial airports worldwide to minimize collisions with wildlife on take-off and landing. Collisions are dangerous to wildlife and crew and costly to aircraft. *Photo by Air Combat Command, U.S. Dept. of Defense.*

Muskrat burrows can collapse under the pressure of livestock, sometimes injuring the livestock but also stimulating additional burrow construction or reconstruction. Thus, damage is often exacerbated at ponds, stock tanks, and waterholes for livestock.

Beavers cut saplings and cut or sometimes girdle large trees, preferring trees such as cottonwood, sweet gum, loblolly pine, and aspen, and use them to build dams that can block water flow and cause flooding (Miller and Yarrow 1994). They sometimes plug road culverts with sticks, build dams under highway bridges and railroad grades, and dig into levees and manmade dams, causing damage. Significant in riparian ecosystems, they can cut and remove numbers of trees and flood and kill large tracts of mature timber. For example, beavers cut and remove Fremont cottonwood (*Populus deltoides*) saplings, trees that are declining in western arid riparian habitat. Breck et al. (2003) showed that access to trees was enhanced in the Green River, Utah, by controlled flooding, increasing their vulnerability to damage by beavers. Beavers sometimes carry diseases such as the protozoan infection *Giardia* spp., which can be transferred to humans, causing diarrhea and gastroenteritis.

Another invasive mammal is the red fox, one of a group of canids—including also coyotes, wolves, and dogs (domestic and feral)—that collectively prey on animals from small reptiles and rodents to large cervids. These animals take livestock, with foxes being a particular problem for poultry. Foxes usually attack the throats of lambs and poultry, sometimes biting the neck or back, and carry the carcass to a den. They open eggs and lick out the contents, leaving the shells beside the nest, and consume first the legs and then the breast of poultry, partially burying the remains. Domestic and feral dog attacks are often characterized by inefficient mutilation of the prey, with broad damage to the flanks, hindquarters, and head. Coyotes often attack the throat and suffocate the prey. Less frequently, they attack from the rear or attempt to pull the prey down from the underbelly (Jansen 1974; Sterner and Crane 2000). If the prey is small enough, the coyote can bite through the skull or the neck. Coyote attacks on pets and small children have increased as humans and coyotes have increasingly shared suburban and urban space (e.g., Gehrt 2007). Coyotes are euryphagic, opportunistic foragers and will damage crops such as watermelons, cantaloupes, grapes, and other fruit. They can carry rabies, sarcoptic manges, and other diseases transmissible to pet dogs and livestock, and they readily prey on game species of wildlife, including wild turkey, antelope, and especially fawns and young poults.

Wolves attack caribou, moose, elk, and cattle, often by damaging muscles and ligaments of the back legs. Anticipated impacts on livestock led ranchers to oppose reintroduction of wolves into Yellowstone National Park. With reintroductions underway, many efforts in North America are now focused on minimizing livestock-wolf conflicts (e.g., Bangs and Shivik 2001). Wolves will attack people, hunting dogs, and domestic pets. McNay (2002) summarized 80 encounters between wolves and people in Alaska and Canada in which the wolves appeared unafraid of people, and wolves have caused human fatalities in both Alaska and Canada in recent years. Of these, 12 wolves were known or suspected to be rabid, and 39 aggressions were by wolves known to be healthy. Sixteen people were bitten by healthy wolves; none of the injuries were life-threatening.

Mountain lions attack deer, elk, and livestock, including sheep, goats, cattle, and horses, in western North America. They are opportunistic and will take available smaller prey. Mountain lions use their powerful jaws to puncture skulls and can break the neck and back of even large prey. These cats often feed on the front quarter of prey first, often also dragging the carcass to a feeding site where it is cached and to which the lion will return repeatedly to feed (e.g., Akenson et al. 2003). Contacts between mountain lions and humans are increasing in North America, as are the number of fatal attacks on both humans and their pets.

Summary

- Wildlife damage occurs widely in North America, caused by both endemic and exotic species that include plants and animals, invertebrate and vertebrate.
- Most damaging plants are exotic invasives; they cause ecosystem problems by altering fire regimes, impacting nutrient and water cycling, and creating genetic pollution.
- Most damaging insects and other invertebrates, such as the gypsy moth and the zebra mussel, are also exotic invasives; they disrupt food production and commercial fishing in aquatic systems and can have widespread impacts on forests and agroecosystems.
- Damaging amphibians and reptiles are both exotic, such as the Burmese python, and endemic, such as the alligator.
- Damaging fish, such as the sea lamprey, are also both exotic and endemic. Often moved by people

between bodies of water, they sometimes outcompete local species and carry diseases.

- Flight and their gregarious nature underlie the damage caused by many birds, such as the exotic starling or the red-winged blackbird, allowing them to travel great distances daily and then congregate in large groups to feed, loaf, or roost.
- Mammals, such as white-tailed deer and coyotes, often cause unseen damage at night or at dusk.

Review and Discussion Questions

1. An important part of wildlife management is restoration of wildlife and its habitat. Often this involves the translocation of wildlife species. As we write, for example, elk are being captured and held in Kentucky for movement and restoration in Missouri, and they have been reintroduced in several eastern states in recent years. What lessons might be drawn from this chapter that should be considered before taking such management actions?

2. What does this chapter suggest regarding relationships between wildlife problems and movement of boats from one body of water to another, e.g., from oceans into lakes of North America? Or from one lake to another in North America? Or from pond to pond?

3. What do you see as the long-term impacts of damaging species on the ecosystems of North America? For example, how about impacts of black or silver carp on the Mississippi River system? How about impacts of invasive weeds on the grasslands and livestock industry of Saskatchewan?

4. The USDA's Animal and Plant Health Inspection Service has recently told teachers and public schools to no longer use giant snails in their classrooms. Why? What environmental impacts is this federal agency trying to avoid?

5. Describe some of the effects of feral pigs on natural ecosystems in Texas or Mississippi. How about their impact on endangered bird species in Hawaii or on islands of North America?

9

Wildlife Diseases and Zoonoses

We explore wildlife diseases and zoonoses and their sources, symptoms, means of transmission, and impacts on humans and their interests, including natural ecosystems; we also explore how humans influence the nature and spread of wildlife diseases and zoonoses.

Statement
Recent increases in wildlife diseases and zoonotic epidemics may be the consequence of increased human transport of wildlife and wildlife pathogens, the movement of people into more intimate contact with wildlife, and natural or human-caused perturbations of ecosystems, with consequent changes in old diseases so they can invade new hosts and areas.

Explanation
Diseases are abnormal conditions that impair the functions of organisms. They are usually associated with sets of signs and symptoms. If affected organisms are nondomesticated, it is a wildlife disease. If a disease can be transmitted from other organisms (either wild or domesticated) to humans or vice versa, it is also a zoonosis.

Diseases are caused by agents. **Agents** are often biotic but might be abiotic, such as radiation or chemicals like heavy metals, toxins, or **prions** (abnormally folded protein bodies, possibly the cause of bovine spongiform encephalopathy or "mad cow disease" in cattle, Creutzfeldt-Jakob disease in humans, chronic wasting disease in cervids, and scrapie in sheep; Barnes 2005). Biological agents are called **pathogens**. Many are also parasites in that they are smaller than organisms they infect and benefit at the expense of infected organisms. Most parasitic pathogens are also **endoparasites** because they live inside infected organisms. Pathogens include **microparasites** (microscopic parasites that can complete life cycles within one organism—e.g., viruses, fungi, obligate intracellular bacterial microparasites called **Rickettsia**, and other bacteria) and **macroparasites** (parasites that can be seen with the naked eye—e.g., protozoa and helminths) that need to move from host to host to complete their life cycles. Examples include *Rickettsia* that cause Rocky Mountain spotted fever and *Yersinia pestis*, bacteria that cause plague. Some pathogens are fungal, such as *Geomyces destructans,* which causes white-nose syndrome and has killed more than a million bats in the northeastern United States and Canada since it was first discovered in New York State in 2006–7.

Hosts are organisms that harbor pathogens. Long-term hosts may be considered **reservoirs**, or **niduses** ("nests" or loci), for the disease (see tables 9.1–9.3). Diseases are transmitted by organisms called **vectors**, or carriers. Vectors may also be ectoparasites if they are smaller than the host and live on the outside of it. These carriers do not cause diseases but rather convey agents from hosts (or reservoirs) to hosts. Vectors are often arthropods or insects such as mosquitoes, flies (single-winged *Dipterans*), **sand flies** (blood-sucking *Dipterans*), biting **midges** (tiny "no-see-ums" or sand-fleas), **lice** (wingless obligate parasites, 3,000 spp.), **fleas** (*Siphonaptera*), ticks, or **mites** (*Acarina*). Movement of vectors can be facilitated by birds or mammals. At times, hosts may themselves serve as vectors. For example, Reed et al. (2003) noted that the spread of the mosquito-borne disease West Nile virus along the eastern coast of the United States coincided with migratory routes of birds.

Vectors carry diseases in various ways. Flies may simply have bodies contaminated by feces or saliva of infected hosts (reservoirs) or by soil. Infected mosquitoes and ticks can inject pathogens directly into bloodstreams of new hosts. Pathogens may accumulate in gelatinous masses in the gut, making it difficult for vectors such as mosquitoes to ingest second meals. Mosquitoes may then proactively expulse (regurgitate) the masses into bloodstreams of hosts in order to clear their gut for more meals (Barnes 2005).

Pathogens and vectors have often coevolved so that pathogens do not harm their vectors. Vectors that carry parasitic pathogens for prolonged periods of time are called **secondary** or **intermediate hosts**. Final targeted organisms are then called **definitive hosts** (Barnes 2005). Definitive hosts are often, but not always, where parasites reach maturity and reproduce. Sometimes, definitive hosts are simply ones considered most important by humans. For example, sludge worms (*Tubifex* spp.) are considered to be secondary hosts for **whirling disease** (a protozoan disease that infects salmon and trout species) even though the protozoa sometimes reproduce inside the worms. In this case, fish are considered the definitive hosts.

Movement of a disease between hosts is called its biological life cycle. The natural portion of the cycle, between wildlife as hosts and vectors, is sometimes called the **sylvatic** (or **enzootic**) **cycle**. Sylvatic cycles

may allow diseases to persist in the wild and hosts to serve as reservoirs. If a disease has only a sylvatic cycle, it is a wildlife disease. If the disease "spills over" to humans, even if the infection "dead ends" without further transmission, the disease is called a zoonosis. Some zoonoses are transmitted between humans.

Biological life cycles can be straightforward or complex. We illustrate using Lyme disease (fig. 9.1). Lyme disease is caused by the spirochete *Borrelia burgdorferi* and occurs in North America, Europe, South Africa, and Australia. This disease appeared or reappeared along with reforestation and movement of people back into suburban settings. The disease was first identified in the United States in Old Lyme, Connecticut, in 1981, but the skin reaction was described in Sweden in 1909.

With Lyme disease, vectors are *Ixodes* ticks—e.g., *I. scapularis*. Tick larval and nymphal stages feed on small animals, preferring the white-footed deer mouse in North America. Pinhead-sized larvae emerge from eggs in soil during summer, attach to a host, feed, and then drop to the ground and molt. The only slightly larger nymphs emerge the next summer, find a host, feed, and drop to the soil to molt. Adults emerge at the end of summer and choose larger mammals such as white-tailed deer as hosts. Adult males die shortly after mating; females die after laying eggs and completing the life cycle (fig. 9.1; Barnes 2005).

If any host is infected with *B. burgdorferi*, the tick becomes a vector, passing the bacteria to its next hosts. Spillover to humans occurs as early as the nymphal stage. The tiny nymphs are hardly visible to the human eye and are difficult to detect on the skin. The spirochetes only begin to replicate in the guts of ticks after the ticks begin ingesting blood, and they must move to salivary glands to be injected, so infected ticks are not contagious for 36 to 48 hours after attaching to humans or other hosts (fig. 9.1; Barnes 2005). Lyme disease symptoms sometimes include a ring-shaped rash that grows to several inches in diameter at the site of the bite. As the ring grows, the center clears, giving the appearance of a target. Initial symptoms are flu-like and include fever, chills, headache, fatigue, and joint and muscle aches. Aggressive treatment with antibiotics is often needed to deal with the disease. Untreated, Lyme disease can lead to chronic arthritis, cardiac and neurologic problems, and in some individuals, death.

Although we think of zoonoses as being transmitted from wildlife to humans, reverse zoonoses also happen. For example, molecular typing studies of *Giardia,* caused by parasitic protozoa carried by beavers, indicate that it sometimes moves along with

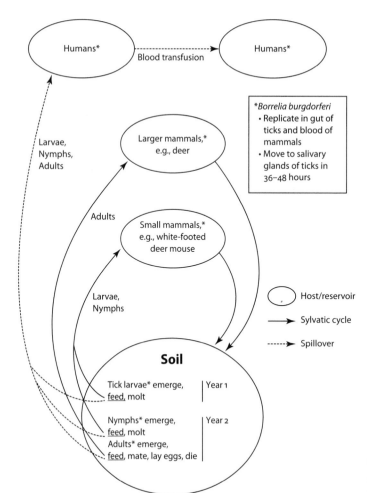

Borrelia burgdorferi
- Replicate in gut of ticks and blood of mammals
- Move to salivary glands of ticks in 36–48 hours

Host/reservoir

Sylvatic cycle

Spillover

Figure 9.1 Life cycle of Lyme disease. *Illustration by Lamar Henderson, Wildhaven Creative LLC.*

human sewage back to beavers. A concern, of course, is that infected beaver populations can then re-infect humans (Thompson, Kutz, et al. 2009).

Outbreaks happen when infectious diseases occur at higher frequencies than expected. If the disease is unknown or has not occurred for a long time, a single case might be considered an outbreak. **Epidemics** occur when infectious diseases spread rapidly among large numbers of people (e.g., the SARS epidemic in Asia during 2003). **Pandemics** are even larger; pandemics occur when spread of a disease is potentially global. Avian influenza pandemics are examples.

Some wildlife diseases have been known since ancient times and are called **lingering**. This does not imply that the diseases are static and relict. Rather, such diseases are dynamic, having survived by changing along with environmental conditions. Lingering diseases re-emerge in changed forms when conditions are right for them. Examples include brucellosis and dog rabies (WHO undated b). Other wildlife diseases

are new and emerging. An **emerging** zoonosis is defined as "a pathogen that is newly recognized or newly evolved, or that has occurred previously but shows an increase in incidence or expansion in geographical, host, or vector range" (WHO undated b). In addition to avian influenza, described below, other emerging zoonoses include bovine spongiform encephalopathy, Nipah virus, and possibly toxoplasmosis (box 9.1).

Diseases respond to environmental changes partly because of the genetic bases of pathogens. Viruses composed of ribonucleic acids (RNAs) mutate frequently and are therefore particularly labile, but deoxyribonucleic acid–based (DNA-based) pathogens also mutate. Pathogens also recombine their nucleic acids to form new types, take nucleic acids from other pathogens already in the host, or sometimes take nucleic acids from the host itself. Pigs and birds have particular bents as "melting pots" where pathogens can meet and exchange strands. While most new combinations fail, a few become emergent pathogens, surviving in new

BOX 9.1 Toxoplasmosis

Toxoplasmosis is a parasitic infection caused by the protozoan *Toxoplasma gondii*. Felids, including domestic and wild cats, are definitive hosts, while many warm-blooded animals serve as intermediate hosts (Frenkel et al. 1970).

Characteristics and status of the disease were reviewed by Jones and Dubey (2010) and are partly summarized here. The disease has three stages—tachyzoites, bradyzoites, and oocysts. Each stage can infect either intermediate or definitive hosts. A tachyzoite penetrates a host's cell and then forms a vacuole that protects the tachyzoite from the host's defenses. It multiplies until the cell ruptures. After repeated divisions, the tachyzoite enters a stage where it forms a cyst, and the cyst grows until it contains up to several hundred bradyzoites. The cyst is usually formed in muscular or skeletal tissue including the brain, eye, or cardiac muscle. The cysts remain dormant without causing problems in most intermediate hosts. When ingested by a felid, however, the cyst is digested by intestinal enzymes, releasing the bradyzoites. They invade extracellular intestinal tissues, forming at least five different asexual types of *T. gondii* and eventually also forming gamonts that begin a sexual cycle. Oocysts are produced and shed from the infected felid for several months.

Jones and Dubey (2010) estimated that with 78 million domestic cats and 73 million feral ones in the United States, and with 1 million oocysts per shedding cat (estimated at about 30% of the total), there are 50 billion oocysts in the environment in the United States. Thousands of cougars and millions of bobcats add to the environmental burden associated with toxoplasmosis, and because of the presence of live prey as well as eviscerated tissues from infected hunted deer and bear, Jones and Dubey suggested a sylvatic cycle for the disease. The potential for the disease is therefore now believed to be cosmopolitan.

Until recently, toxoplasmosis was viewed mostly as a wildlife disease transmitted by carnivores eating contaminated prey. The disease was also seen as a zoonotic, but the "spillover" was believed to be mostly due to infected domestic cats.

In 1994 an outbreak of toxoplasmosis occurred in a town in Canada that was traced to contaminated drinking water. Infected cats, both wild (*Felis concolor*) and domestic, were found near the water supply (Aramini et al. 1998). An outbreak of toxoplasmosis also caused high mortality in sea otters along the coast of California. The source of the protozoa was believed to be contamination of seawater from oocysts in runoff water containing urine and feces of domestic and wild cats (Dabritz, Gardner, et al. 2007; Dabritz, Miller, et al. 2007). Sea otters probably got infected by feeding on marine invertebrates (Miller et al. 2008). The disease is now considered waterborne as well as terrestrial.

hosts, vectors, or environments. Rapid expansion of the disease may follow. For example, "swine flu" virus, H1N1, originated in North America in April 2009 and quickly became the 2009 influenza A / H1N1 pandemic that by June 15, 2009, had involved 76 countries, 36,000 infections, and 163 human deaths (WHO undated c). H1N1 was composed of two swine, one human, and one avian strain of influenza (Belshe 2009).

Frequency of wildlife diseases (including zoonoses) appears to be increasing (e.g., Woolhouse and Gowtage-Sequeria 2005). Factors facilitating diseases appear to be the same as those already described for other categories of wildlife damage: a burgeoning human population with increased food demands, intensified agriculture, expanded human habitation and contact with wildlife, and increased human densities in urban areas (e.g, McMichael 2004); increasing transport of people, organisms, and products around the world for business or recreation, including ecotourism; and a changing climate that facilitates mutations of pathogens and alters population dynamics of pathogens, vectors, and hosts (e.g., Cutler et al. 2010).

For example, intensified agriculture changed how feed was processed in the United Kingdom, allowing prion-contaminated feed to be given to cows and leading to the emergence of a previously unknown bovine spongiform encephalopathy in cattle (Mahy and Brown 2000). The transport of infected rodents from Ghana into the United States, and their subsequent comingling in pet shops, allowed monkeypox virus to "species-jump" (see below) into pet prairie dogs, subsequently infecting over 70 people (e.g., Hutson et al. 2007).

A **species jump** occurs when a pathogen overcomes a species barrier and infects a new organism. Woolhouse et al. (2005; table 9.1) provide examples and describe underlying epidemiological theory. The likely importance of an outbreak is based partly on its **basic reproduction number**, R_0. R_0 is the ratio of secondary cases (number of individuals in the new population that are infected from primary cases) to primary cases (number of individuals infected by the jump). Suppose, for example, a pathogen jumped a species barrier and infected two individuals (i.e., two primary cases). Suppose the primary cases subsequently infected six more individuals. The R_0 would then be 6:2, or 3.

Woolhouse et al. (2005) stated that if $R_0 < 1$, then transmission potential is low and the disease is unlikely to pose a great threat to the new hosts. Examples include Ebola, monkeypox, and most avian influenza viruses. If, however, $R_0 > 1$, there is risk for a major epidemic or pandemic. Examples include avian influenza type A and SARS-coV viruses. Woolhouse et al. (2005) point out that when $R_0 > 1$, most new infections come from within the new population. When $R_0 < 1$, however, spread of the disease into the new population depends on repeated and continued species jumps, which is unlikely.

Native pathogens operate in all ecosystems and seres, serving critical functions in detritus circuits,

killing susceptible organisms, and breaking down dead organisms to release and recirculate minerals and nutrients. Ecosystems have coevolved with native pathogens, resulting in pathogens that are neutral or beneficial. For example, native pathogens can facilitate tree decline and therefore degrade tree composition and overall diversity of forest ecosystems. Pathogens work in concert with abiotic factors such as mineral deficiencies, wind, and lightning to remove susceptible trees. Such pathogens include viruses, bacteria, fungi, nematodes, parasites such as mistletoe, and insect defoliators, as well as bark and wood borers. Hagle et al. (2003), among others, point to over 1,300 pathogens, diseases, and insect pests of conifers in the northern and central Rocky Mountains, most of which are native and of little or no concern. Native pathogens usually become problematic when conditions change, favoring new pathogenic forms.

Exotic pathogens, however, can disrupt normal ecosystem function by affecting individual species and organisms. There is not enough time for co-evolution of defenses against new pathogens, and such defenses may not already exist. Susceptibility to American chestnut blight and Dutch elm disease are examples. CISEH (undated) provides at least 185 additional rec-

TABLE 9.1   *Examples of pathogens considered to have emerged via a species jump*

Pathogen	Original host	New host	Year reported
TRANSMISSIBLE SPONGIFORM ENCEPHALOPATHIES			
BSE/vCJD	Cattle	Humans	1996
VIRUSES			
Rinderpest	Eurasian cattle	African ruminants	Late 1800s
Myxoma virus	Brush rabbit / Brazilian rabbit	European rabbit	1950s
Ebola virus	Unknown	Humans	1977
FPLV/CPV	Cats	Dogs	1978
SIV/HIV-1	Primates	Humans	1983
SIV/HIV-2	Primates	Humans	1986
Canine / Phocine distemper virus	Canids	Seals	1988
Hendra virus	Bats	Horses and humans	1994
Australian bat lyssavirus	Bats	Humans	1996
H5N1 influenza A	Chickens	Humans	1997
Nipah virus	Bats	Pigs and humans	1999
SARS coronavirus	Palm civets	Humans	2003
Monkeypox virus	Prairie dogs	Humans	2003
BACTERIA			
Escherichia coli O157:H7	Cattle	Humans	1982
Borrelia burgdorferi	Deer	Humans	1982
FUNGI			
Phytophthora infestans	Andean potato	Cultivated potato	1840s
Cryphonectria parasitica	Japanese chestnut	American chestnut	Late 1800s

Source: Table 1 in Woolhouse et al. 2005, with permission.

ords of invasive and exotic pathogens in North America, affecting both terrestrial and aquatic species.

Effects of exotic pathogens on ecosystems can be both subtle and complex but have profound consequences. For example, Borer et al. (2007) suggested that generalist invasive viral pathogens (e.g., barley and cereal yellow dwarf viruses) contribute to the invasion and domination of California's perennial grasslands by exotic annual grasses. Over 20 million acres of California's native perennial grasses have been replaced by exotic annual grasses and forbs since European settlement. The exotic annual grasses have been successful, even though they are inferior to native perennial species in competing for resources.

Borer et al. (2007), modifying a Lotka-Volterra competition model to accommodate diseases and seasons, found that disease-free native perennials could not be successfully invaded by healthy annuals. However, presence of viruses changed susceptibility. Because the native perennials are a necessary reservoir for the viruses, the simulations predicted an equilibrium in which native perennials were dominated by invasive annuals. Aphids serve as vectors for reinfection of perennials, and annual immigration of aphids from perennial lawn grasses or irrigated grass crops such as barley are needed for the equilibrium to persist.

Plant diseases of interest to wildlife damage management probably number in the thousands. There are also over 230 wildlife diseases that infect animals (NBII undated), some of which are also zoonotic. We provide selected listings in tables 9.2–9.3 and examples below.

Examples

PLANT DISEASES. Chestnut blight (*Cryphonectria parasitica*) was first observed in 1904 infecting chestnut trees in the Bronx Zoo, New York City. Released from ecosystems that kept it in check in Asia, the fungus spread rapidly throughout forests of the eastern United States, so that by 1940 the chestnut was no longer a dominant species in the forest canopy. Chestnuts have since been mostly replaced by oak and hickory. Chestnut blight also spread over Europe; however, chestnuts there showed resistance and recovered. *C. parasitica* infects upper branches of trees but does not kill root collars or root systems. Consequently, native chestnuts still survive in parts of the United States as new saplings emerging from old root systems. Most become infected and die before they can produce viable seeds.

Diamond et al. (2000) analyzed mast (edible seeds and fruits of woody plants) production in the eastern United States before and after the presence of chest-nut blight. Even though chestnuts were replaced by two other mast-producing species, amounts and quality of masts were sufficiently reduced to impact animals. The researchers pointed particularly to white-tailed deer, black bear, wild turkey, gray squirrel, and ruffed grouse (*Bonasa umbellus*) as wildlife that depend on mast production for survival.

Dogwood anthracnose (*Discula destructiva*) is a fungal infection of dogwoods (*Cornus florida*). It may have been introduced into the United States on infected horticultural stock from Asia in the 1970s. Symptoms appeared in dogwoods in Washington, DC, New York State, and Connecticut at about the same time and spread throughout the Appalachian mountains in the 1980s. Oswalt and Oswalt (2010) compared population samples taken throughout the Appalachians in the 1980s with those taken in the 2000s. Losses ranged from 25% to 100%. The fungus was reported in Germany in 2002 (Stinzing and Lang 2003), Italy in 2004 (Tantardini et al. 2004), and Switzerland in 2006 (Holdenrieder and Sieber 2007). Infected plants form girdling cankers at leaf nodes and then exhibit twig dieback. After successive years of dieback, shoots develop at the plant base. The shoots are susceptible to the fungus, and cankers form around them and sometimes around entire trunks, eventually killing the dogwood.

Dogwood trees remove large amounts of calcium from soil. When leaves drop, calcium becomes part of the litter zone. Dogwoods are therefore considered a primary calcium pump for deciduous forest ecosystems. Jenkins et al. (2007) studied loss of calcium in several ecosystem types affected by dogwood anthracnoses. Oak-hickory forests appear most likely to be affected, with a 42% decline in the amount of calcium cycled, losses not completely offset by overstory increases in cycling. Nation (2007) found calcium-dependent land snails at higher densities in litter from flowering dogwood than from American beech (*Fagus grandifolia*) and blackgum (*Nyssa sylvatica*), indicating that effects of the fungus may cascade beyond dogwoods to other forest species.

Oak wilt (*Ceratocystis fagacearum*) is a fungal infection found only within the United States. It was first recognized in Wisconsin in 1942. Oak wilt may be the most destructive disease of oak in the United States, having particularly severe impacts on trees in the Edwards Plateau and other parts of Texas. Damage in Texas is estimated at over a billion dollars. The disease is an economic concern for both forest industries and individual homeowners, where individual removal and treatment of oaks costs upward of $20,000 per incident (Wilson 2005).

TABLE 9.2   *Selected wildlife diseases*

Disease	Agent	Vector/transmission	Host/reservoir
PLANTS			
chestnut blight	fungus	spores in wind	chestnut trees
dogwood anthracnose	fungus	spores in wind, rain	dogwoods
Dutch elm disease	fungus	elm bark beetle	Dutch elm trees
eelgrass wasting disease	fungus	unknown—warm temperatures?	eelgrass
Florida torreya mycosis	fungus	unknown	Florida torreya
oak wilt	fungus	insects, root transfer	oaks in U.S.
pondberry stem dieback	fungus	unknown—beetle?	pondberry
powdery mildew, grape	fungus	spores in wind	European horse chestnut, Indian beam tree
root fungus	fungus	infected soil	various species, e.g., jarrah
sudden oak death syndrome	fungus	infected air, water, plants	some oaks, rhododendron, huckleberry
ANIMALS			
African swine fever	virus	soft ticks	wild boar
aujeszky's disease (pseudorabies)	virus	oral, nasal contact	wild boar
avian pox	virus	mosquitoes, direct contact	Hawaiian honeycreepers
bass virus	virus	contaminated water	largemouth bass
blue tongue	virus	biting flies	wild ruminants, livestock
bovine viral diarrhea, alpha	virus	direct, indirect contact, fluids	cervids
herpesvirus, catarrhal fever	virus	contact with birthing sheep	cattle, ungulates
canine distemper	virus	contaminated fluids, aerosol	many carnivores
duck plague	virus	direct contact; in water	waterfowl
epizootic haemorrhagic disease (EHD)	virus	biting flies	deer, blesbuck, aridad, bighorn sheep, elk, pronghorn
hemorrhagic septicemia, type IV-b (VHS)	virus	contaminated water	finfish
hemorrhagic disease	virus	direct contact, wind, dry air	rabbit
myxomatosis	virus	insects, direct contact	rabbit
salmon anemia, infectious	virus	direct contact, sea louse?	Atlantic salmon
swine fever, classical	virus	direct, indirect contact	wild boar
avian malaria	protozoa	mosquitoes	Hawaiian honeycreepers, other birds
crayfish plague	protozoa	infected crayfish	crayfish
whirling disease	protozoa		sludge worms, salmon, trout, whitefish
anaplasmosis	bacteria	tick	cattle, deer, sheep, antelope, wildebeest, blesbuck, duike
avian botulism, type C	bacteria	contaminated carcus, maggots	waterfowl, shorebirds, gulls
avian cholera	bacteria	direct, indirect contact	waterfowl
furunculosis, fish	bacteria	direct contact; in water	trout, other fish
mycoplasm conjunctivitis	bacteria	direct contact, aerosol, eggs	various passerine species
upper respiratory tract disease, tortoise	bacteria	direct contact	desert tortoises
aspergillosis	fungus	inhalation	birds worldwide
bat white-nose syndrome	fungus	direct, human facilitated?	bats
chytridiomycosis, frog chytrid fungus	fungus	distributed by infected pets	frogs, other amphibia
chronic wasting disease	prion	contaminated food,water, soil	cervids

Oak wilt may have originated in Central America, South America, or Mexico. Infected trees wilt and die quickly. Patches of infected trees can be seen in forested areas by observing discoloration of canopy tops. The disease is transferred by spores carried from tree to tree by insect vectors. For example, sap beetles are attracted to aggregation pheromones of other beetles (Bartelt et al. 2004) and fruity smells of mats of fungal spores for breeding and feeding. Contaminated beetles then pass spores onto wounds of other trees (Cease and Juzwik 2001). Intertwined root systems of trees also transfer the disease, even between species of oak. Tree removal is often needed. Trenching around trees sometimes prevents transfer of the disease between root systems.

WILD ANIMAL DISEASES. Crayfish plague (*Aphanomyces astaci*) arrived in Italy in 1860, probably in the ballast of a North American ship (Ninni 1865). This

TABLE 9.3 *Selected zoonoses*

Disease	Agent	Vector/transmission	Host/reservoir
avian influenza, highly pathogenic (HPAI, H5N1)	virus	direct contact	poultry, wild birds, incl. geese, swans, raptors, gulls, herons, pigs, cats, tigers
Colorado tick fever	virus	ticks	porcupines, chipmunks
ebola	virus	body fluids	bats
encephalitis, European	virus	tick, contaminated cow/goat milk	tick, rodents, cows, goats
encephalitis, Japanese	virus	mosquito	water birds, pigs
encephalitis, LaCrosse	virus	mosquito	chipmunks, tree squirrels
encephalitis, Powassan	virus	ticks	spotted skunk
encephalitis, Russian	virus	ticks, contaminated cow/goat milk	ticks, rodents, cows, goats
equine encephalitis, eastern	virus	mosquitoes	riparian birds, horse, dogs
equine encephalitis, western	virus	mosquitoes	passerines
equine encephalitis, Venezuelan	virus	mosquitoes	horses, rodents
hantavirus (cardio)pulmonary syndrome (HPS, HCPS)	virus	contact w/ wildlife, contaminated material	mice, rats
hemorrhagic fever w/ renal syndrome (HFRS)	virus	contact w/ wildlife, contaminated material	mice, rats, voles
hendra virus	virus	contact w/ horses	fruit bat (Pteropid), horses
monkeypox	virus	animal blood or bite	black-tailed prairie dog, rodents, primates
Newcastle disease	virus	direct contact	pigeons, waterfowl, other birds
nipah virus	virus	contact w/ wildlife, contaminated material	fruit bat (Pteropid), pigs
rabies	virus	saliva of infected animals	raccoon, raccoon dog, wolf, fox, bats, dogs, skunks, mongoose
severe acute respiratory syndrome (SARS)	virus	aerosol, contact w/ infected animal	horseshoe bats (Rhinolophidae)
swine influenza, H1N1	virus	contact w/ pigs	swine
West Nile encephalitis	virus	mosquitoes	birds
West Nile virus	virus	mosquitoes	birds, other vertebrate
anaplasmosis, human monocytic	bacteria	ticks	small mammals
anthrax	bacteria	contact with hoofed animals	hoofed animals
avian botulism, A, and B	bacteria	ingest toxins, uncooked meat	waterfowl, shorebirds, gulls
avian tuberculosis	bacteria	contaminated feces, food, water	birds, mammals
bovine tuberculosis	bacteria	contact	cervids, canids, bovids, others
brucellosis	bacteria	ingest infected material	elk, mule deer, buffalo, cattle, canids, goats, pigs
cholera	bacteria	contaminated water	contaminated water, shellfish
ehrlichiosis, human monocytic	bacteria	ticks	white-tailed deer
leptospirosis	bacteria	contaminated water, food, soil	livestock, commensal rodents, dogs, many wildlife
Lyme disease	bacteria	ticks	mice, deer
plague	bacteria	fleas	rodents, prairie dogs, ground squirrels
Q-fever	bacteria	ticks	many wildlife
rocky mountain spotted fever	bacteria	ticks	small mammals
salmonellosis	bacteria	contaminated food, water, soil	many reptiles, birds, mammals
sylvatic plague	bacteria	direct contact or fleas	rodents, canids
tularemia	bacteria	ticks, biting flies	birds, mammals, rabbits (type A), rodents (type B), hares (both), prairie dogs (pets)
babesiosis	protozoa	ticks	many mammals
giardiasis	protozoa	fecal contaminated water	beaver
malaria	protozoa	mosquitoes	vertebrates
toxoplasmosis	protozoa	contaminated food or water	felids, sea otter, birds, mammals
visceral leishmaniasis	protozoa	sandfly	dogs, other mammals
aspergillosis	fungus	inhalation	birds worldwide
coccidioidomycosis	fungus	inhalation of fungus	sea otter, soil
histoplasmosis	fungus	inhalation	bat, bird guano–enriched soil

(continued)

TABLE 9.3   (*Continued*)

Disease	Agent	Vector/transmission	Host/reservoir
acanthocephaliasis	parasite	infected water	sea otters, birds, mammals
chagas disease	parasite	blood-sucking insects	armadillos, opossum, other mammals
schistosomiasis	parasite	infected water	snails
trichinellosis	parasite	infected meat	many birds, mammals, incl. rodents, pig, mountain lion, bear
bovine spongiform encephalopathy	prion	infected food	cattle

disease is believed to be caused by a protist that only infects crayfish. It is endemic to North America, infecting signal crayfish and Louisiana swamp crayfish (*Procambarus clarkii* and *Orconectes limosus*). On exposure to the oomycetes that cause the disease, North American species respond vigorously with immune defenses, often showing no symptoms (Unestam and Weiss 1970).

In Europe, the noble crayfish (*Astacus astacus*), the white-clawed crayfish (*Austropotamobius pallipes*), the slender-clawed or Turkish crayfish (*Astacus astacus*), and other European crayfish are popular food and important prey for fish such as trout, pike, perch, and chub (Westman and Savolainen 2001). Because these European crayfish offered little immunity, plague spread rapidly through crayfish populations of northern Italy, quickly devastating them, as it moved on to crayfish populations in France and Germany and then to other European countries (Alderman 1996). The plague created both the economic loss of crayfish as a valued food and the fishery loss of important species in the food chains.

European responses included replacing noble crayfish with the larger and comparably flavored signal crayfish from North America. Because they were asymptomatic, signal crayfish were assumed to be free of plague. Signal crayfish harbored plague, however, and vectored further impacts on European crayfish populations. At the time of writing, problems continued in Europe. Furthermore, tests on crayfish in other parts of the world, including Australia and the Pacific Islands, indicate high vulnerability to the plague should these crayfish come into contact with the disease (Unestam 1976).

Frog chytrid fungus is a cause of declining global populations of amphibians (e.g., Fisher et al. 2009). For example, the fungus contributed to loss of 40% (20 species) of frogs in the Monteverde Cloud Forest Preserve in Costa Rica (Pounds et al. 1997) in the late 1980s. At about the same time, the fungus was found in amphibians disappearing from wet tropical areas in Australia (McDonald 1990). The fungus has since been associated with declines elsewhere, including diurnal frogs, genus *Atelopus*, in South and Central America, and *Bufo* and *Rana* species in the Sierra Nevada, California (Lips et al. 2006; Pounds et al. 2006).

Largemouth bass virus (*Ranavirus lmbv*) was first isolated in Lake Weir, Florida, in 1991 and was associated with a kill of about 1,000 fish in the Santee-Cooper Reservoir, South Carolina, in 1995 (Grizzle et al. 2002). It has since been found in lakes and ponds in 18 other southeastern U.S. states. The origin of the virus remains unknown.

Bass often show no symptoms until they are stressed by factors such as hot weather (Grant et al. 2003), poor water quality, or excessive handling by anglers. Symptoms then include floating on or near the surface and difficulty swimming or remaining upright. The virus may affect swim bladders of fish, which sometimes exude colored substances or appear overinflated (Goldberg 2002).

Although many bass may be infected, die-offs usually impact only small portions of populations. Other fishes and some amphibians may also carry the virus, but only largemouth bass die from infections. Schramm et al. (2004) looked at temperatures of holding containers in fishing tournaments and presence of bass virus and developed a model that predicted likelihood of survival of bass caught and released. They concluded that the virus affected post-release survival.

Plasmodium relictum is a filarial protist that causes avian malaria. The disease infects erythrocytes of birds, damaging them and causing anemia. Given the importance of oxygen for metabolism and flight, impacts on birds are often lethal unless they have developed some tolerance to the disease. The protist is not spread directly between birds but requires vectors such as mosquitoes. *P. relictum* has been found in birds in England, France, New Zealand, and several mainland U.S. states.

P. relictum spread quickly through Hawaii's bird fauna after the southern house mosquito arrived in a ship's water barrels in 1826. The disease infected birds, including honeycreepers and Hawaiian crows (*Corvus hawaiiensis*), at lower elevations where mosquitoes could survive, and it became a factor in the extinctions of at least ten bird species (Atkinson et al. 1995). Some honeycreepers in forests at lower elevations are developing tolerance to the disease, so the reservoir of avian malaria is increasing. Freed et al. (2005) have argued that this increases pressure for transfer of the disease to birds at higher elevations.

Chronic wasting disease (CWD) is a transmissible spongiform encephalopathy found in cervids, including mule deer (*Odocoileus hemionus*), white-tailed deer, elk, and moose. The syndrome was first identified in 1967 in mule deer in a research facility in Colorado and subsequently in free-ranging deer and elk in Colorado and Wyoming. CWD may be endemic to Colorado, Wyoming, and Nebraska. It has since been found in free-ranging deer in at least 22 other states (recently in a Missouri captive deer facility) and the provinces of Alberta and Saskatchewan, perhaps brought along with deer reintroductions.

Infected deer lose weight over weeks or months and experience excessive salivation, difficulty swallowing, excessive thirst, and excessive production of urine. Most die within months, sometimes of aspiration pneumonia. The disease is highly transmissible, probably from animal to animal by direct contact or through contaminated food or water.

Hunters and others who eat venison are concerned that CWD might be transmitted to humans. Ranchers are worried that infected free-ranging deer or elk might contact or contaminate food or water that would transmit the disease to livestock. Experimental studies show that some transmission may be possible under laboratory conditions. However, transmission to noncervid mammals has not been observed under natural conditions and appears unlikely. The U.S. Centers for Disease Control and Prevention (CDC) recommend that venison from CWD-infected cervids not be eaten and that all venison be well cooked (Belay et al. 2004).

ZOONOSES. The influenza A virus genome is made of eight segments of negative-stranded RNA. Two of the eleven proteins coded by RNA are used to classify influenza viruses in wild birds and poultry. Sixteen subtypes of one glycoprotein, hemagglutinin (HA), and nine subtypes of another, neuraminidase (NA), have been found. The name of the particular influenza is determined by the combination of HA and NA subtypes, e.g., H5N1 being composed of the fifth subtype of hemagglutinin combined with the first subtype of neuraminidase (fig. 9.2). Influenza A has been identified in many mammalian and avian taxa, but the natural reservoir is believed to be wildfowl and waterbirds. Found mostly in a low-pathogenic (LPAI) form, H5 and H7 subtypes may become highly pathogenic (HPAI, formerly "fowl plague"; Clark and Hall 2006; Olsen et al. 2006).

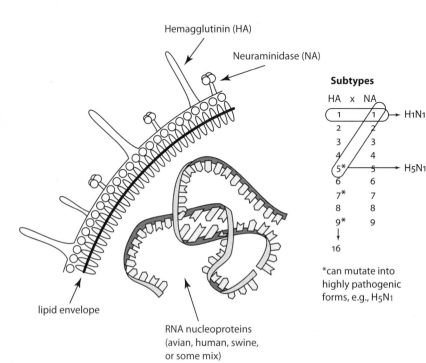

Figure 9.2 Avian influenzas are named according to subtypes of hemagglutinin and neuraminidase. *Illustration by Lamar Henderson, Wildhaven Creative LLC.*

Olsen et al. (2006) describe a series of Eurasian outbreaks of H5N1. An HPAI outbreak of this virus occurred on poultry farms and markets in Hong Kong in 1997, resulting in the first documented case of human H5N1 influenza and associated fatality. The H5N1 outbreak recurred in 2002. It was found in waterfowl in two parks in Hong Kong and in other wild birds. By 2003 it had resurfaced again, this time raising havoc with the whole of the poultry industry in Southeast Asia. In 2005, the virus was found in migratory birds in Qinghai Lake, China, and subsequently affected wild birds broadly. That epidemic alone reduced by about 10% global numbers of bar-headed geese (*Anser indicus*). More recently, the virus has been found in Asia, Europe, the Middle East, and some countries in Africa. In Europe, it has affected mute swans (*Cygnus olor*) and whooper swans (*C. cygnus*) as well as other wild waterfowl and some raptors, gulls, and herons. The disease has caused mortality in over 60 species of wild birds. It has been transmitted to at least 175 humans, causing 95 deaths, and has been found in pigs, cats, tigers, and leopards (Olsen et al. 2006). In the past decade, H5N1 outbreaks have occurred in Asia, Russia, the Middle East, Europe, and Africa; H5N2 in Italy and Texas; H7N1 in Australia, Pakistan, Chile, and Canada; H7N4 in Australia; and H7N7 in the Netherlands.

Another zoonotic virus, Nipah virus is an emerging disease that causes encephalitis (inflammation of the brain) in infected people. Fruit bats (*Pteropodidae*) are a reservoir, henipaviruses (a category of virus that includes Nipah virus) having been found in fruit bats ranging from Australia (*Pteropus* spp.) to Africa (*Eidolon* spp.). Bats appear asymptomatic. The disease is transmitted from animals such as pigs to humans as well as from humans to humans. Since it was recognized in pig farmers in Malaysia in 1999, there have

been 12 outbreaks, all in Southeast Asia, resulting in hundreds of human deaths (e.g., Chua et al. 2000).

In Bangladesh, an outbreak of Nipah virus was related to consumption of raw date palm juice that was contaminated with urine from fruit bats (Luby et al. 2006; fig. 9.3). Chua et al. (2002) and Chua (2010) evaluated factors associated with the epidemic in Malaysia and concluded that the emergent disease was driven by climatic as well as anthropogenic factors. Deforestation for pulpwood and industry resulted in fewer native trees for pteropid bats. The problem was exacerbated by an El Niño–caused drought, which, along with slash-and-burn-induced haze, reduced flowering and fruiting of forest trees. Fruit bats moved into cultivated fruit trees grown near confined pig operations. The bats then had access to pigsties where they hung to eat fruit, sometimes dropping fruit or urinating, spreading potential sources of the virus.

Rabid animals were described in Mesopotamia as early as the twenty-third century BC (Fu 1997). Centuries later, Louis Pasteur described the disease as viral (before the term "virus" was clearly defined) and developed an attenuated preparation from rabbits that was used, with some risk, to treat patients bitten by rabid dogs.

Rabies is caused by rhabdoviruses (rod-shaped viruses) of the genus *Lyssavirus*. Rhabdoviruses infect plants and animals, including insects, fish, and mammals. These viruses are made of single-, negative-stranded RNA surrounded by a matrix protein, in turn surrounded by a bilayered envelope, with glycoproteins shaped into spikes on an outer surface. The RNA contains the codes for five proteins, enough information for replication of the virus in the cytoplasm of a cell. The viruses are transmitted by contact with the

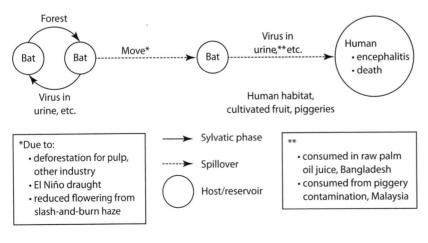

Figure 9.3 Nipah virus has spilled over from bats to humans. Human activity and environmental change has forced movement of bats closer to humans, where consumption of items contaminated with infected bat urine has led to human deaths. *Illustration by Lamar Henderson, Wildhaven Creative LLC.*

saliva of an infected animal—e.g., by bites, scratches, or licks on open skin or mucous membranes. In humans, symptoms appear at first to be flu-like, for weeks to years, and then they progress rapidly to paralysis, anxiety, confusion, paranoia, hallucinations, delirium, and death. Paralysis includes the throat and jaw, resulting in an inability to swallow to quench thirst and consequent hydrophobia (fear of water) and collection of saliva in the mouth until it overflows, hence "foaming."

Vaccination of pets, careful monitoring for presence of diseased animals, and safe and effective treatment immediately after exposure to rabid animals have greatly reduced the human incidence of rabies in Europe and North America. The cost for such efforts is high, over $300 million per annum in the United States, mostly paid by pet owners. In regions of the world where pets have been widely vaccinated, most human cases are now transmitted by wildlife, these wild cases sometimes collectively called "sylvatic" rabies. In Europe, fox and raccoon dogs serve as reservoirs. In the United States, primary hosts include raccoons along the eastern seaboard, skunks and foxes in the south-central states, skunks in the north-central and southwestern states, foxes in Alaska, mongoose in Puerto Rico, and bats broadly (Slate et al. 2002; fig. 9.4). A recent outbreak in coyotes in Texas was controlled.

Often caused by dog bites, rabies is rampant in Africa and Asia. There, over 55,000 people die annually (WHO undated d). The World Health Organization (WHO) has declared September 28 Rabies Day in an attempt to increase global awareness of the continued presence and importance of the disease.

West Nile virus was first isolated in the West Nile Province of Uganda in 1937, and it occurred in epidemics in parts of Europe during the 1950s, 1960s, and 1990s. An outbreak occurred in humans in New York City in 1999, possibly a consequence of international travel by infected persons or transport of infected birds. West Nile virus infects many vertebrate species, but infected animals are asymptomatic in the normal range of the virus—Africa, the Middle East, western Asia, and Europe. This virus is transmitted by mosquitoes (*Culex* spp. mainly, but also *Aedes* spp. and *Anopheles* spp.).

Most infected people never develop symptoms. Those who have them experience fever, headache, fatigue, skin rash, and occasionally swollen lymph glands and eye pain. In the severe form of West Nile virus, the central nervous system is infected and meningitis or encephalitis may occur. In the United States, about 1,700 cases were reported to the CDC in 2009–10, with about 69 fatalities.

"Beaver fever," or giardiasis, is a mammalian zoonotic disease transmitted by a flagellated protozoan that causes mild to severe stomach distress in people. Giardiasis can be transmitted from wildlife such as

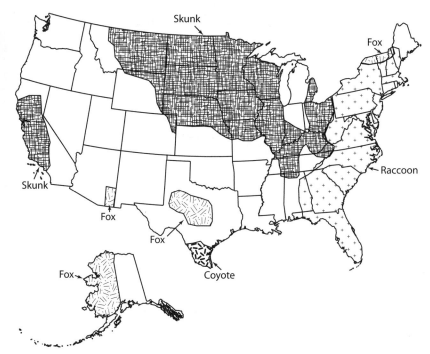

Figure 9.4 Distribution of sylvatic rabies in the United States. Redrawn after Slate et al. 2002, fig. 1, with permission. *Illustration by Lamar Henderson, Wildhaven Creative LLC. With permission of the Vertebrate Pest Council.*

beavers to people by fecal contamination of drinking water. For example, clear water of mountain streams may appear pristine and safe and yet contain giardia cysts if beavers upstream are infected. About 19,000 cases were reported annually to the CDC in the United States during 2006–8 (Yoder et al. 2010). Children aged 1 to 9 years and adults 35 to 44 had the greatest incidence of the disease, with peak onset from early summer through early fall, probably associated with increased outdoor activities such as camping and swimming.

Histoplasmosis is a fungal infection caused by *Histoplasma capsulatum*. The fungus grows in soil enriched with bird or bat guano and has been found in litter of poultry barns, in bat roosts in caves and houses, and under bird roosts. Histoplasmosis occurs globally, but it is most common in North and Central America. In the United States, histoplasmosis mostly occurs in the Mississippi and Ohio River valleys. Spores can reside for years in guano. The spores become airborne when disturbed—e.g., with movement of soil for development of a subdivision—and can then cause infections upon inhalation. Histoplasmosis most often affects the lungs of infected people, who often think they have a cold. Particularly for those with emphysema or compromised immune systems, the disease can progress into a disseminated form that affects other organs (Kauffman 2007). For example, Reddy et al. (1970) found the disseminated disease in 25 of 530 patients who had been diagnosed with active histoplasmosis. In that study, all untreated cases of disseminated histoplasmosis were fatal. In an outbreak involving over 100,000 residents of Indianapolis, Indiana, in 1978–79, 46 had progressive disseminated infection, and 15 died (Wheat et al. 1981).

Summary

- Wildlife diseases are abnormal conditions that impair functions of nondomesticated organisms; if they infect humans, they are also zoonoses.
- All wildlife diseases have sylvatic cycles wherein vectors move diseases between wildlife hosts; zoonoses also spill over to humans as, e.g., Lyme disease carried by ticks to humans, field mice, and deer.
- RNA- and DNA-based pathogens mutate and sometimes mix with other pathogens in hosts, thereby adapting to changing environmental conditions, and sometimes allowing species jumps.
- Frequency of wildlife diseases and zoonoses appears to be increasing, probably due to burgeoning human populations and increased contact with wildlife.
- Native pathogens serve important ecosystem functions and are of little or no concern; exotic pathogens have had no time for co-evolution with affected species and can disrupt normal ecosystem function.
- Examples of wildlife diseases include chestnut blight, dogwood anthracnose, oak wilt, crayfish plague, frog chytrid fungus, largemouth bass virus, avian malaria, and chronic wasting disease.
- Examples of zoonoses include influenza A, Nipah virus, rabies, West Nile virus, giardiasis, toxoplasmosis, and histoplasmosis.

Review and Discussion Questions

1. If wildlife diseases are a normal part of ecosystems, why would they be of interest to wildlife damage managers? Explain with specific examples.
2. What unique problems are posed by wildlife diseases of aquatic organisms? How, for example, does one get a pathogen out of a pond, a stream, a lake?
3. Do you think it is just by chance that many wildlife viruses are made of RNA? How might RNA relate to survivability of wildlife diseases? Are DNA-based diseases at a disadvantage? Share your thoughts along with some specific examples.
4. Might the notion of propagule pressure have an application in the study of wildlife diseases? Explain with examples. Conversely, might the notion of the basic reproduction number, R_0, be applied to invasive species? Explain, using examples.
5. Do wildlife diseases impact threatened or endangered species? How about, for example, the African wild dog? Or the black-footed ferret? Explain.

PART IV • METHODS

In this section, we describe methods used to solve problems caused by wildlife, organizing them into physical, chemical, and biological techniques.

10

Physical Methods

Physical methods may be used to exclude, restrain, or kill damaging species; frighten, intermittently repel, or persistently repel a species; reduce natality of populations; or alter habitats or ecosystems to reduce the presence of or damage caused by wildlife (table 10.1).

METHODS THAT RESTRAIN OR KILL

Statement
Damaging wildlife may sometimes be physically restrained to prevent damage, to gather information, or to take other management actions, or damaging wildlife may be killed using physical methods.

Explanation
Restraining devices include cage traps, foothold traps (both open and enclosed), body-grip traps, cable devices (i.e., snares, both powered and nonpowered), net traps, and net guns. Use of traps is guided by standards based on factors such as effectiveness, humaneness, economy, and safety (e.g., "Best Management Practices for Trapping in the United States," Association of Fish and Wildlife Agencies undated; Schemnitz 1996).

Cage or box traps are designed so that animals can enter but not exit them. They may have triggers that release when the animal moves against them, closing one-way doors, or the doors may be controlled remotely. Or, traps may have complex entrance designs that facilitate getting in but prevent animals from getting out. When closed, traps may send electronic signals to the trappers. Traps or cages may be baited with food, conspecifics or their effigies, or other attractants, such as pheromones.

Cage traps are used on land or in water. They may be used for capture and release, relocation, or capture and euthanasia. Some traps resemble boxes with walls and drop-doors made of wire, solid wood, plastic, netting, or metal. Others, such as the Hancock or Bailey beaver traps, resemble suitcases. They can be mouse-sized or smaller, like minnow traps, or culvert-sized, like Stephenson box traps (McBeath

TABLE 10.1 *Physical methods used in wildlife damage management*

Type	Example
EXCLUSIONARY	
bud caps	Breathe Easy Bud Cap
fencing	high tensile electric woven wire
netting, grids	Kevlar wire grids
spikes, sticky substances	"Ag spikes," Tanglefoot
tube barriers	Vexar tubing
HABITAT, ECOSYSTEM ALTERATION	roosts, Clemson Beaver Pond Leveler
KILLING	
snap trap	conibear trap, Victor 4 Way
REPELLENT	
aircraft, boats	Fixed-winged; remote control
bioacoustical	Av-Alarm
effigies	vultures
flags, tapes	fladry, Mylar
laver, strobe lights	Lord-Ingerie Lem 50
pyrotechnics	firecrackers, (bangers, screamers); propane cannons
scarecrows	"Scary Man"
RESTRAINT	
cable devices	Collarum, neck restraint
cage traps	Australian crow trap
foothold traps	"raptor" pole trap
net guns	net guns
net traps	fish trammel nets
SHOOTING	
blowpipes	Blowpipe B16
handguns	pistol, revolver
rifles, hunting	Winchester model 1894
shotguns	pump action, Remington 870
tranquilizer gun	dart gun, e.g., Dan-inject projector rifle
STERILIZATION	addling, surgical

1941) or Clover (1954) traps, made for animals such as feral pigs, deer, and bears.

A foothold trap can be of the long-spring, coil-spring, or jump-trap type with a spring underneath (fig. 10.1). Traps range in size from No. 0 for small rodents and some birds to No. 1 for a muskrat- or skunk-sized animal to No. 4.5 for a mountain-lion-sized animal. Foothold traps have two opposing jaws attached to a baseplate. The jaws may be offset (i.e., have a small space between the gripping surfaces), double, laminated, or padded, or they may have additional springs (i.e., "four-coiling"). The baseplate has a pan and a dog (trigger) attached to it. The trigger is released when pressure on the pan exceeds a threshold, and the springs snap the jaws closed. The threshold can be adjusted, and the pan may have stops that limit the distance between it and the trap jaws.

Foothold traps may be modified for specific uses. For example, these traps are sometimes used to restrain raptors as part of a set called a "pole trap." The jaws may be wrapped in cloth and the coil springs heated to weaken them, to further reduce chances of injury to the bird. The traps are set on top of posts but designed to allow the restrained bird to drop to the ground, where it may be more relaxed after capture. For example, Butchko (1990) used pole traps to protect endangered California least terns from predation by raptors.

Nets are also used to restrain and capture wildlife. Nets to capture fish include **seines** (e.g., beach seine nets, purse seine nets), dip nets, **trawl nets** (long bags or sock-shaped nets that are pulled through water), **gill nets** (designed so fish or reptiles of a particular range of sizes are caught when they try to pass through the net), **enmeshing** nets (not as selective as gill nets), and **trammel nets** (used for species not easily caught in gill nets, such as flatfish or sturgeon).

Net traps can be shaped in hoops and hand thrown; for example, hoop nets designed by Ronconi et al. (2010) were used to capture sea birds, including great and sooty shearwaters (*Puffinus gravis* and *P. griseus*) and red-necked and red phalaropes (*Phalaropus lobatus* and *P. fulicarius*). Specialized types such as **fyke nets** have wings for use in lakes and internal cones that direct fish into a collecting area. Fyke nets can be further modified for turtles and other reptiles. Net traps can be fired from cannons or launched with rockets so that they quickly cover wildlife that has been lured to a baited site. Cannon or rocket nets have been used to capture and restrain many types of birds and mammals, including deer and bighorn sheep.

Net guns are handheld and are sometimes fired at wildlife from aircraft. O'Gara and Getz (1986), for example, used net guns to capture offending golden eagles.

A **cable device** (a **snare**) is a steel cable with a loop on one end. The loop draws closed when an animal attempts to pass through it, put a limb through it, or withdraw from it. Cables are made of wire strands, and sizes vary. For example, a seven-by-seven cable is made of seven bundles, with each bundle containing seven wires. A nonpowered cable depends on the

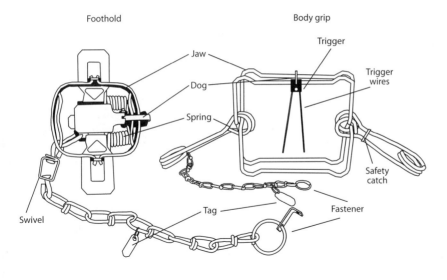

Foothold

Body grip

Jaw

Dog

Spring

Trigger

Trigger wires

Safety catch

Swivel

Tag

Fastener

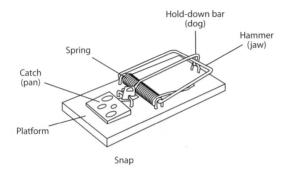

Hold-down bar (dog)

Hammer (jaw)

Spring

Catch (pan)

Platform

Snap

Figure 10.1 Different-looking traps often share common structures, such as jaws, dogs, springs, and triggers. *Illustration by Lamar Henderson, Wildhaven Creative LLC.*

forward movement of the animal to close. A powered cable has a spring or other power source to place the cable on a limb of the animal or close the loop of the cable (see, e.g., Collarum Neck Restraint or the Wildlife Services' Turman snare; Shivik et al. 2005). Cables can have relaxing locks, breakaway hooks, loop-stop ferrules, in-line swivels, and/or anchor swivels. Relaxing locks allow closure of the loop when the animal pulls but no additional closure when the animal stops. Break-away devices release animals that pull with greater strength than the targeted animal. A loop-stop ferrule prevents a loop from closing past a preset diameter. A maximum loop stop prevents too large an animal from entering the loop. A minimum loop stop prevents the loop from closing around the animal's foot. Cable devices can be set horizontally or vertically, and they may include a spring-activated throw arm to capture the animal's leg.

Restraining devices may also be specially designed for specific needs. An example is the "EGG Trap," a form of spring trap used particularly in the United States to restrain raccoons.

Mechanically powered killing traps are designed to kill quickly. They are often powered by one or more springs. Examples are snap and body-grip (e.g., conibear) traps. Snap traps include "four-way" traps (so called because a rodent will be caught regardless of which direction it approaches the trigger from) and "museum specials," designed to withstand the vigor of frequent use, both made for mouse- or rat-sized animals. Most snap traps break the back or neck of the animal, which dies quickly. Conibear traps are often used for aquatic animals and are made in different sizes, such as 110 (muskrat size), 220, and 330 (beaver size). Other kill traps include gopher and mole traps that are set in burrow openings (see fig. 10.1).

Restraining systems include foothold or cable devices, swivels, and anchors. Swivels allow movements that keep the anchor or cable from breaking and help to reduce injury. One swivel is usually placed on the base plate of a foothold trap, and another two or more are placed along the anchoring system. Anchors can be stakes or cable stakes, but they need to be strong enough to hold the largest animal that can be caught.

In-line shock springs are sometimes also used to cushion lunges of the captured animal and reduce chance of injury or escape. A submersion trapping system might be used to kill an offending animal that lives in aquatic habitats or near water, such as a beaver or muskrat. The system would be designed with a restraining device (e.g., a body-grip trap, suitcase, or cable device) and cables to hold the animal underwater so that death occurs relatively quickly.

The restraining system also includes the "set." The quality of the set depends largely on the experience of the practitioner. The set attracts intended animals and discourages others. The set may include baits such as food, scents, or lures as well as specific configurations. A myriad of baits and attractants are used successfully. Grains such as corn, milo, wheat, and oats, sometimes dyed in colors and soaked to swell and look like berries, have been used to attract granivorous and frugivorous birds. Live mice have served to visually attract predatory birds. **Jesses** (cables with loops, usually made of monofilament plastic) may be attached to cages to catch raptors. A mix of peanut butter and oatmeal is a common bait for rodents (e.g., Schemnitz 1996). Punctured cans of dog or cat food and predator urine, real or synthetic, with glycerin sometimes added as a preservative, are often used to attract predators. Rotted meat or eggs also effectively attract predators and scavengers. Synthetic fermented egg was designed as a predator attractant (and deer repellent; Bullard et al. 1978). Pheromones such as musks and beaver castor are sometimes used to attract predators and other mammals. The set may also include sieved dirt, branches, or other vegetation to camouflage the trap or direct an animal toward the pan.

Pistols, rifles, or shotguns may be used to shoot and kill offending animals, from the ground or from aircraft. Shooting can also be part of a remote delivery system used to deliver agents such as anesthetics, immobilizing agents, contraceptives, or immunogenic compounds. Blowpipes (powered or not) and dart guns are also used, the agent delivered in a matrix such as a "biobullet" that breaks down once inside the animal (e.g., Kreeger 1997).

Aerial gunning is performed with either light fixed-wing aircraft or helicopters. This type of gunning is sometimes used to manage predators such as coyotes in the western United States and wolves in interior Alaska, on both public and private land.

Restraining and killing systems carry risks. One is injury to wildlife from restraining devices such as leghold traps or cable devices. Another risk is capture-related myopathy, a muscle disease characterized by damage to muscle tissue caused by physiological changes due to extreme exertion, struggling, and stress. Some groups of animals, such as cervids, seem particularly vulnerable (Beringer et al. 1996). Reducing visual and auditory stimulation in the postcapture environment by not looking directly at wildlife, covering the cage, avoiding loud noises, administering selenium and vitamin E, and injecting sodium bicarbonate (to neutralize ketone body buildup) counters capture myopathy (e.g., Businga et al. 2007).

Humaneness of trapping systems depends on the ethical nature and consideration of the practitioner. With rare exceptions, professional practitioners have specialized in wildlife damage and its management out of passion for wildlife, including its humane treatment, and use restraining and killing systems with a deep-rooted concern and compassion for individual wildlife and its future (e.g., Powell and Proulx 2003; chapter 14).

Example

Called a "Williams Carp Separation Cage," this carp trap was designed and tested by Alan Williams, a lock master, for removal of carp from waterways in Australia (fig. 10.2). The traps are set on the upstream side of vertical slots, bottlenecks through which migratory fish must pass. In prototype trials, about 83% of carp that passed through the slots and into the traps jumped into a containment area where they were subsequently harvested. Native fish passed through the trap (Stuart et al. 2006). In a trial of a modified design, the trap removed about 80 tons of carp from a single location between late 2007 and 2009, up to almost three tons per day (Stuart and Conallin 2009). The traps may have applications in North America with other Asian carp that tend to jump and in separating sea lampreys from jumping fish, such as rainbow trout (Stuart et al. 2006).

METHODS THAT EXCLUDE

Statement

Exclusionary methods are often favored because they are nonlethal, but they must be maintained to retain their effectiveness.

Explanation

Exclusionary methods physically separate wildlife from a resource that humans want to protect. Exclusionary devices include barriers such as fencing and netting, tubing, wire grids, and bird spikes.

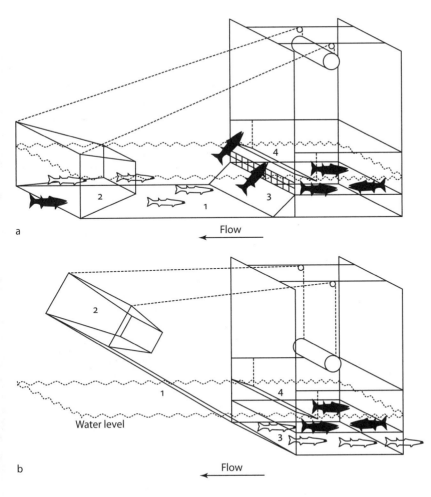

a

Flow

b

Water level

Flow

Figure 10.2 The Australian Williams Carp Separation Cage takes advantage of jumping behavior to separate invasive carp from native fish, removing up to two tons of carp per day. The cage in (*a*) operating and (*b*) bypass modes. *Redrawn from Stuart et al. 2006, with permission of Taylor & Francis Group, www.informaworld.com.*

Barriers such as tubes and fencing have been designed in a myriad of shapes, sizes, and materials for exclusion of animals ranging from small fish to large elephants and for wildlife that swim, walk, climb, or fly. For example, barriers made of water are used to help prevent movement of sea lampreys upstream and into the U.S. Great Lakes. On land, barriers can protect an individual plant or animal, and they can protect herds, entire crops, and even entire natural ecosystems such as national parks.

Tubing is sometimes used to protect seedlings from herbivores such as gophers, mice, rabbits, mountain beavers, beavers, and deer (Marsh et al. 1990). For example, Durite and Tiller protection netting provide protection from pocket gophers, mountain beavers (*Aplodontia rufa*), and lagomorphs for newly transplanted seedlings (e.g., Arjo 2003). Such tubing may be used in small gardens or in major reforestation programs, as in the U.S. Pacific Northwest. Tubing is placed over the transplant and held in place with bam-

boo or plastic stakes. The tubing is biodegradable and may be impregnated with a chemical repellent.

Bud caps can reduce browsing by deer, presumably because the deer do not see the bud (Barnacle 1997). Caps can be as simple as paper or plastic netting placed over the tops of the transplants and held in place with one or more staples. Caps are sometimes made by farmers using local materials, but commercial versions are available.

Fencing can be strictly physical or reinforced with repellent flavors, visual cues (such as flagging on fences used in **fladry** lines), or electrical charge. Fences can be permanent or temporary and portable. Materials such as wood, concrete, plastic, or metal can be made into fencing as well as posts and corner braces. The material may be solid, woven, or braided. Fences may be placed vertically or at various angles. Fencing may be set partly underground to prevent the passage of burrowing animals or may include overhangs to prevent successful climbing. The overhangs may be solid or may give when an animal attempts to move over them. Fences can have aprons that extend outward on either side. A socket, a battery, or the sun can be used to charge fences with electricity. The charge may be lethal or repellent.

Fence designers need to take into account where the fencing is supposed to function. Flat terrain is usually the easiest, whereas hilly terrain requires special consideration. For example, one side of a fence will be closer to the earth when traversing a hill, sometimes necessitating a taller fence. Rocks and other landscape features also need to be considered. Special designs may be required for exclusion over streams or for fences that open onto bodies of water. Electric fences require moist soil to function most effectively. Severe storms and flooding can damage fences and reduce their effectiveness. Design of gates and corners needs to be given careful thought, because these are often weak points exploited by wildlife.

Designers of fencing should also consider the physical ability, behavior, and motivation of the targeted species as well as the purpose of excluding the wildlife. We illustrate with fences intended to keep out white-tailed deer. VerCauteren et al. (2006a), for example, found that a fence at least nine feet tall was needed to exclude white-tailed deer in rough terrain. When motivated, deer can exhibit nontypical behaviors. For example, a single-strand electric fence may be sufficient to protect an orchard when other food deer eat is abundant, but a multiwired fence may be required when alternative food becomes unavailable. A fence that is 50–60% effective may be sufficient and economically advantageous in reducing deer damage to corn from 10% to 5%. However, if the intent is to prevent transmission of diseases to livestock or prevent deer collisions with motor vehicles (e.g., Mastro et al. 2008), no deer intrusions can be permitted (VerCauteren et al. 2006a).

High-tensile electric fences are increasingly used for effective exclusion of deer. These fences do not physically exclude deer but rather depend on avoidance reinforced with electric shock. The fences are usually six feet high, with parallel wires that vary from a foot apart at the top to eight inches apart in the middle. The wires are either positively charged or alternate positive and negative. Slant designs are common in the eastern United States because they adapt to hilly terrain. These high-tensile fences have been used successfully in New Zealand for over 40 years (VerCauteren et al. 2006a).

Single-strand electric fences for deer are often used where portable temporary fencing or relatively inexpensive exclusion will provide satisfactory protection. Seventeen-gauge steel wire can be used, but "poly-fences" that use polytape or polyrope with interwoven wires are more visible (e.g., Seamans and VerCauteren 2006). Peanut butter or other attractants can facilitate effectiveness of the fence designs by attracting deer to them (Hygnstrom and Craven 1988; see box).

Thompson, Jonkel, et al. (2009) provide a variety of electric fence designs that can do everything from protecting garbage cans, bird feeders, and apiaries from bears to excluding livestock from coyotes, wolves, and mountain lions. Their designs include permanent fencing as well as portable fences for campsites, coolers, or game carcasses. In one design, car tires are laid between two sections of cattle fencing (the lowest fencing is placed on the ground) and the garbage can is placed on top. The system is wired so that a bear receives a shock when it touches the can. Similar designs are used to shock bears that touch metal bird feeders. Breck et al. (2006) invented a shocking device, using batteries and an automobile vibrator coil/condenser, that protected simulated bird feeders from black bears at test sites in rural Minnesota.

Barrier fences have also been designed to prevent movement of smaller animals, such as rodents, rabbits, and raccoons. These are much shorter than deer fences and sometimes have a curtain or are placed partly underground to block digging. Designs may or may not be electrified. Ahmed and Fiedler (2002) found both lethal and nonlethal designs that kept rodents out of rice paddies in the Philippines. Reidinger et al. (1985) suggested that nonlethal electrical fencing might "train" rodents to establish territories along the fence; the trained rodents would then protect their territories, excluding conspecifics from the paddies.

Long and Robley (2004) evaluated effectiveness of fence designs used in Australia to exclude certain animals (red foxes, feral cats, goats, pigs, and rabbits) from entire conservation areas. They found that effective feral pig and goat fences could be fabricated with one or two offset electric wires; feral rabbit fences were 900-mm wire netting with apron; wild dog and dingo fences were made with netting with aprons but varied considerably in overall design; and fox and feral cat designs also varied in design, but the most effective ones had floppy tops (Catalogue of fence designs undated).

Barrier fences can also constrain movements of reptiles, including snakes. Examples are the habu (*Trimeresurus flavoviridis*) and the brown tree snake. The habu is a poisonous sedentary snake that is a problem

Scott E. Hygnstrom

Scott Hygnstrom is a professor and extension wildlife damage specialist for the School of Natural Resources, University of Nebraska at Lincoln. Dr. Hygnstrom teaches wildlife damage management and oversees undergraduates and graduate students within the fisheries and wildlife major. He has graduated over 30 students who have gone on to professional careers in wildlife damage management for federal, state, and private institutions.

Dr. Hygnstrom served as co-editor for the 1994 edition of the handbook *Prevention and Control of Wildlife Damage*. He is working on a revision of this text as well as a book focused on wildlife damage problems in southern Africa. Dr. Hygnstrom conducts research and has published extensively in the area of wildlife damage management. Recent studies have included ecology of elk in Nebraska, efficacy of frightening devices for managing deer damage in cornfields, ecological studies on prairie dogs, and studies on chronic wasting disease in Nebraska. Dr. Hygnstrom developed and coordinates the Internet Center for Wildlife Damage Management. The site gets about 15 million hits per year from about 1.5 million visitors in 140 countries.

Dr. Hygnstrom received his B.S. degree in biology from the University of Wisconsin–River Falls, his M.S. in natural resources–wildlife from the University of Wisconsin–Stevens Point, and his Ph.D. in wildlife ecology from the University of Wisconsin–Madison. The recipient of the Conservation Education Award from The Wildlife Society in 2010, he earlier received the Communication Award from the Berryman Institute for Wildlife Damage Management. He is past chair of the Wildlife Diseases and Wildlife Damage Management Working Groups of The Wildlife Society as well as past president of the society's Nebraska chapter. His professional interests include wildlife habitat and population restoration and management, methods of damage management, biocontrol, wildlife diseases, and human dimensions of wildlife and ecology of vertebrates, including ungulates, rodents, and predators.

in Japan. Electric fencing has been used to create snake-free zones and to keep snakes from entering villages on Tokunoshima and Amamioshima and electric substations in Okinawa Prefecture (Campbell 1999). For the brown tree snake, Campbell found a five-wire nylon netting fence most effective as a barrier.

Netting is also used worldwide to exclude smaller mammals or birds from structures or crops. Cloth or plastic, netting may be treated to resist degradation by sunlight and weather. Sometimes the netting is incorporated into cage designs for small gardens or high-cash-value crops. The netting might protect individual plants or trees or orchard or vineyard rows or have suitable infrastructure to be pulled over entire high-value crops, such as cherries or other stone fruit (e.g., Dellamano 2006). Pochop et al. (2001) used plastic netting to drastically reduce nesting by ring-billed gulls (*Larus delawarensis*) on Upper Nelson Island in Washington State. The gulls feed on fry of chinook and other salmon (*Oncorhynchus tshawytscha*) that pass through turbines on the Columbia and Snake Rivers, become disoriented, and float to the surface momentarily. Grid systems of wires have also been used to protect the fry from the gulls (Steuber et al. 1995).

Grid systems of wires can be erected over bodies of water to exclude birds such as geese or ducks. Grid lines are often set 20 feet apart and suspended about 3 feet above the water. Monofilament and other types of string or wire can be used, but **Kevlar cord**, a para-aramid synthetic fiber, is a favorite because it is relatively easy to anchor and tighten and resists degradation by sunlight. Netting erected just 12 inches above the ground can be effective in preventing movement of birds such as geese, ducks, or gulls between the water and shorelines, encouraging the birds to find more suitable habitat. Terry (1984) described the use of grid systems to reduce access of waterfowl to ponds at airports.

Spikes have also found worldwide applications, excluding birds from perching, loafing, or roosting surfaces. Manufactured from either plastic or metal materials, spikes can be attached, for example, to overhangs of doorways or window sills, where the birds or their droppings are unwanted. Spikes have been used to keep birds from perching on antennae, signs, and ledges of structures at airports (Seamans et al. 2007).

Exclusionary devices have problems and carry some risks. Barriers must be maintained to retain their effectiveness. Metal parts corrode and need to be replaced. Storms and wind can damage barriers, either directly or with falling debris. Fire can destroy fencing. Vegetation can grow under electric fences and cause short circuits. Vandalism is a common problem and can be expensive to monitor and repair (Long and Robley 2004). Fence lines sometimes exclude unintended species and can alter migratory routes and interrupt important wildlife behaviors, such as reproduction. Further, fence-line contact between wildlife and livestock may facilitate transmission of diseases such as chronic wasting disease (Wyckoff et al. 2009) and bovine tuberculosis (Rhyan et al. 1995). Double rows of fences have been suggested as a means of eliminating this concern.

Example

Predator fences have been used in Australia to keep dingoes and other feral dogs from livestock since the 1880s. The Dingo Fence extends about 5,320 km (3,306 mi) from Jimbour, Darling Downs, to Eyre Peninsula, Great Australian Bight. The fence is composed of the Great Barrier Fence (Wild Dog Fence, 1,553 miles long) in Queensland, which has a full-time maintenance staff that patrols it weekly; the Queensland Border Fence; the South Australian Border Fence; and the Dog Fence in South Australia (about 1,383 miles long). Mostly wire mesh, the fence is about 6 feet high with an additional foot underground and metal posts about every 27 feet. Fitzwater (1972) recounted the situation of Beltana Pastoral Company, whose property was bisected by the dingo fence. The firm could run sheep only on the protected side of the fence; when sheep were placed on the unprotected side after a drought, the company suffered about a 3% loss due to predation.

SYSTEMS THAT REPEL

Statement

Physical methods can be used to repel wildlife. Some startle damaging wildlife with sudden noise or lights. Others use visual or auditory cues to communicate to wildlife that an area is dangerous and should be avoided.

Explanation

Repellent methods include frightening devices, pyrotechnics (self-contained reactions that can produce heat, fire, explosions), bright or flashing lights, and human activity. Liquid propane gas exploders have been used for years to frighten damaging wildlife, including deer and birds. Some models discharge at random intervals, and some momentarily inflate pop-up scare-

crows (such as ScaryMan) that add to the overall effect. Other pyrotechnic devices include ropes of firecrackers or individual firecrackers—e.g., 12-gauge exploding shell crackers and "bangers" and "screamers" that can be fired from specially designed 15-mm or 17-mm pistols. Pyrotechnics can be effective in dispersing birds from waterways or roosts, but habituation is a problem (Booth 1994).

Laser lights have been surprisingly effective in dispersing some bird roosts. Ones tested are class II (low power) or class III (moderate power). Baxter (2007) used two Lord-Ingerie Lem 50 lasers to completely remove 33,000 gulls (*Larus* spp.) nightly from ponds at a military airfield in England. Gulls moved back daily, but they were dispersed without habituation for at least 26 nights of use. Chipman et al. (2008) included lasers in 53% of their operations as part of an overall hazing program for dispersing American crow roosts in New York. Glahn et al. (2001) tested a Desman model FLR 005 class IIIB laser that directed red light with a 12-mm diameter at the source to see if it would disperse double-crested cormorants from night roosts. They also tested a Dissuader laser used for security; it was a class II diode-type laser with a 76-mm diameter at its source. Both failed to evoke avoidance in laboratory trials but effectively dispersed cormorants in field trials. The lasers offer a quiet, selective, and effective alternative to pyrotechnics for use in refuges or wetland habitat where cormorants tend to roost (Glahn and Blackwell 2000).

Aircraft might be used to haze offending wildlife. Fixed-wing aircraft are often used over open, flat, or gently rolling terrain, whereas helicopters are used over bushy, timbered, broken, or mountainous areas. Use of aircraft involves maneuvering close to the ground at slow airspeeds, and the aircraft may be specially designed for that purpose and the pilots specially trained. Handegard (1988) reported use of fixed-wing aircraft to haze blackbirds damaging sunflowers in 6 districts and 7,000 to 10,000 square miles in North Dakota; these aircraft were used to harass depredating birds with low-level flying and shotguns. Although the overall efficacy has been questioned, the approach spreads damage among farmers and may allow plants to compensate for some damage by increasing the size of remaining sunflower heads. Remote-control model aircraft and boats have been used to haze birds at airports, agricultural and aquaculture areas, and landfill sites (Saul 1967; Littauer 1990; Yashon 1994).

Effigies, or likenesses, can be effective for varied periods of time. Effigies for repelling vultures have already been described. Stickley and King (1995) described the effectiveness of ScaryMan effigies in dispersing overwintering double-crested cormorants at a catfish fingerling complex in Mississippi. They found a "sudden and extreme drop in the number of cormorants flushed on the trial site in the first week of use" (p. 91). Morrison and Allcorn (2006) found the ScaryMan effective as part of an integrated approach to disperse gulls (*Larus* spp.) from a Coquet Island reserve in Northumberland, England. Gulls competed for nesting space with terns (*Sterna* spp.) and were becoming an increasing problem on the island. In another case, Saul (1967) used effigies (gull carcasses preserved in formalin, wings outstretched) to remove gulls from the international airport in Aukland, New Zealand. Dejong and Blokpoel (1966) had less spectacular results in Holland, reporting the need to move effigies frequently to retain their effectiveness.

Okarma and Jędrzejewski (1997) reported that fladry—ropes with large rags—hung along forest paths were used to circle wolves for hunts by Polish kings in the fifteenth century, a practice picked up for hunting wolves in Russia in the 1800s. These researchers used fladry and netting to capture wolves for study in the Białowieża Primeval Forest in Poland. A concern is attenuation, because avoidance can be a temporary neophobic response. Musiani et al. (2003), however, found that fladry-based barriers protected ranches in Idaho and Alberta, Canada, from wolf depredation for up to 60 days. The researchers reported at least 23 approaches to barriers before 4 wolves crossed the barrier and killed cattle. A second concern, still largely unresolved, is the relatively high labor cost for installing and maintaining fladry systems.

Mylar tape, a metallic plastic tape that is red on one side and silver on the other, has been used to repel birds. This tape is available in widths of less than an inch to several inches. It can be stretched from post to post to create a fencelike barrier or hung from string or posts at intervals. When in place, wind causes the tape to vibrate so that it hums and reflects light in patterns. Some practitioners believe that, from above, it gives the appearance of fire. Bruggers et al. (1986) first reported success with Mylar tape in reducing bird activity in the Philippines, India, and Bangladesh. Dolbeer et al. (1986) subsequently reported that the tape repelled blackbirds from millet, sunflowers, and sweet corn in the United States. However, the tape failed to protect blueberries from bird damage in field trials in New York State. Mylar tape has subsequently been shown to be effective for keeping Canada geese out of areas where they are not wanted, such as farmers' fields (e.g., Heinrich and Craven 1990), and herring gulls (*Larus argentatus*) and ring-billed gulls from certain areas, such as loafing places (e.g., Belant and Ickes 1997). The tape may be used as part of an integrated deterrent program. For example, Marcus et al. (2007) used Mylar tape, along with other frightening devices, to redirect selection of nest sites by endangered interior least terns (*Sterna antillarum athalassos*) and threatened piping plovers (*Charadrius melodus*) away from gravel mines (where they would have poor nesting success) in Nebraska.

Bioacoustical systems take advantage of distress or alarm calls of wildlife and have been used successfully in many situations and countries. Actual recordings of calls are used, as are electronic simulations, such as those generated by the Av-Alarm or Bird-X systems. Holcomb (1976) tested Av-Alarm for effectiveness in protecting ricefields from red-billed quelea in Africa in 1975 and 1976. He found damage was lowest nearest the speakers and increased in a linear fashion as one moved away from the speakers, up to 450 feet. Gildorf et al. (2004) found that a bioacoustical system was unsuccessful in reducing deer damage to silk-stage corn in Nebraska. These researchers believed that the large fields and height of the corn afforded safe areas for the deer during periods of activation; the deer would take corn after the calls ended.

Example

Stevens et al. (2000) described a radar-activated system designed to keep waterfowl from large ponds contaminated with potentially lethal wastes from a power plant. The system was designed as "demand performance" in that detection of birds by radar initiated an integrated hazing program that included acoustic alarms, pyrotechnics, and chemical repellents. Because hazing was a direct response to presence of waterfowl, habituation was not a problem. The system was effective but expensive to install and required skilled personnel for maintenance and operation. However, failure of the system could result in substantial fines under the U. S. Migratory Bird Treaty Act, so the system was viewed as cost effective.

ADDLING AND PHYSICAL STERILIZATION

Statement

Physical methods can be used to reduce birth rates of populations or lower survival rates of young.

Explanation and Examples

Physical methods are sometimes used to reduce or eliminate fertility. Methods include physical addling and surgical sterilization. Meaning "loss of development," **addling** in the strictest sense is destruction of eggs by shaking. Addling has come to mean destroying eggs by any physical or chemical means—puncturing, freezing, or coating with vegetable oil. Addling is often used to manage goose populations in urban and suburban areas (e.g., McMurtry 2002). It has also been used to reduce growth of invasive mute swan populations in Rhode Island, with over 10,000 eggs addled since 1978 (Allin undated). Nest destruction is a closely related activity and might be favored when eggs are not present (e.g., in controlling damage by barn swallows) or if the eggs are newly laid.

Bromley and Gese (2001a) sterilized whole packs of coyotes and found reduced predation on sheep compared with coyote pairs having pups. The approach was cost effective and had no significant effects on territoriality or related behaviors (Bromley and Gese 2001b). Reduced fertility, however, has been viewed as impractical for many wildlife situations, for instance, management of large deer populations (e.g., Merrill et al. 2006).

PHYSICAL ALTERATION OF HABITATS AND ECOSYSTEMS

Statement

Habitats or ecosystems can be physically altered, thereby reducing their carrying capacities and making them undesirable for offending wildlife.

Explanation

To make habitats unattractive to wildlife, roosts might be pruned, thinned, or removed; an ecosystem pushed to an earlier sere; dams removed physically or with pyrotechnics; water levels manipulated; or food, water, or shelter otherwise made unavailable to offending species. The desired consequence of such manipulations is reduction in wildlife damage, including wildlife diseases and zoonoses. However, such habitat manipulation can also negatively impact nontarget species, and such effects must be considered.

Pruning or removal of vegetation that serves as sites for bird roosts has been a common worldwide practice for management of bird damage, for example, for management of red-billed quelea in Africa or the dispersal of nuisance house crows (*Corvus splen-*

dens) in Singapore (e.g., Peh and Sodhi 2002). Manipulation of water levels can sometimes discourage the presence of wildlife such as muskrat or beaver. Raising winter water levels in ponds almost to flooding and then lowering levels in the summer can sometimes force muskrats to find more suitable locations (e.g., Link 2005). This action may also flood muskrat burrows in wintertime and make them more vulnerable to predators in the summer. Removing food sources—for instance, vegetation adjacent to beaver ponds—can sometimes also force beavers to abandon a pond.

Airports, landfills, and sewage treatment facilities are landscapes where the presence of undesirable wildlife is reduced or eliminated by physical alteration. Here, the notion is to make unavailable a factor critical for wildlife, such as food, water, or shelter. Airports and landfills work from management plans that consider impacts and management of wildlife. Standard practices at landfills can be modified so that trash is covered regularly with soil or a repellent coating. Unnecessary ponds can be removed from airports, and necessary ponds can be covered with wire grids or chemical repellents. Vegetation that provides cover for undesired wildlife can be modified or removed. Fences can prevent movement of wildlife into critical areas, such as air operation areas of airports. All attractants, such as food or solid waste, are cleared from operational areas, including those surrounding runways (Cleary and Dolbeer 2005).

Example

Developed by Gene Wood at Clemson University, the "Clemson Beaver Pond Leveler" uses a PVC pipe placed within a beaver dam to drain a beaver pond; it works by having holes in the portion of the pipe inside the beaver pond and screening to prevent debris from plugging water intakes and access by beavers to the pipe on the pond side. An elbow and vertical standpipe riser at the exit end of the pipe allow regulation of pond level. With this regulation, the pond can be maintained at a level where it is beneficial for people and wildlife and will not cause problems such as flooding standing timber, roads, or farmland (Miller and Yarrow 1994). The Leveler has been modified to permit the passage of brook trout, allowing the device to be utilized as a bypass for these fish through beaver dams. The Leveler and similar designs have found use worldwide (fig. 10.3).

Summary

- Physical methods are used to restrain or kill animals, to frighten or repel them, to reduce

Figure 10.3 A pond leveler with culvert fence, designed to discourage beaver from building dams. *Photo by Michael Callahan.*

populations of offending animals by reducing births, and to alter habitats so as to make them undesirable for offending wildlife.

- Restraining devices include cage traps, foothold traps, net traps and guns, and cable devices. Systems include devices, swivels, and anchors to reduce injury to captured wildlife and to prevent breakage, and the "set" includes baits, attractants, and placement.
- Killing traps, such as rat or mouse snap traps, are designed to kill quickly.
- Shooting can be done with pistols, rifles, shotguns, blowpipes, or dart guns, from the ground or from aircraft.
- Exclusionary methods separate wildlife from things humans want to protect and include fences, netting, tubing, wire grids, and bird spikes.
- Physical methods that repel include those that startle damaging wildlife with the use of sudden sounds or lights, temporarily stopping damage, and those that use the senses to

communicate that an area is dangerous and should be avoided.

- Physical methods used to reduce populations of offending wildlife include addling and surgical sterilization.
- Physical methods used to alter habitats so they are undesirable to offending wildlife include pruning, manipulating water levels, and limiting available food, water, or shelter.

Review and Discussion Questions

1. Are traps lethal or nonlethal devices? Explain your answer, using specific examples.
2. Explain the basis for the effectiveness of the "Williams Carp Separation Cage." How have these ideas been incorporated in the design of some barriers for sea lampreys? How might they find other applications for damaging species in North America?
3. How important is the experience of the practitioner in using traps or snares to catch a coyote? Explain, please.

4. What techniques are used to ensure that a trap intended to catch a starling catches or holds only starlings? How about a snare set to catch a coyote or a conibear trap set to catch a muskrat?

5. Do you think a pyrotechnic firecracker can be an effective tool for frightening geese away from a pond they might want to use as a spring nesting site? Will the firecracker be effective in removing a pair of geese that are already nesting? Explain.

11

Pesticides

We review here chemical methods that are used in wildlife damage management.

Statement

Pesticides have been discovered that sedate, repel, kill, or sterilize offending wildlife; these pesticides have been formulated, evaluated, and registered for use following public policies and regulations particular to local to national governments worldwide.

Explanation

"**Pesticide**" means literally "to kill a pest." In regulatory practice, the term includes any mix of substances (or devices) that affects behavior of pest plants or animals. The U. S. Environmental Protection Agency (USEPA) defines a pesticide as "any substance or mixture of substances intended for preventing, destroying, repelling, or mitigating any pest" (USEPA undated). Pesticides may be chemicals such as **warfarin** (an anticoagulant poison), genetic parts of organisms, or whole living organisms such as viruses, bacteria, or protozoa. Pesticides made from living organisms or their genetic parts are called **biopesticides**.

Pesticides may also be described according to targeted organisms: e.g., **antibiotics** for bacteria; **avicides** for birds; **fungicides** for fungi; **herbicides** for plants; **molluscicides** for snails; **piscicides** for fish; **predacides** for predators; and **rodenticides** for rodents and other small mammals.

Pesticides presently used in wildlife damage management include chemical attractants and repellents; immobilizing (stupefying) agents; antifertility agents, including antibodies designed to immunize against fertility; biopesticides, including bacterial or viral agents used for biological control (see chapter 12); and toxicants, including herbicides (table 11.1). Because of their biological or pharmacological activity, pesticides, as **active ingredients,** are the components of greatest importance for registration and legal use of pesticidal methods.

Most pesticides are formulated into baits designed to selectively attract targeted animals and encourage them to feed. **Baits** may include mixes of liquids, solids, or powders formed into any of a variety of

TABLE 11.1   *Pesticides used in wildlife damage management*

Mode/Target	Active Ingredient
Antifertility	
Birds	nicarbazin
Mammals	gonatropin releasing hormone (GnRH)
	porcine zona pellucida (PZP)
Immobilization	
Birds	α-chloralose (also toxicant, birds, mammals)
Mammals	propiopromazine HCl
Repellent	
Birds	4-aminopyridine
	anthraquinone
	methiocarb
	methyl (or dimethyl) anthranilate
	polybutene
Mammals	capsaicin
	egg acrylic
	semiochemicals (e.g., dried blood, urine)
	thiram
Reptiles	cinnamon, clove, and anise oil
Toxicant	
Birds	3-chloro-4-methylanine HCl (DRC-1339)
	oil (addling)
	sodium lauryl sulfate
Fishes	antimycin A
	niclosamide
	4-nitro-3-(trifluoromethyl) phenol HCl
	rotenone
	saponins
Mammals	aluminum phosphide
	anticoagulants
	bromethalin
	cholecalciferol
	magnesium phosphide
	para-aminopriophenone (PAPP)
	polybutene
	sodium cyanide
	sodium fluoroacetate (Compound 1080)
	sodium nitrate, carbon (gas cartridge)
	sodium nitrite
	strychnine
	zinc phosphide
Plants	glyphosate (Rodeo)
	other herbicides
Reptiles	acetominophen
	methyl bromide

shapes and sizes. Baits may be complex and include whole organisms. For example, dead neonatal mice containing acetaminophen (Tylenol) are used to deliver the toxicant to brown tree snakes in Guam and Saipan (Shivik et al. 2002). Along with the active ingredient pesticides, baits include **inactive ingredients** that serve other functions (box 11.1).

Delivery to targeted wildlife can be performed in various ways. Liquid baits may be put in a station, as with warehouses for managing rodents, or delivered with a sprayer, as for managing weeds. Powders may be injected, as with some zinc phosphide products designed to be placed behind drywall. Rodents get this powder onto their fur and ingest lethal amounts while grooming. Solid baits may be broadcast by hand or with specialized equipment, as with rodenticides applied in orchards, reforestation or island habitat restoration programs, or pasture crops such as alfalfa (e.g., Witmer et al. 2007). Sometimes baits are delivered by aircraft over large areas, such as in baiting programs to immunize raccoons for rabies in the United States or the red fox and raccoon dog in Europe or to manage invading species such as rodents or predators on islands (e.g., Innes et al. 1995).

Baits may be put into specially designed containers at baiting stations. Locally available materials often serve as bait stations. In the Philippines, for example, coconut husks are split in two and used to house baits for rodent control. The containers may be designed not just to hold the formulation but to keep it from the weather, to attract and allow selective access by the targeted wildlife, and to keep other wildlife out (fig. 11.1). Draw stations, such as carcasses from prior predator kills, may be used to attract predators and scavengers to a general area. Selective attractants may then be used to entice targeted species to other locations, where there are baited devices. For example, an M-44 mechanical ejector containing sodium cyanide and gauze soaked in a selective attractant may be placed under a rock near a draw station. The attractant may be designed to induce bite-and-pull behavior in canids, making the M-44 specific for targeted predators and not attractive to other species.

Rifles and blowguns have been used to deliver immunizations and antifertility compounds to small populations—e.g., the feral horse population on Assateague Island, Maryland, and elephants in small refugia in Africa. Bacteria and viruses are being studied as carriers for active ingredients such as antifertility compounds (see chapter 17).

Pesticides are registered at the national level in most countries, and many nations require approval at lower governmental levels as well, such as state or municipality (see chapter 16). In 2006, the U.S. Food and Drug Administration (USFDA) gave the USEPA responsibility and authority for registration of all

BOX 11.1 Inactive Ingredients in Baits

Inactive ingredients can include a **carrier**, the matrix in which all of the ingredients are mixed; a **binder** or sticker, used to ensure that the active ingredient adheres to the carrier; an **attractant**, which selectively attracts intended animals; an **enhancer**, which encourages consumption of sufficient amounts of the formulation; a **masking agent** that overshadows or otherwise hides flavors in baits that might induce avoidance; and a **preservative** that prevents degradation of the bait and protects it from the weather.

Carriers can be liquid or solid. Liquid matrices are sometimes used, for example, in warehouses or arid environments where water might be particularly attractive to rodents or other wildlife. Carriers might be matrices that slow the release of volatile ingredients such as dimethyl anthranilate (a bird repellent). Often, carriers are food familiar to and preferred by targeted animals. Thus, ferns are a food preferred by mountain beavers and also serve as carriers for pesticides. Cereals, starch, pastes, and meats are other common carriers. Waxes sometimes serve as weather-resistant carriers for baits, are well tolerated by some wildlife, and are included in some commercially available rodenticides.

Binders not only ensure that active ingredients stick to carriers but can also weatherproof formulations. For example, oils and fats are often used as binders for zinc phosphide (a rodenticide) baits in tropical regions. The oils keep ambient acidic moisture from the rodenticide until it is in the gut of the animal, where acids convert the rodenticide into phosphine gas.

Attractants, such as apple or peanut-butter flavorants for deer, tend to be aromatic and favored by targeted animals. Favorite attractants for rodents and predators include chicken, beef, and fish flavors. In aquatic systems, byproducts of decomposition, such as the amino acid alanine, can attract scavengers such as catfish. Synthetic fermented egg, used in the United States for national surveys of coyote populations, can also be an attractant in predator baits. Highly attractive flavors have an important disadvantage; if unfamiliar, they tend to induce both neophobia and learned aversions, facilitating bait-shyness (Reidinger and Mason 1983).

Enhancers encourage consumption once the animal has found the formulation. For rodents and other herbivores, carbohydrates (putatively connoting safe food) such as sugars often encourage optimal consumption. Fatty materials may be more effective for omnivores and predators.

Masking agents, or sometimes microencapsulation, may be used to hide flavors that targeted animals find disgusting and avoid. For example, wildlife often perceive bitter flavors, such as those in some toxicants and antifertility agents, as potentially poisonous and avoid them (Reidinger and Mason 1983). Preservatives such as EDTA or insect repellents may also be added to prevent consumption by unintended wildlife, for example, ants.

All inactive ingredients contribute to the final flavor of a bait and, in turn, its salience in inducing behavioral avoidance such as neophobia or bait-shyness.

drugs used in wildlife damage management, so that the USEPA is now responsible for registration of all wildlife pesticides used in the United States. Pesticide regulators in other countries include the Australian Pesticides and Veterinary Medicines Authority (APVMA) in Australia, the Environmental Risk Management Authority (ERMA) in New Zealand, and the European Commission (Environment) for the European Union.

Examples

Each registered pesticide has a label that provides detailed information on the product and terms and conditions for its use (e.g., acetaminophen; NWRC

undated; fig. 11.2). It specifies that protective equipment and clothing must be worn when handling, mixing, or applying acetaminophen. The label also gives safety requirements and recommendations, environmental hazards (in this case, to avoid contaminating water), and storage and disposal requirements. It provides information on whom to consult, such as the Government of Guam Division of Aquatic and Wildlife Resources, and lists specific threatened or endangered species that may be of concern in Guam and the Commonwealth of the Northern Mariana Islands. The label specifies that actual use is restricted to employees of the U. S. state and federal governments, the government of

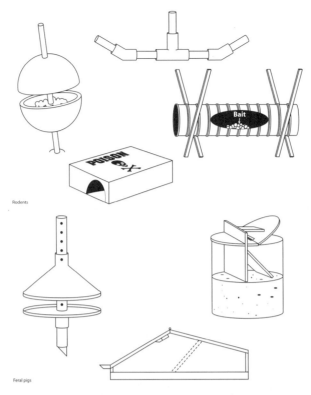

Rodents

Feral pigs

Figure 11.1 Variations in bait stations designed for rodents and feral pigs (*top*). A bear attempts to get bait from a BOS feeder (*bottom*). *Illustration by Lamar Henderson, Wildhaven Creative LLC. Photo by Michael Avery, APHIS/WS/NWRC.*

<table>
<tr>
<td>

ENVIRONMENTAL HAZARDS

Do not apply directly to water, or to areas where surface water is present or to intertidal areas below the mean high water mark. Do not contaminate water when cleaning equipment or disposing of equipment washwaters.

STORAGE AND DISPOSAL

Do not contaminate water, food, or feed by storage and disposal.

Pesticide Storage: Store only in a closed container in a dry place that is inaccessible to children, pets, and domestic animals.

Pesticide Disposal: If the product cannot be disposed of through use in accordance with product label directions, contact your Territorial Pesticide or Environmental Control agency, or the Hazardous Waste representative at the nearest EPA Regional Office for guidance.

Container Disposal: Nonrefillable container. Do not reuse or refill this container. Triple rinse container (or equivalent) promptly after emptying. Offer for recycling, if available. Otherwise, puncture and dispose of in a sanitary landfill or incinerator, or, if allowed by state and local authorities, burn. If burned, stay out of smoke.

</td>
<td>

ENDANGERED SPECIES CONSIDERATIONS

NOTICE: It is a Federal offense to use any pesticide in a manner that results in the death of an endangered species. Before undertaking any control operations with the product, consult with local, State, and Federal wildlife authorities to ensure the use of this product presents no hazard to any endangered species.

To reduce the hazard baiting may present to non-target species, consultation and concurrence will be obtained from Federal and local wildlife management authorities (e.g., U.S. Fish and Wildlife Service (FWS) Ecological Services, and Government of Guam Division of Aquatic and Wildlife Resources) prior to use in threatened or endangered species habitat (including habitat of the Mariana crow, Guam rail, or other reintroduced species).

Threatened or endangered Species – Guam:
Bat, Little Mariana fruit (*Pteropus tokudae*)
Bat, Mariana fruit (*Pteropus mariannus mariannus*)
Broadbill, Guam (*Myiagra freycineti*)
Crow, Mariana (*Corvus kubaryi*)
Kingfisher, Guam Micronesian (*Halcyon cinnamomina cinnamomina*)
Mallard, Mariana (*Anas oustaleti*)
Moorhen, Mariana common (*Gallinula chloropus guami*)
Rail, Guam (*Gallirallus owstoni*)
Sea turtle, green (*Chelonia mydas*)
Sea turtle, hawksbill (*Eretmochelys imbicata*)
Sea turtle, leatherback (*Dermochelys coriacea*)
Sea turtle, loggerhead (*Caretta caretta*)
Swiftlet, Mariana gray (*Aerodramus vanikorensis bartschi*)

Endangered Plants: Iagu, Hayun (*Serianthes nelsonii*)

</td>
<td>

ENDANGERED SPECIES CONSIDERATIONS (continued)

Threatened or endangered Species – Commonwealth of the Northern Mariana Islands:
Crow, Mariana (*Corvus kubaryi*)
Mallard, Mariana (*Anas oustaleti*)
Megapode, Micronesian (*Megapodius laperouse*)
Monarch, Tinian (*Monarcha takatsukasae*)
Moorhen, Mariana common (*Gallinula chloropus guami*)
Rail, Guam (*Gallirallus owstoni*)
Sea turtle, green (*Chelonia mydas*)
Sea turtle, hawksbill (*Eretmochelys imbicata*)
Sea turtle, leatherback (*Dermochelys coriacea*)
Sea turtle, loggerhead (*Caretta caretta*)
Swiftlet, Mariana gray (*Aerodramus vanikorensis bartschi*)
Warbler, nightingale reed (*Acercephalus luscinia*)

Endangered Plants: Iagu, Hayun (*Serianthes nelsonii*)

Amended 11/2009

</td>
</tr>
</table>

Figure 11.2 A pesticide label issued by the U.S. Environmental Protection Agency.

Guam, or the Commonwealth of the Northern Mariana Islands, and only employees trained in control of brown tree snakes or persons working directly under their supervision. The directions provide sufficient information so that a trained person can refer to it for guidance—for example, "manually insert one 80-mg or two 40-mg acetaminophen tablets into the throat of a dead mouse pup (approximate age: 10–15 days)."

HERBICIDES

Statement

Herbicides are used to control damage caused by weeds such as exotic and invasive plant species, to make habitat unsuitable for offending animals, and to break sylvatic cycles of wildlife diseases or zoonoses.

Explanation

Herbicides may be specific, in that they kill only targeted plants; selective, in that they kill only types of plants, such as broadleafed ones; or general, in that they kill any plant. These chemicals can depend on contact or be systemic, wherein they are translocated throughout a plant and serve also as a soil sterilant, preventing any plant survival for a period of time. Herbicides can be pre-emergent, preventing germination or early growth, or post-emergent, killing plants after they have germinated.

Whether to use herbicides, and what types to use, depends on the specific offending plants and the situation—i.e., managing exotic weeds versus altering habitat to reduce the presence of offending wildlife or diseases. The reader is referred to county and state cooperative extension guidelines and other local sources for recommendations, because many herbicides are available.

Examples

The herbicide glyphosate (Rodeo, 53.8%, CAS No. 38641-94-0) is a post-emergent, nonselective herbicide formulated for use in aquatic environments. Rodeo is often delivered by aircraft and is used to control invasive cattails (*Typha* spp.) and other vegetation in marshy areas where blackbirds roost, particularly in the prairie pothole regions of North America where sunflowers are grown (Linz and Homan 2011). Open potholes (ponds formed from glacier scours, important for wildlife and waterfowl production in North America) provide less favorable roost sites for blackbirds than potholes with thick cattails. Conversely, open potholes are preferred by desirable wildlife, such as ducks and other waterfowl. About 1,400 hectares of cattails are treated annually in North Dakota. The areas treated are those with the most severe damage from blackbirds, and they represent less than 1% of the estimated yearly total area of cattail stands in the state.

Spotted knapweed (*Centaurea maculosa*) is an exotic forb that has aggressively invaded semiarid habitats of western North America. Two gall flies (*Urophora affinis* and *U. quadrifasciata*) were introduced to control seed production of the knapweed. While the galls reduced seed production overall, the flies were unsuccessful in suppressing knapweed populations, so both the knapweed and the flies increased in abundance. The gall flies have become an abundant exotic food source for deer mice (*Peromyscus maniculatus*), which have doubled in areas with gall flies. The increases in the number of field mice in turn increased the prevalence of sin nombre virus, the pathogen causing hantavirus pulmonary syndrome.

Pearson and Fletcher (2008) aerially sprayed the broadleaf herbicide Tordon (Dow Agrosciences) to treat experimental areas and remove knapweed (and gall flies). Comparing treated with untreated areas, they monitored mouse abundance. They found that the herbicide treatment removed sufficient exotic gall flies by taking away their food base, so that deer mice populations were reduced to levels occurring before the invasion of knapweed. This process also reduced the risk of hantavirus pulmonary syndrome to baseline levels.

DRUGS, INCLUDING IMMOBILIZING AGENTS AND TRANQUILIZERS

Statement

Chemical drugs sedate offending wildlife, so they can be captured and removed from an area or calmed, to reduce self-inflicted injuries while being retained.

Explanation

Immobilizing agents, or stupefying agents, render wildlife temporarily unresponsive to environmental stimuli, so they can be readily captured, or to reduce resistance and injury when they are captured. Two of these, alpha-chloralose (CAS No. 15879-93-3) and propiopromazine hydrochloride (CAS No. 7681-67-6), are registered for use in the United States under Investigational New Animal Drug (INAD) permits.

Alpha-chloralose is used as an anesthetic, hypnotic, immobilizing agent, sedative, and toxicant. It is utilized by trained and certified "applicators" of Wildlife Services, USDA, to capture and remove nuisance waterfowl and other birds—pigeons, ravens (*Corvus* spp.), and sandhill cranes (*Grus canadensis*)—often from urban or suburban settings. It can be used in recreational and residential areas and near swimming pools, shoreline residential areas, golf courses, and resorts. This chemical is mixed in a single bread or corn bait for waterfowl and in corn baits for pigeons. From October 2004 through September 2005, 443 g of alpha-chloralose were used to capture mostly Canada geese, with a mortality rate under 5% (O'Hare et al. 2007). In some European countries, alpha-chloralose is used to immobilize and kill birds and rodents (Woronecki et al. 1990). The chemical is registered in Australia and New Zealand to manage some bird species. As in the United States, use is restricted to certified applicators.

Propiopromazine hydrochloride is being considered and tested in the United States and Australia as the active ingredient in a tranquilizer tab (rubber nipple) attached where caught wildlife will instinctively bite at traps and other restraints. The compound sedates—without loss of consciousness—coyotes, feral dogs, wolves, and other wildlife (Savarie et al. 2004; Hunt and McDougall 2005).

Example

Smith (2004) compared capturing geese in roundups using funnel traps versus baiting with alpha-chloralose in Reno, Nevada. Removal efforts focused on public areas such as parks and airports. Funnel traps with wings extending 100 yards could be assembled by 4 people in about 15 minutes; extra time was needed to dock a boat if water hazards were a concern. Use of alpha-chloralose required prebaiting with bread. Both traps and alpha-chloralose were effective in removing geese. Alpha-chloralose offered some overall advantages, because geese could be removed without regard to molting stage and because trap-shy geese could be effectively captured. Stress from handling in sedated geese was low, and use of the immobilizing agent attracted little public interest.

CHEMICAL REPELLENTS

Statement
Chemical repellents use the sensory systems of wildlife to irritate or frighten them, thereby averting offending behavior. Learned effects need to be reinforced or they attenuate quickly, whereas unlearned effects can be long-lasting.

Explanation
We list common repellents (see table 11.1) and describe their uses here.

Anthraquinone (CAS No.84-65-1) is a naturally occurring substance that causes gastric illness in learned flavor aversions (Avery et al. 1997; Werner and Provenza 2011). It was first evaluated as a bird repellent in the 1940s and was tested as a means of repelling blackbirds from rice seed in the 1950s. Anthraquinone is the active ingredient in products used for bird repellency in the United States (e.g., Flight Control, Avipel, and Airepel) and New Zealand (Avex; Spurr and Coleman 2005a). A critical step for effective use on corn was developing a formulation that encouraged sufficient consumption to stimulate learned aversions by birds such as Canada geese, red-winged blackbirds, and ring-necked pheasants (*Phasianus colchicus*; Werner et al. 2009).

Four-aminopyridine (CAS No. 504-24-5) induces an alarm response in birds that eat it. These birds move erratically and emit calls that frighten other birds. Birds ingesting the compound usually die. Other birds leave or avoid the area. The chemical is the active ingredient in Avitrol (Avitrol Corporation) grain baits in the United States and Australia (Eason et al. 2010), for use with birds such as pigeons and blackbirds that can be legally killed.

Capsaicin (e.g., capsicum oleoresin, CAS No. 404-86-4), the spicy chemical in hot capsicum peppers, is the active ingredient in some repellents (e.g., Hot Sauce) and one of several active ingredients in others (e.g., Deer Off, Havahart, Woodstream Corporation). The products are sold as repellents for wildlife including deer, predators, and birds. Effectiveness of capsaicin as a deer repellent appears at least partly dependent on its concentration (Andelt et al. 1994). Red

pepper is the active ingredient in some sprays that are used to protect humans from attacks by dogs and predators such as bears.

To repel snakes, Wildlife Services, USDA, recommends the use of cinnamon, clove, or eugenol oil (APHIS 2003). Aerosols of one or more of these oils at 2% of the total formulation, sprayed directly at the head of a snake or as a fumigant, will repel brown tree snakes from cargo or other confined spaces (Clark and Shivik 2002). This application is exempted from U.S. federal registration because it poses minimal risk to users and the environment.

DMA (synthetic dimethyl anthranilate, CAS No. 85-91-6) and MA (natural methyl anthranilate, CAS No. 134-20-3) irritate the trigeminal nerve (which innervates the common chemical sense) of most birds and repel them (e.g., Mason et al. 1991; Avery 1992). Birds subsequently avoid this unlearned irritation. DMA and MA are grape flavorants used in flavored drinks and as purple colorants for labels on meats. The compounds are listed on the USFDA's GRAS (generally recognized as safe) list. DMA and MA, formulated for controlled release, are used to repel birds from areas such as airports, golf courses, landfills, turf farms, and lawns. The natural form, MA, is less expensive than the synthesized form, DMA, and is the active ingredient in products such as Migrate, Goose Repellent, and Fog Force. Effectiveness varies, and some applications need to be repeated after rainfall. DMA and MA break down quickly under ultraviolet light, often retaining effectiveness for less than a week. Economical use therefore depends on damage carrying high risks and costs (e.g., Spurr and Coleman 2005b).

Methiocarb (CAS No. 2032-65-7), or Mesurol, is an insecticide and molluscicide that has also been tested and used in many countries as a bird repellent (e.g., Bruggers et al. 1981). It serves as an illness-inducing agent in learned flavor aversions (Reidinger and Mason 1983). The product can be hazardous to wildlife such as fish, and its use in the United States is limited for that reason. One use of methiocarb, conditioning ravens to avoid eggs of endangered birds and turtles, is maintained by Wildlife Services, USDA. Methiocarb is registered in more than 60 countries as an insecticide and molluscicide. It is registered in New Zealand, South Africa, and some European countries as a bird repellent for use on emerging seeds and nonfood crops and as a snail and slug spray (e.g., Spurr and Coleman 2005a).

Polybutene (CAS No. 9003-29-6) is transparent and sticky and can be applied in a liquid or solid formulation to discourage birds from roosting on perches or ledges. Products are sold under such trade names as Bird-X Bird Repellent Liquid and Bird-B-Gone Transparent Bird Gel. Some formulations may also have other active ingredients, such as capsaicin (to give them a "hot foot" if they alight on the surface). Glue boards using polybutene are also sold to capture snakes and commensal rodents and rats under trade names such as J. T. Eaton's Stick-Em.

There are many other mixtures of compounds designed to repel offending wildlife. These include combinations of dried blood, lipids, putrid compounds including real and synthetic fermented egg, and predator urine. Some include human hair and soaps. Effectiveness varies broadly, but in general most have at least temporary neophobic effects, and frequent movement and rotation of compounds can sometimes extend protection.

Examples

Smith et al. (2008) reviewed data from 83 encounters in which humans used sprays to stop undesired behaviors of bears (brown, black, and polar). All encounters were in Alaska between 1985 and 2006. In 72 cases, sprays were used to defend the people from the bear, whereas in the 11 other cases, people sprayed an object or an area to protect it. Each spray had capsaicin as its active ingredient. Most (69%) of the encounters were with brown bears, and only 3% were with polar bears. The researchers sorted negative behavior into aggressive, defensive, or nuisance, and they separated food searching from other activities, such as curiosity.

Regardless of brand or canister size, effective spraying distance (often advertised as 15–20 feet), or nature of behavior or activity, spraying stopped undesired behaviors in 92% of the encounters, and 98% of the people involved were uninjured. Of those injured, only one required stitches for minor lacerations, and none were hospitalized. Although wind influenced accuracy, sprays always reached bears. In two cases, the spray nearly incapacitated the person as well as the bear. Overall, Smith et al. (2008) concluded that the capsaicin-based repellents served as an effective alternative to shooting.

Ward and Williams (2010) treated yew plants (*Taxus cuspidata*) with ten repellents at two locations in Connecticut for two growing seasons and measured deer browse. The authors followed product recommendations and kept detailed records of application costs. Products included Chew-Not, Deer Off, Deer-Away, Big Game Repellent, Plantskydd, Bobbex, Liquid Fence, Deer Solution, Hinder, and Repellex. The researchers developed a protection index based on dry weight of needles and plant size at the end of the second growing season. Untreated controls served as the base for complete (100%) damage.

This research team noted that regardless of mode of action, products applied most frequently were also most effective. The most expensive repellent, Bobbex, was also the most effective. Hinder performed about as well and cost much less. Plants protected by Repellex, Deer Solution, and Plantskydd had damage roughly equivalent to untreated yews. The authors concluded that no repellent completely protected yews from browse damage, but some provided protection equivalent to fencing. They suggested that choice would have to be based on a trade-off between effectiveness and costs, plus ability to reapply the chemicals at the intervals recommended by the manufacturers.

TOXICANTS

Statement

Toxicants are effective and selective tools, important for wildlife damage management.

Explanation

Uses of toxicants are carefully regulated by governmental agencies. Many can be utilized only by specially trained persons, sometimes also of designated agencies. We list common toxicants (see table 11.1) and describe their uses here.

AVICIDES. The only avicide presently registered in the United States is Compound DRC-1339 (CAS No. 7745-89-3). DRC-1339, chemically known as 3-chloro-4-methylanine, and some closely related compounds, was the 1,339th compound tested by the Denver Research Center (now the NWRC) as a potential avicide. It is the active ingredient in products sold as Starlicide. DRC-1339 is slow acting and lethal in a single feeding. This chemical deposits uric acid in kidneys and blood vessels, causing necrosis and impaired circulation. Death is from uremic poisoning and congestion of major organs.

DRC-1339 formulations include a 98% a.i. (active ingredient) concentrate to mix in grain, French fries, eggs, or meat baits. Its primary uses in the United States are to reduce consumption and contamination of livestock feed by starlings and blackbirds; to control starlings, pigeons, crows, and grackles in structures; and to control starlings, pigeons, and grackles

Figure 11.3 Loading a hopper with anticoagulant baits for distribution by helicopter in Hawaii. *Photo courtesy of Justin Fischer, APHIS/WS/NWRC.*

in staging areas. DRC-1339 is also used with bread-cube baits as a gull toxicant along coastal breeding areas and in egg or meat baits for the control of ra-vens, crows, and magpies and for livestock protection. A premixed pelleted bait called Starlicide Complete is available for restricted use by certified applicators to control starlings and blackbirds at feeding stations near livestock and poultry. DRC-1339 is registered for use in New Zealand but not in Australia (Lapidge et al. 2005).

Sodium lauryl sulfate (CAS No. 151-21-3), a wet-ting agent exempted from federal registration in the United States, is also a lethal control agent for man-aging blackbird roosts. Roosts are sprayed with wa-ter containing the wetting agent when weather and

conditions are appropriate. The compound allows wa-ter to saturate feathers, so birds die of hypothermia at ambient temperatures under 41° F. In field trials dur-ing 2004–7 in Missouri, birds died quickly, as soon as 30 minutes after exposure (APHIS 2008).

RODENTICIDES. Rodenticides can be **chronic** (require repeated feedings, mostly for anticoagulants) or **acute** (effective in single doses). Warfarin (CAS No. 81-81-2), the active ingredient in baits such as COV-R-TOX, Co-Rax, D-Con, Mouse Pak, RAX, and Waran, was the first anticoagulant rodenticide. Discovered (with research funded by the Wisconsin Alumni Research Foundation) in the 1920s as the sweet-smelling com-pound with anticoagulant properties in decomposing sweet clover hay, the name warfarin is an amalgam of the funding organization and the active ingredient, coumarin. The compound found heavy use worldwide, mostly for control of commensal rodents in urban areas and on farms. Its use as a blood-thinning medication began in the 1950s, and today warfarin remains the most widely used anticoagulant medication in the world. Other coumarin-based compounds are brodifacoum and bromadiolone (with these "second generation" compounds, single feedings are lethal). Chlorophacinone (registered for mountain beaver control in Oregon and Washington; Arjo 2006), valone, and pindone are also first-generation (multiple feedings required) compounds, but these are based on indandione rather than coumarin. Diphacinone is a second-generation indandione-based anticoagulant.

Genetic resistance to warfarin by Norway rats was first discovered on a dairy farm in Scotland (Boyle 1960). Subsequently, genetic resistance was documented in rural areas of Europe, in major cities around the globe, and with both first-generation rodenticides such as pival and diphacinone and second-generation anticoagulants such as difenacoum and bromadiolone (Jackson and Ashton 1986). Blood samples are sometimes taken and monitored for clotting time to detect genetic resistance. The geographic focus of resistance can then be contained or eradicated using a different chemical or control method (Greaves 1986).

While anticoagulant poisons have mostly been used to manage commensal and agricultural rodents, the compounds have also been used to manage other mammals, including muskrats, monkeys, and feral pigs. Uses now include eradications of invasive rodents on islands for conservation purposes (fig. 11.3). Wildlife Services, USDA, maintains three such registrations, two for brodifacoum (see below) and one for diphacinone. Because brodifacoum has secondary hazards, diphacinone is increasingly the compound of choice for these conservation applications.

Brodifacoum (CAS No. 56073-10-0) is a second-generation anticoagulant rodenticide that has seen broad application worldwide as the active ingredient in rodenticide baits such as D-Con II, Enforcer, Final, Havoc, Jaguar, Klerat, Ratak, Ropax, and Talon. It is effective against warfarin-resistant rodents. Several formulations include a wax base that helps resist the effects of weather. These compounds must be used outside in ways that limit access to nontargeted wildlife, because of secondary hazards to raptors and other predators and scavengers.

Cholecalciferol (CAS No. 67-97-0) is Vitamin D3, a human health supplement when taken in low doses. In higher doses, however, cholecalciferol mobilizes calcium. The calcium then mineralizes organs, having a lethal effect within three to five days. Cholecalciferol is the active ingredient in products such as $Agrid_3$, Rampage, and Quintox, rodenticides that have found particular use against anticoagulant-resistant rodents. The compound can be hazardous to nontarget animals such as pet dogs, which limits its use in many countries.

Acute rodenticides are fast acting but can cause bait shyness. These types of rodenticides might be formulated as fumigants, powders, or food baits. **Prebaiting** (using the bait without the active ingredient) is sometimes done to encourage consumption. Examples of acute rodenticides follow.

Aluminum phosphide (CAS No. 20859-73-8) is formulated as a tablet that converts to phosphine gas in a wet acidic environment. Trade names include Phostoxin, Phosphume, and Phostex. This tablet, used for burrowing rodents, is placed in a burrow to fumigate it. While secondary hazards are rare because the gas quickly disperses, primary hazards to unintended wildlife that may also be in the burrow, such as pigmy owls, are a concern.

Gas cartridges, with such trade names as Giant Destroyer, Smoke'Em, Gopher Gasser, Revenge Mole, Gopher Smoke Bombs, and Dexol Gasser, are also fumigants. Sodium nitrate (CAS No. 7631-99-4) and charcoal are the active ingredients. Wildlife Services, USDA, maintains registrations for large and small cartridges. Small cartridges are used for ground squirrels, pocket gophers, woodchucks (marmots), and prairie dogs. Larger cartridges are for coyotes, red foxes, and striped skunks in rangeland, crop, and noncrop areas. Cartridges are lit with a fuse and placed into a burrow, and they burn very rapidly and inefficiently. The consequence is generation of lethal carbon monoxide.

Zinc phosphide (CAS No. 1314-84-7) is one of the most widely used rodenticides in the world (Eisemann et al. 2003). Trade names in the United States include Arrex, Denkarin Grains, Gopha-Rid, Phosvin, Pollux, Ridall, Ratol, Rodenticide AG, Zinc-Tox, and ZP. It acts by converting to phosphine gas on ingestion, activated by the acidic nature of the stomach. Because it acts quickly, often killing rodents within a few feet of a bait station, it is a favorite of farmers. Zinc phosphide is used to reduce damage by rodents such as mice, voles, rats, prairie dogs, ground squirrels, muskrats, and nutria to cereal crops, pastures, forage crops, orchards, rangeland, and field borders. This chemical is effective in a single dose but is believed to have a strong taste to wildlife, thus requiring prebaiting to minimize bait shyness.

Strychnine (CAS Nos. 57-24-9, 60-41-3) is used as a rodenticide in the United States and also as an avicide in some other countries. Its use is restricted to burrows in some countries, including the United States, because of concerns for nontarget hazards. Strychnine remains undigested in affected animals, which then become hazards to scavengers and other wildlife. By restricting use to below ground, these hazards can be minimized. Wildlife Services maintains four registrations, all for managing pocket gophers.

PISCICIDES. Piscicides are chemicals poisonous to fish and include rotenone (CAS No. 83-79-4); antimycin A (CAS No. 642-15-9, marketed as Fintrol); saponins (amphipathetic glycosides, a group of chemicals abundant in some plants that have soaplike qualities); sea lamprey larvicide [4-nitro-3-(trifluoromethyl) phenol, TFM, CAS No. 88-30-2]; niclosamide (CAS No. 50-65-7); and ethanolamine salt of niclosamide (Bayluscide). Piscicides are used in hatcheries, ponds, lakes, and streams to eradicate or reduce the presence of undesired dominant, parasitic, diseased, or invasive fish. Entire ponds or lakes are sometimes treated with nonselective piscicides—e.g., rotenone, to eliminate a community of fishes—and then stocked with more desirable species.

Rotenone is sold as Devcol, Liquid Derris, Tubatoxin, and Parderil. It is both a piscicide and an insecticide, used worldwide to control some insect pests and remove unwanted, often invasive, fish species from ponds and streams (e.g., Rayner and Creese 2006). It interferes with the electron transport chain in mitochondria. This compound is absorbed through the gills of fish, making them particularly sensitive to it (Lockett 1998). Indigenous people in some Southeast Asian countries found that if they crushed the roots of certain plants and placed them in water, fish would become paralyzed, float to the surface, and be easily harvested. The plants contained rotenone. Of the four piscicides registered for use in the United States, rotenone is by far the most commonly used compound (Finlayson et al. 2000).

Sea lamprey larvicide (Lamprecide, TFM, 4-nitro-3-difluoromethylphenol) is registered in the United States for control of the sea lamprey. It is applied as part of an integrated program of control in the tributaries of the Great Lakes (Lennon et al. 1971).

At least 28 other countries use piscicides to control undesired fishes (Lennon et al. 1971), often along with other methods, such as trapping, in an integrated pest management program.

PREDACIDES. Predacides are used to kill predators including coyotes, skunks, raccoons, and foxes. Two, sodium cyanide and Compound 1080, are used in the United States and Australia, and three others—para-aminopropiophenone (PAPP), sodium nitrite, and theobromine with caffeine—are under development and evaluation (see chapter 17).

Sodium cyanide (CAS No. 143-33-9) produces the nerve gas hydrogen cyanide when it enters the mouth or stomach of an animal, causing death within a few minutes. This compound is the active ingredient in the capsule placed in the holder on top of the M-44 mechanical ejector. The M-44 ejector is tube-shaped and about seven inches long by one inch in diameter. It is made of a base containing a loaded spring that is anchored in the ground, the capsule and holder, and an ejector mechanism that includes a spring-driven plunger. Gauze containing an attractant for coyotes is placed over the capsule above the ground. A chemical may be added that stimulates bite-and-pull behavior in canids. When pulled, tension on the spring is released and the plunger pushes up into the capsule, ejecting the sodium cyanide into the mouth of the coyote. Death ensues quickly. Because cyanide gas dissipates rapidly, there is little concern for secondary hazards. Wildlife Services maintains two federal restricted-use registrations for M-44 cyanide capsules, one for managing coyotes, foxes, and wild dogs nationwide where state conditions and registrations allow, and one for managing the arctic fox in the Aleutian Islands.

Cyanide is registered for control of possums (Trichosurus vulpecula) in New Zealand and registered experimentally for control of foxes in Australia (Eason et al. 2010). It is considered the most humane compound for control of the possum when delivered in an optimized system (Gregory et al. 1998).

Sodium fluoroacetate (Compound 1080, CAS No. 62-74-8) is a competitive inhibitor of acetate in the citric acid cycle and so has potential toxicity to kill virtually any living organism. It affects energy production in cells and organs, and death is usually due to ventricular fibrillation. Compound 1080 is particularly toxic to canids. It is less toxic to herbivores, omnivores such as rats, and humans, but it has been used as a rodenticide (e.g., Atzert 1971). Nonetheless, in humans one teaspoonful is toxic (e.g., McIlroy 1994).

In the United States, the only current use of Compound 1080 is in the Livestock Protection Collar. This collar is attached to the neck (which is usually the site of attack) of a lamb or goat that will be tethered for accessibility to an attacking coyote. On attack, the rubber bladder containing the toxicant is punctured and the liquid is squirted into the mouth of the attacking animal. Death usually ensues within five hours. The Collar is relatively expensive and is viewed as a last resort in removing an offending coyote that has eluded such methods as trapping and M-44s.

Compound 1080 has been used successfully to eradicate rodents and predators from islands. For example, it has been utilized with the Arctic fox on Kiska Island to protect the endangered Aleutian blue goose (Palmateer 1987) and has found recent interest as a compound of choice to eliminate invasive species on other islands (Veitch 2001; Burbidge and Morris 2002; Nogales et al. 2004). Compound 1080 has a relatively low toxicity to species of particular concern in Australia and New Zealand (King et al. 1989), including wombats (*Vombatidae*), quolls, and Tasmanian devils. It is therefore used widely in both countries for the control of feral animals, including pigs, dogs, foxes, and rabbits (Rammell and Fleming 1978). About 80% of the world's use of Compound 1080 is in New Zealand, but this chemical has also served as a rodenticide and a predacide in many countries in Africa, Asia, and the Americas.

REPTILICIDE. Snakes, including brown tree snakes, are particularly sensitive to acetaminophen (CAS No. 103-90-2), the active ingredient in pain relievers such as Tylenol. Acetaminophen was registered in 2003 as a restricted-use compound for lethal control of brown tree snakes in Guam and the Commonwealth of the Northern Mariana Islands. A tablet is placed in the throat of a dead mouse pup as bait, and this, in turn, is placed in a bait station or broadcast by hand or airplane. Baits broadcast by airplane have small parachutes attached to them and many get caught in trees, where they are eaten by brown tree snakes. This approach is used as part of an integrated pest management strategy for control of the snake on these islands (Savarie et al. 2001). Acetaminophen has also been an effective toxicant for juvenile Nile monitors and Burmese pythons, and it may play a role in managing these invasive species in the United States (Mauldin and Savarie 2010).

Examples

Downtown Omaha, Nebraska, was selected by starlings as the site for a major winter roosting area

(Thiele et al. 2012). Several parking garages and high-rise office complexes became roosting areas, as did a new mall complex with groves of older trees. The birds benefited from warm air associated with vents and ducts, and the roost size expanded to the tens of thousands. Droppings accumulated on parked cars under roosting areas and rained onto people. Business at the new mall suffered because of the mess and concerns for disease. The City of Omaha asked for assistance from federal agencies, and Wildlife Services, USDA, took a lead role in developing a management plan.

In the first phase of the plan, the damage practitioners captured, marked, and released blackbirds. About 6,000 blackbirds, mostly starlings, were tagged with streamers that could be readily identified from a distance. The practitioners sought tagged blackbirds at feeding sites, beginning just outside of Omaha. They first found tagged blackbirds at a cattle feedlot about 20 miles from Omaha. Over time, they found 6 cattle feedlots with blackbirds, at distances up to 60 miles from the city, and were able to show some movement of individual tagged birds between roosts over years. The feedlot operators were also concerned about the birds.

The damage practitioners treated several feedlots with Starlicide bait, after meeting regulatory legal requirements and informing and involving the public. The overall population and size of the winter roost in Omaha were greatly reduced. Building owners added exclusionary devices and screening to make their buildings less suitable for roosting. The city removed larger trees from the mall area and replaced them with new landscapes less usable as roosts. Pyrotechnics and other repellent devices were utilized to disperse roosts when birds attempted to form them. The program was continuing at the time this chapter was written but has already been considered a success.

ANTIFERTILITY COMPOUNDS

Statement

Antifertility compounds offer the potential of reducing growth of offending populations by reducing birth rate. Finding antifertility agents with the characteristics needed for wildlife damage management, particularly multiyear effectiveness with free-ranging wildlife, has been difficult, but several products are now available for commercial and restricted use.

Explanation

Ideal antifertility agents would have certain characteristics, including single-dose effectiveness that lasted one or more seasons or continued acceptability of delivery compounds; reversibility, wherein wildlife could return to full reproductive function after the period of treatment; species specificity, in that only targeted wildlife would be contracepted; deliverability to free-ranging wildlife; and economic benefits that clearly outweighed management costs. Such an ideal antifertility agent has been elusive and may not be necessary for some uses. For example, delivery to individuals rather than large, free-roaming populations may be sufficient for managing confined wildlife populations in refuges and zoos (e.g., Rutberg 2005).

Antifertility agents have been sought for many years, with varied success (e.g., Fagerstone et al. 2010; see box). At the time we write this chapter, only Ovotrol P and Ovotrol G are registered and commercially available in the United States as antifertility agents for control of wildlife reproduction. Ovotrol P is registered for control of pigeon reproduction, and Ovotrol G for goose reproduction. Both products have nicarbazin as the active ingredient, which affects the membrane separating the egg white from the yolk, making conditions unsuitable for growth of the embryo.

GonaCon is also registered in the United States for use as the first immunologically based antifertility agent. It blocks GnRH production and is registered for inhibiting deer reproduction, but it is available only as a restricted-use compound. Its registration has been hailed as a major advance by leading experts in wildlife damage management (e.g., Warren 2011); however, delivery and effectiveness in population management of free-ranging deer remain difficult and questionable.

Egg oil, specifically corn oil, is used as an addling agent, sprayed on gull and goose eggs to prevent them from hatching. It is also available for use in the United States under restricted conditions determined by federal and individual state laws. Corn oil is exempt from the registration requirements of the USEPA.

Examples

A sterile-male-release program has been in progress since 1991 to help control the spread of sea lampreys in the U.S. Great Lakes (Twohey et al. 2003). The Great Lakes Fishery Commissions maintain sea lamprey traps that collectively catch about 25,000 male sea lampreys each year. Instead of being destroyed, these males are sent to the Hammond Bay Biological Station of the U. S. Geological Survey, where they are treated with bisazir, a chemical sterilant. The lampreys are retained for two days until the compound

Kathleen A. Fagerstone

Kathleen A. Fagerstone is currently the manager of the Technology Transfer Program, National Wildlife Research Center, Wildlife Services, Fort Collins, Colorado. Previously she was manager of the Product Development Research Program. In these positions, Dr. Fagerstone sees that all data requirements are met for over 20 federally registered vertebrate pesticide products, including those needed for bird, rodent, predator, and snake management. Data for many registrations are also broadly available for use by private- and public-sector industries and practitioners.

She has also overseen broadly ranging research that moves new ideas, such as repellents and immunocontraceptive vaccines, toward registration and practical use in wildlife damage management. Thus, Dr. Fagerstone has been instrumental in meeting registration requirements for products ranging from immobilizing agents containing alpha-chloralose or propiopromazine hydrochloride to contraceptives such as GonaCon, which uses gonadatropin-releasing **hormone**.

Dr. Fagerstone often serves as a liaison between private-sector registrants and the federal research program—for example, forming data-gathering consortia or summarizing registration progress at national and international meetings. She coordinates work with scientists from other countries; for example, she has worked with scientists in Australia and New Zealand who are exploring use of sodium nitrite as a toxicant for feral pigs.

Dr. Fagerstone received her B.S. in zoology from Colorado State University, Fort Collins, and her Ph.D. from the University of Colorado, Boulder. Her personal research interests include risk assessment and small-mammal ecology, particularly prairie dogs, ground squirrels, and black-footed ferrets. She is active in professional societies and has served as president of the Colorado chapter of The Wildlife Society, associate editor for *Journal of Wildlife Management*, and chair of the Wildlife Damage Management Working Group of The Wildlife Society and the Vertebrate Pest Council.

fully clears from their bodies, and then they are returned to open water on the St. Mary's River as part of a whole-lake program being tested on Lake Superior.

The idea is that sterile males behave reproductively as normal at spawning sites and, by sheer numbers, overwhelm the effectiveness of nonsterile males. The anticipated impact is a greatly reduced reproductive rate for the sea lamprey populations in targeted areas. A problem is cost of manufacture and use of bisazir, about $0.50 per treated lamprey (Bergstedt and Twohey 2007).

Summary

- Pesticides are substances or devices, including antibiotics, avicides, fungicides, herbicides, molluscicides, piscicides, predacides, and rodenticides, that affect the behavior of pest plants or animals. ·
- Pesticides are active ingredients in formulations, mostly baits, that also include inactive ingredients, such as carriers, binders, attractants, enhancers, masking agents, and preservatives.
- Pesticides are delivered by broadcasting, by spraying, by placement in bait stations, with specialized devices such as M-44s, or with rifles or blowguns.
- Pesticides are registered and have labels that describe allowable and restricted uses as well as conditions for use.
- Many herbicides are available to manage weeds, to make habitat unsuitable for offending animals, or to break sylvatic cycles of wildlife diseases or zoonoses. Alpha-chloralose is an immobilizing agent used to capture offending wildlife such as geese and pigeons; propiopromazine is being tested as a sedative for canids caught in restraining devices.
- Repellents that use sensory systems to irritate or frighten offending wildlife include anthraquinone, 4-aminopyridine, capsaicin, mixtures of oil scents, DMA and MA, methiocarb, and polybutene.
- Toxicants include avicides such as DRC-1339; chronic rodenticides such as warfarin, brodifacoum, and cholecalciferol; acute rodenticides such as aluminum and zinc phosphide, gas cartridges, and strychnine; piscicides such as rotenone and sea lamprey larvicide; predacides such as sodium cyanide and sodium fluoracetate (Compound 1080); reptilicides such as acetaminophen; and antifertility compounds such as Ovotrol P, Ovotrol G, GonaCon, and egg oil.

Review and Discussion Questions

1. If you extracted the odors from human hair, distilled and concentrated them, and then set a vial of the extract in a field of alfalfa to keep out deer, is the extract considered a pesticide? Would it require registration by the USEPA?
2. How might herbicides be used to break the sylvatic cycle of a zoonosis? Give a specific example.
3. How important is rotenone as a toxicant in wildlife damage management? Are there suitable alternative piscicides to, for example, clear a pond of undesired fish before stocking it with desired ones? Would it be better to just drain the pond?
4. Is bait shyness learned from the flavor of an active ingredient (e.g., warfarin or zinc phosphide), from the composite flavor of a formulation, or both? How might baits be designed to minimize bait shyness? Give an example.
5. Some argue that wildlife contraception as a management tool takes too long to be practical for farmers and ranchers who need quick solutions. Deer continue feeding and coyotes continue killing during the time it takes contraception to reduce populations. Do you agree? Do you disagree? Explain your view, including specific references.

12

Biological Methods

In this chapter we review biological methods used to manage wildlife damage. We include immunologically based disease control—e.g., rabies control—here because the antigens are live or attenuated viruses or bacteria and depend on biological responses to foreign materials in blood. Bacterial- or viral-based delivery systems of chemical pesticides are simultaneously chemical and biological methods. None are developed enough for routine use, so we describe those in the chapter on future directions (chapter 17).

Statement

With biocontrol, management is accomplished by introducing a living agent such as an alternate plant or prey, pathogen, parasite, herbivore, predator, or interspecific aggressor. Success with biocontrol offers alternatives to chemical pesticides or reduced amounts of chemicals.

Explanation

Biological control (i.e., **biocontrol**) uses interactions among organisms to limit damage or the presence of damaging species. Biological methods include predator and prey interactions, parasite and host interactions, disease and host interactions, and interspecific aggression for territorial or social defenses, such as guard animals attacking predators of livestock. Biocontrol also includes **lure** (or **decoy**) crops as alternate foods and **diversionary feeding.**

Biocontrol has had particular success in managing wildlife damage caused by weeds and insects. For example, use of the insecticide producing the bacterium *Bacillus thuringiensis* (Bt), as well as Bt-engineered plants (plants that have Bt genes incorporated into their genetic makeup to provide protection from insects), are examples of biocontrol familiar to people who garden or farm. Parasitic wasps and preying mantises are also familiar biocontrol agents. When seeking a biological agent, it is better to enlist an organism already available in the ecosystem, such as Bt, than to import one from a distant land. An organism already adapted to the ecosystem, and vice versa, probably reduces opportunities for ecological backlash.

Biocontrol includes classical and inundative methods for introducing biocontrol agents into ecosystems, classical methods being most commonly used. With **classical methods**, one or more agents is found and, after the appropriate guidelines are followed, the agent(s) is/are released one or more times. Over subsequent generations, the agent becomes established and gradually manages the pest. An example is the release of several parasitoid wasps in the United States for management of the emerald ash borer. With **inundative methods**, large numbers of agents (sometimes using sterile ones as a tactic) are released periodically throughout a single season. The sheer numbers overwhelm the problem population. Inundative methods usually require mass producing the agents. An example is the massive release of sterile males to manage sea lampreys in the Great Lakes.

Biological control of weeds has been applied in aquatic ecosystems (e.g., for common grass carp) as well as terrestrial ones. McFadyen (2000) listed 41 weeds managed successfully with biocontrol as the only or the preferred method. Included are introduced flower-and-seed-feeding weevils to manage purple loosestrife and herbivorous insects to manage leafy spurge (Anderson et al. 2000). With both examples, however, biocontrol imparts only partial control, and success depends on the method being part of a broader integrated management program (see chapter 17).

Biocontrol may be only partly successful or successful in one location but not in another (McFadyen 2000). For example, the moth cactoblastis (*Cactoblastis cactorum*) was released in Australia in 1926 to limit the spread of prickly pear cactus (*Opuntia inermis* and *O. stricta*). The cacti were themselves introduced around 1839 as horticultural curiosities and quickly moved into arid areas as problematic feral plants. Cochineal insects (*Dactylopius* spp.), including four species that came along with the first European settlers, were also used. The moth and other insects controlled the cacti, a success story in biocontrol (Zimmermann et al. 2000). The cactoblastis moth was also introduced into the Caribbean for biocontrol. It is believed to have subsequently moved into Florida, where it was discovered in 1989, causing concerns for unintended impacts on cacti of Florida, the southwestern United States, and Mexico (e.g., Johnson and Stiling 1998).

The viruses Myxoma and rabbit hemorrhagic disease (RHD, calicivirus) have been used to control the European rabbit in Australia and New Zealand, the first time biocontrol was used successfully against a mammalian pest. Here, however, the effectiveness of Myxoma in Australia was limited to about 40 years (1950–90), after which resistant rabbits began repopulating the country. Rabbit hemorrhagic disease has effectively reduced these rabbit populations after the animal was illegally (New Zealand) or erroneously (Australia) introduced in 1995 (see the Examples section below).

Biocontrol has been used successfully to break the sylvatic cycle of rabies in wildlife, first in Europe and later in North America. Reservoirs of rabies, such as foxes, dogs, raccoon dogs, and raccoons, are provided with antigens that result in the production of antibodies and immunity from the disease (see the Examples section below).

Guard animals have been used successfully worldwide to protect domestic livestock from predators such as coyotes and wolves (e.g., Braithwait 1996; Rigg 2001; USDA 2002; fig. 12.1). Guard animals are often dogs (**livestock protection dogs**, or **LPDs**) but can also be cattle, goats, llamas (*Lama glama*; e.g., Franklin and Powell 1994), mules, or donkeys (e.g., Walton and Field 1989). Rigg (2001) lists over 44 varieties of LPDs and at least 28 countries where LPDs are used. He points to mongrels as effective guarding dogs—for example, mongrels are used by the Navajo. LPDs may see additional uses, such as protecting livestock from diseases, conserving wildlife (Rigg 2001; Gehring et al. 2010), and protecting high-cash-value crops from herbivory (see the Examples section below).

Depending on the feeding preferences of damaging species, lure crops or diversionary feeding can be utilized. These are, therefore, forms of biocontrol and have many uses in wildlife damage management. In one scenario, a crop or feed that is particularly attractive to pests is used to lure them away from a more expensive crop or from a situation where they can cause damage to humans or their property (e.g., black bears at a campsite), thereby reducing human injury or economic losses. The approach is sometimes used by wildlife refuge managers to keep waterfowl in refuges rather than in neighbors' fields. Crops sought by waterfowl are planted in the refuge and the waterfowl are allowed to feed in the lure fields unmolested. Sometimes, neighboring fields where waterfowl feeding patterns are already established are purchased for refuge purposes. Scare devices are sometimes used in surrounding fields to further encourage use of the lure crop (e.g., Fairaizl and Pfeifer 1987). Cummings et al. (1987) suggested decoy plantings as a potentially effective means to reduce blackbird damage to commercial sunflower fields, and Linz et al. (2004) resurrected the idea in the form of "Wildlife Conservation Sunflower Plots" that would provide habitat for some

Figure 12.1   A guard dog protecting goats. *Photo courtesy of Kurt C. VerCauteren, APHIS/WS/NWRC.*

49 species of fall-migrating birds, reduce damage to commercial sunflowers, and garner support from both agriculturists and conservationists.

In other scenarios, pests are attracted to lure crops, where they may be captured and removed or poisoned. For example, the trap barrier system uses rice planted earlier than the main crop as a lure crop. In some designs, rats wanting to access the rice need to swim across a moat, where they are trapped and removed (e.g., Singleton et al. 1998). Sullivan and Klenner (1993) used food to divert red squirrels (*Sciurus vulgaris*) from damaging crop trees in lodgepole pine (*Pinus contorta*) forests, and Witmer and Ver-Cauteren (2001) reported use of cracked corn or soybean to divert voles (*Microtus* spp.) from drilled seeds on no-till cropland.

Lure crops and diversionary feeding raise concerns, especially when they are used to manage damage by game species such as bears (e.g., Ziegltrum 2008). Lure crops can be viewed as not only providing a food choice to damaging wildlife but also as supplemental feeding. The Wildlife Society (Inslerman et al. 2006; TWS 2007) has expressed concern that unregulated supplemental feeding by the public involving game species could concentrate wildlife in unnatural densities, furthering both stress and disease transmission. Supplemental feed could also impact home ranges and territories or direct attention and resources away from habitat and population management that might provide more long-term solutions to wildlife issues such as disease outbreaks and eradication efforts. "A fed bear results in a dead bear" is a scenario often suggested by wildlife agencies because of bear habituation to artificial food sources, e.g., garbage and bird feeders.

TWS encourages federal, state, and academic institutions to monitor the full spectrum of impacts on wildlife affected by baiting and supplemental feeding As an example, Rogers (2009) studied over eight years of diversionary feeding of bears at U.S. Forest Service campgrounds and residences. Only one bear, a transient sub-adult male, was removed. This researcher looked at other locations and found fewer house breakins, attacks, and bear removals where diversionary feeding was used. He also found that food-conditioned bears did not jeopardize public safety, and he encouraged reevaluation of policies regarding use of diversionary feeding.

Some attempts at biocontrol have failed spectacularly. Introductions of monitor lizards, giant cane toads, and the Asian mongoose on Pacific islands to manage rats are examples already described (see chapter 4). Primarily because of concerns for ecological backlash, biocontrol (through species introductions) remains a controversial method (e.g., Louda and Stiling 2004). This is true not only for the introduction of exotic agents but also for live bacteria or RNA-based viruses that can mutate quickly (see part III). Correctly or not, people tend to see agents of biocontrol as easily adapting to new environments, in ways not anticipated by experts, and becoming pests. Given such controversy, why do intensive and expensive searches for new agents continue? At least partly, says McFadyen (2000), because new biocontrol agents still offer the potential for long-term solutions to wildlife problems. Despite this promise, people are often unaware that biocontrol agents were part of successful solutions and tend to distrust these agents or remember failures rather than successes.

Governments, international organizations (e.g., FAO 1997), and individual experts (e.g., Balciunas 2000) have developed guidelines and regulations intended to minimize opportunities for ecological backlash when using biocontrol. In general, these guidelines ensure that potential benefits exceed potential risks, that the public is informed throughout the project, that all regulations and policies are followed, and that results and impacts are monitored so that adaptive practices can be used.

Transfer of successful methods between countries can also be complex, because exotic agents may be involved and assistance may be needed for technical aspects of the program, e.g., implementation of releases, program monitoring, and rearing of agents. Bilateral agreements are often involved, and sometimes more than two countries need to cooperate. Despite these obstacles, there have been many successful transfers (see the water hyacinth example below).

Examples

VOLES. Sullivan et al. (2001) explored using diversionary food to protect lodgepole pine (*Pinus contorta*) seedlings from voles (three *Microtus* spp. and the red-backed vole, *Clethrionomys gapperi*) in old-field habitats near Summerland, British Columbia, Canada, and in forested areas. The intent was to provide food that was more attractive than seedlings but not so nutritious as to trigger reproduction in the voles. The

researchers made "logs" of alfalfa pellets, wood pellets, or bark mulch mixed with wax and sunflower oil. These scientists found that bark mulch, followed by alfalfa pellet logs, reduced damage to seedlings in both old-field and forested habitats, with voles feeding 2.6 to 2.8 times more on seedlings in untreated than in treated sites (biologically, not statistically, significant, $P = 0.09$). No increase in reproduction of the voles was noted. These results support supplemental feeding as a potentially effective method for reducing vole damage.

WATER HYACINTH. In the 1970s, three species of weevil (*Neochetina bruchi*, *N. eichhorniae*, and *Sameodes albiguttalis*) that feed on water hyacinth were introduced by USDA researchers to manage the weed in North America. The results, as part of an integrated program that includes use of herbicides, have been successful. Australia, partly based on the experience in the United States, imported one weevil (*N. eichhorniae*) and a moth (*Niphograpta albiguttalis*) and released them in the 1970s. Later, in the 1990s, the Australians released *N. bruchi*. The releases were also successful, and the Australians then provided assistance to other countries having problems with water hyacinth. Weevils released in the Sepik River lagoons of Papua New Guinea cleared formerly choked waterways in less than five years. Clogged waterways in Lake Victoria across Kenya, Uganda, and Tanzania were cleared in fewer than three years. Australians also helped Benin, South Africa (Cilliers 1991), Nigeria (Farri and Boroffice 1999) and Thailand to control water hyacinth. In Rwanda, weevils from Uganda were introduced with help from the United States (Agaba et al. 2009). The British assisted with weevil introductions on the Shire River in Malawi.

In spite of at least 22 introductions in countries since the first releases in the United States, biocontrol is not a panacea for managing water hyacinth. The weevils have slow life cycles compared to the rapid growth of water hyacinth. Herbicides and insects or pathogens are also needed to manage the hyacinths. The method is best viewed as an effective part of a broader integrated pest management program (Jimenez 2003).

LPDS. VerCauteren et al. (2008) provided experimental evidence that livestock protection dogs might be used to protect livestock from wildlife diseases. They evaluated whether 4 LPDs would keep potentially infected white-tailed deer at least 15 feet (in radii) from cattle, thereby protecting them from bovine tuberculosis. For this study, they chose enclosures with high deer densities, about 35 deer/mi^2 and 93 deer/

mi², and Great Pyrenees dogs. The dogs were bonded with cattle and trained for about one year. Fenced paddocks were set up within the larger enclosures, and deer became accustomed to the areas. Shelters were built within the paddocks, and electric fencing was provided to contain calves. An invisible electric fence (invisible so that it would not influence fence-line contact between deer and calves) was used to contain the LPDs, which were trained to stay within the fence. VerCauteren et al. (2005) found the dogs to be highly effective in keeping deer from hay bales within the paddocks and also keeping deer from approaching the calves.

MYXOMATOSIS. **Myxomatosis** was first seen in rabbits in Uruguay in the late 1800s. In the European rabbit, the disease causes skin lumps and puffiness, sometimes causes conjunctivitis and blindness, and reduces resistance to secondary bacterial infections such as pneumonia. Infected rabbits become listless, lose appetite, and die within two weeks. The disease has less severe effects on New World rabbits (*Sylvilagus* spp.). The idea of introducing the Myxoma virus into Australia's rabbit population came from a Brazilian scientist in 1919. The Australian government initially considered the introduction too dangerous. However, a prominent Melbourne pediatrician urged reconsideration and, following 14 years of tests, the disease was introduced in 1950. The rabbit population declined 99% within 2 years, and the livestock industry received enormous economic benefits. By the end of the 1950s, though, resistance to the disease had emerged, myxomatosis became less effective, and the rabbit population gradually began to increase.

RHDV. **Rabbit hemorrhagic disease virus** (RHDV, or rabbit calicivirus) was considered as a second biological control agent in Australia. The virus causes rabbit hemorrhagic disease, a highly contagious illness specific to rabbits both domestic and wild. The disease was first discovered in China in the 1980s, spread to Europe by 1988, and subsequently appeared in Australia, Cuba, Mexico, New Zealand, and the United States. RHDV was evaluated for efficacy and nontarget hazards on Wardang Island, off the coast of Australia, and pros and cons were heavily debated. In 1995 the disease escaped the island, making it to the mainland on its own; the pros and cons became moot points.

In 1997, the government of New Zealand determined that it would not allow introduction of RHDV. However, the disease was deliberately introduced on the South Island by individuals who opposed the decision. The New Zealand government attempted to contain the disease while farmers, using diseased rabbits and blenders, furthered the spread of the disease. Unfortunately, the releases were not only illegal and but also poorly timed, allowing immune responses by rabbits under two weeks old and the rapid growth of a resistant population.

The disease has nonetheless heavily impacted rabbit populations in both countries, particularly in humid areas. Effects, though, were not as dramatic as those observed after the introduction of myxomatosis. The natural resistance of young rabbits and the presence of a less virulent natural form of RHDV that conferred some natural immunity to some rabbit populations lessened the overall effectiveness of the method. Today, Australian scientists are considering use of the viruses as live vectors to deliver antigens, to induce sterility in rabbits.

ORAL RABIES VACCINATION. An epidemic of rabies had spread from Poland to the eastern part of Europe at 12 to 36 miles per annum, beginning in 1939. By the 1950s, most cases of rabies in Europe were sylvatic rather than urban, mainly due to red foxes rather than pets (pets were mostly immunized). Raccoon dogs, introduced from Asia into western Russia for hunting around 1920, subsequently moved into other parts of Europe and became a wild species important for transmitting rabies. The epidemic reached Switzerland in 1967 and parts of central Europe by the 1970s. At this time, American scientists at the U.S. Centers for Disease Control showed that red foxes could be immunized with attenuated virus vaccines delivered orally. The European and Canadian communities applied these concepts as a broad-scale biological method to reduce or eliminate rabies in wild red foxes and raccoon dogs (Vitasek 2004).

Switzerland, in 1978, was the first country to attempt mass vaccination, and in doing so freed much of the country from sylvatic rabies within four years. Other countries followed, including Austria, Belgium, France, Luxembourg, Finland, the Netherlands, and then central and eastern European countries (Vitasek 2004). During the first years, baits were made of chicken heads with capsules containing the vaccination (mostly SAD Bern strain), placed by hand. Subsequently, manufactured baits were developed and alternative vaccinations became available. The Bavarian Method, in which hunters were enlisted to help place baits, was developed and used in some countries; with other techniques, baits were broadcast by helicopter and fixed-wing aircraft. European countries found that by coordinating efforts and timing of treatments with season (e.g., raccoon dogs hibernate, making treatment less fruitful during that time), optimal effectiveness for cost of treatment could be achieved. These countries also learned to expect an increase in the size of fox populations as one consequence of effective rabies control, and they adjusted baiting densities to maintain effectiveness. The campaigns were intensified in the late 1980s, following a resurgence of sylvatic rabies in some countries. Sylvatic rabies then declined precipitously, pushing the eastern boundary of the epizootic to the eastern border of Poland and the western edge of Italy (Anonymous 2002).

Over 40,000 cases of sylvatic rabies were diagnosed in Ontario, Canada, between 1958 and 1988 and over 3,500 cases in 1986. In 1989, the province of Ontario began a program of oral vaccination of foxes, the main sylvatic reservoir. The plan was to continue vaccination for at least one complete rabies cycle, about four years. At its peak, over 500,000 baits were distributed by aircraft in one season. At the time this chapter was written, more than 12 million baits have been distributed in southern Ontario. Baits were made of small plastic packets that contained a mixture of fat and wax—i.e., oleo stock, a strengthening wax, and mineral or fish-liver oil—and chicken flavorant. Tetracycline served as a marker. The vaccine was an attenuated but live ERA strain of rabies embedded in a blister packet of bait. Over 7 years, an average of 357 cases per annum of sylvatic rabies was reduced to 0, and the program was declared a success (MacInnes et al. 2001). In addition, the incidence of rabies in skunks parallelled the reduction in foxes, a consequence of treatment predicted from the onset.

Oral rabies vaccination was begun in the United States in 1994, with field trials in Virginia, Pennsylvania, and New Jersey (Rupprecht et al. 1995). A national management program (Slate et al. 2005) gradually emerged, in which oral vaccinations were distributed by aircraft in strategic locations called ORV (oral rabies vaccine) zones, to prevent movement of rabies from states where the virus already existed (fig. 12.2; Slate et al. 2002). In the United States, only Raboral V-RG is registered for use. Raboral V-RG is a vaccinia-based virus; vaccinia is the large, DNA-based poxvirus that was used to eradicate smallpox. Raboral V-RG was made using recombinant techniques in which the only rabies genes incorporated into the vector virus were those for the glycoprotein outer surface. This virus cannot cause rabies and is less sensitive to heat than other oral vaccines. Vaccinia is effective in producing antibodies against several variants of the rabies virus. For raccoons, the vaccine is dyed pink, placed in a plastic sachet resembling a fast-food packet

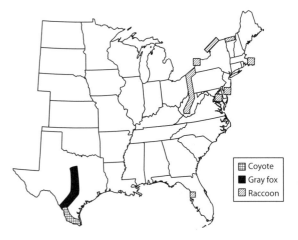

Figure 12.2 Strategic placement of oral rabies vaccine in zones to prevent movement from states where the virus already occurs. *Redrawn from Slate et al. 2002, fig. 3, with permission. Illustration by Lamar Henderson, Wildhaven Creative LLC. With permission of the Vertebrate Pest Council.*

of ketchup, and coated with a mixture of fishmeal and oil. Baits are distributed by ground and aircraft (Slate et al. 2002, 2005). In 2005, almost 12 million packets were delivered by hand and aircraft to strategic locations in 18 states (Nelson 2005). Using a rapid immunochemical testing procedure, 91 of 2,738 animals collected in areas after treatment were found to still be rabid (Nelson 2005), and about 25% of over 4,000 serum samples were positive for rabies antibodies. So, although the approach is already a limited success, continued field efforts are needed, as are continued improvement in methods.

Summary

- Biocontrol uses interactions among organisms to limit wildlife damage or the presence of damaging species. It uses both classical and inundative methods.
- At least 41 weed species have been managed successfully using biocontrol as the only or the preferred method.
- Biocontrol using oral vaccination has been successfully used to break the sylvatic cycle of rabies in Europe and North America.
- Over 44 varieties of livestock protection dogs are used in at least 28 countries to protect livestock from predators. Other livestock protection animals include cattle, goats, llamas, mules, and donkeys.
- Myxoma virus and rabbit hemorrhagic disease virus have been used to manage feral rabbit damage in Australia and New Zealand.

- Lure crops or diversionary feeding are sometimes used to entice damaging species away from more expensive crops or from a situation where they can injure humans or their property. The Wildlife Society has expressed concerns about use of artificial or supplemental feeding by the public because of potential adverse impacts on game and nontarget wildlife species.

- Spectacular failures using biocontrol have included introduction of monitor lizards, the mongoose, and cane toads on islands to control rats or other damaging species. Such failures are often remembered longer than the successes.

Review and Discussion Questions

1. Based on our previous discussions regarding exotic invasive species, why might it be preferable to find an endemic rather than a foreign biological control agent?

2. Based on our previous discussions regarding ecosystems, why might it be necessary to import a foreign biological control agent to manage an invasive exotic pest?

3. Why do you think The Wildlife Society has expressed concerns about public use of supplemental feeding, such as diversionary food or lure crops, in managing wildlife damage? Explain why the Society is especially concerned about game and nontarget wildlife species.

4. What has been the impact of oral rabies vaccination on fox and raccoon dog populations in Europe? Can the same effects be expected when oral disease vaccinations are applied to other species and diseases? Explain.

5. Do you think an oral rabies vaccination program can be developed for management of bat rabies in North America? Why or why not? How would such a program differ from the one currently in place for control of raccoon rabies?

PART V • **HUMAN DIMENSIONS**

In this section we explore the human aspects of an effective management action, including its economic, political, legal, religious, cultural, and social dimensions.

13

Economic Dimensions

Natural resource managers live in a world of limited resources and competing needs, where opportunities to accomplish an important action are sometimes foregone in order to accomplish another even more important one. How the manager decides to allocate resources is often subject to public scrutiny and debate. Given the range of public views on decisions involving natural resources, universal acceptance of any decision is unlikely. One can therefore anticipate vocal and sometimes legal challenges on any management action. Thus it behooves the practitioner to base any action on sound judgment, using the best available information and analytical techniques. Fortunately, economics offers concepts and methodologies that can help the wildlife damage management practitioner make informed choices (trade-offs) among management options.

DAMAGE ASSESSMENT

Statement
Most meaningful economic assessments of wildlife damage require statistically based sampling and analytical methods.

Explanation
With sampling methods, assessments of damage (or some variable related to damage) are made in portions of the total area (or population), and then damage found in the portions is used to interpolate or extrapolate an estimate of damage for the entire area (or population). For example, the practitioner might be able to assess only 1% of a forest damaged by North American porcupines, because it would be too costly to examine every tree. This might mean the practitioner measures damage in about 100 samples of 50 trees each. From the samples, he or she then extrapolates damage for the forest. Quadrat sampling and minimum distance are two methods used to assess wildlife damage (box 13.1).

With sampling, the potential for errors—systematic (e.g., miscalibrated mechanical damage counters that consistently add 10 to the total count) or random—becomes a factor. Even with appropriate

In **quadrat sampling**, a grid of specific
dimensions—a 10-by-10 grid of cornstalks, for
example—is used to sample damage, starting
with a randomly determined coordinate. In this
example, all ears damaged by birds might be
counted. Depending on need, surface area of
damage on each ear might also be recorded. A
partially damaged ear of field corn might have
economic value in that it could still be fed to cattle.
However, for sweet corn, a partially damaged ear
would be counted as a total loss, since people do
not normally buy damaged corn. Quadrat sampling
is based on assumptions that can often accommodate damage patterns common with wildlife—e.g.,
random clumped distributions—and still provide
reliable results. The method tends to be labor
intensive and expensive, however (Engeman 2002).

With **minimum distance sampling**, a point is
randomly selected within a field and the distance
is measured between that point and the nearest
damage. Then the distance is measured between
that damage and the second-nearest point of
damage, then between the second and third points,
and so on. The procedure is repeated a predeter-
mined number of times, and an estimate of damage
is garnered from these data. The method is based
on assumptions of symmetry, but these have been
modified so that minimum distance sampling
offers quality as well as greatly reduced labor and
cost (Engeman 2002).

methods and no errors, variability is part of sampling.
For example, suppose the practitioner examines all
100 tree plots for porcupine damage. It is unlikely that
all plots will yield exactly the same number, say, eight
damaged trees, because porcupines are not uniform
in either distribution among trees or inclination to
damage them.

Depending on the overall distribution of damage, stratification may be needed. For instance, if a
field of corn is being assessed for bird damage, and
most of the damage occurs along fence rows and
edges, the areas of more intense damage may need
to be sampled separately from the rest of the fields.
Sometimes sampling occurs at different levels; e.g.,
a survey of coyote damage to livestock might be

done at county, state, and national levels. Each level
represents a layer for sampling. Costs of sampling
vary with layer, and variability at one layer contributes to the next, so that biases and errors are carried
through all of the higher layers. Layers contributing
the greatest variability should be sampled most heavily (Engeman 2002).

When it is not practical to measure damage itself,
another variable must be found that can be quantita-
tively related to the damage. Thus, missing lambs
might be used to indicate predation on sheep. Often,
population sizes of damaging species are assumed to
be related to levels of damage. For example, population measures of muskrats in constructed wetlands
of sewage treatment facilities might be used as a measure of the level of damage that can be expected to the
wetland vegetation as well as a measure of the effectiveness of any control program that has been attempted. In these cases, a clear relationship must be
established between populations and damage. At that
point populations, or some index of the populations,
can be used to estimate damage.

A similar approach is used with survey methods,
but here respondents, in sampling the affected population, are asked to judge aspects of damage. For instance, selected people from the affected population
might be asked to describe their impressions of damage. Because sampling is involved, variability and errors associated with all sampling techniques need to
be considered. In addition, biases associated with human perceptions of damage and dishonest responses
should also be considered. For example, a farmer who
believes any federal activity is a waste of money and
who wants to avoid a possible federal response may
deny having damage, even though he has experienced
a substantial amount. Conversely, a farmer might bias
his responses upward because he believes this will get
more attention for a problem. Or, a farmer might simply see damage differently from how it would be interpreted by researchers.

Statistical analysis of the sample results is used to
provide information on damage for the area (or population) sampled. Results are often presented as mean
(average) or **median** (the middle value when all are
ranked from lowest to highest) levels of damage, along
with some measure of variation (e.g., standard devia-
tion or standard error of the mean). For instance, results of the analysis of porcupine damage may have
shown that mean damage to the forest was $8.2 \pm 0.5\%$
of the trees at the 95% confidence level (i.e., at least 95
out of 100 field assessments would yield means within
0.5% of 8.2).

COSTS OF DAMAGE AND ITS MANAGEMENT

Statement

Cost of damage provides one measure of its importance and can be used in determining the level of management that is warranted. Assessments of cost are available for major areas of wildlife damage, such as animal-vehicle collisions and invasive species.

Explanation

Costs of damage are gathered by tallying the numbers of goods damaged and assessing their value. Goods include agricultural commodities such as crops or livestock, properties, human health, and other wildlife. For example, the U.S. National Agricultural Statistical Services surveys ranchers annually to derive a national estimate of sheep and goat losses to predation (NASS 2010).

Costs can be direct or indirect. The direct value of a good is not necessarily what it sells for in the local market—i.e., its **market value**—but what a person is willing to pay for the good (e.g., Schuhmann and Schwabe 2002). Nonetheless, market value is often easy to obtain, and it serves as a minimum value. For agricultural crops including timber, livestock, and fish, for example, price at market reflects its actual consumptive value.

The direct value of damaged property; of human health, health threats, and health impacts; and of livestock or pets can sometimes be estimated from insurance payments or medical expenses. If the event involved injuries or fatalities to humans, livestock, or pets, then medical or veterinary expenses and potentially funeral and burial expenses need to be included. If the event involved livestock, loss of market value should be considered in addition to veterinary or other medical expenses. Zinsstag et al. (2007) found that cost effectiveness for management of some zoonoses (e.g., mass vaccination for brucellosis in Mongolia and vaccination of dogs for rabies in Chad) was realized only when total societal costs were considered.

Indirect costs must also be considered, and these can be greater than direct costs. For humans infected with a zoonosis or injured in an animal-vehicle collision, indirect costs may include a shortened productive life span, loss of productive work, health care, and disability compensation (e.g., Havelaar 2007). Indirect costs also occur with crops. For example, wild rodents gnaw between the internodes of canes of sugar, often causing only minor amounts of direct damage and sugar loss (e.g., Martin et al. 2007). However, the wounds open the cane to secondary bacterial infections such as red rot, which sours the entire cane. If a crop or livestock is insured against damage by wildlife (including compensation programs—see below), cost of insurance must be included. Sometimes crops or livestock are purposely not grown or raised in an area because of concerns for wildlife damage, known to economists as a **forgone opportunity cost**. To assess this cost, one would have to estimate the income that might have been generated had the area been cropped or had livestock been raised there.

Some plants can compensate for damage, an indirect capability that can reduce costs of damage. For example, Williams (1974) found that rat damage to coconuts was compensated for by increased size of remaining nuts in groves in Fiji, a factor that reduced costs of damage to coconuts. Rice plants sometimes compensate for rat damage to panicles early in the growing season by producing more panicles. If panicle compensation occurs before harvest, the new grains contribute to overall yield. Often, however, the new grains are still immature at harvest, and farmers are reluctant to conduct a second harvest unless substantial (over 40%) rat damage has occurred (e.g., Islam and Hossain 2003).

Values of wildlife include those considered consumptive (e.g., catching fish) and nonconsumptive (e.g., watching birds). Consider, for example, the cost of an endangered least tern egg taken by a raven. The cost goes beyond the egg's consumptive value (estimated as the market cost of a chicken egg) to include the potential contribution of the egg to the overall survival of the species.

Market values can be obtained in countries where wildlife is legally harvested and brought to market for sale or where there is little concern or no legal regulation, for such species as commensal rats or some common birds. Wildlife regulations have also been used to estimate value. Here, costs in fines for illegal take (Engeman, Shwiff, Smith, et al. 2002), importation, or sales give an indication of value (Sterner 2009). Engeman, Shwiff, Constantin, et al. (2002) use $100 per marine turtle, the Florida state minimum fine, to calculate cost effectiveness of methods for reducing predation by raccoons and armadillos.

Estimates of the value of wildlife might also be based on **willingness to pay**—for a hunting, fishing, or hiking trip, for example. Humans ascribe some value to wildlife simply because it is there, known as **existence value** (Krutilla 1967). Sometimes people have neither the time nor the opportunity to use wildlife at the moment but want it there for use at some other time—i.e., an **option value** (Weisbrod 1964).

For some people, it is important to know that wildlife is present for use by others now (altruistic value) or for future generations (**bequest value**; Schuhmann and Schwabe 2002).

Willingness to pay has been used to assess the value of a broad range of wildlife, including eagles, grizzly bears, deer, and waterfowl (Schuhmann and Schwabe 2002). This concept has also been utilized to assess the willingness of villagers in the eastern Caprivi Region of Namibia to pay for technologies, e.g., fencing, to reduce damage and attacks by wildlife such as elephants (Sutton et al. 2004) and to model economic considerations of community-based wildlife damage management in Zimbabwe (Muchapondwa 2003). With this method, people are asked either how much they are willing to pay for a specific change in access to the wildlife or whether they are willing to pay a specific amount for the change.

Ecosystem services (Anonymous 2005) have also been used to assess the economics of wildlife damage. For example, Frost and Bond (2008) used ecosystem services to assess Communal Areas Management Programmes for Indigenous Resources (CAMPFIREs), now found throughout parts of southern Africa. In CAMPFIREs programs, local communities manage natural resources. In the analyses, wildlife damage is mitigated by revenue sharing between the safari operators and the local communities.

In addition to costs from damage, costs to manage damage need to be assessed. Costs can be substantial, sometimes precluding management. Economic analyses comparing relative costs of management options may help the manager decide which, if any, action to take. For example, Hygnstrom and VerCauteren (2000) compared the costs and effectiveness of five different burrow fumigants for managing black-tailed prairie dogs (*Cynomys ludovicianus*) in colonies in central Nebraska. The researchers found that the products were about equally efficacious but that application costs were higher for pressurized fumigants.

As with damage costs, costs of management are indirect as well as direct. Direct costs include those for personnel, equipment, travel, materials and supplies, permits or licenses, insurance or compensation, and overhead and administration. Management costs can be actual or anticipated and can be collected and tabulated from fiscal budgets and accounts. Costs actually accrued help to evaluate economic benefits of completed management actions.

Indirect costs might encompass such items as loss of wildlife or delays in traffic due to management activities at an urban bird roost. Overall ecosystem impacts, often indirect, need to be considered and can be difficult to value. Oppel et al. (2010) listed the following indirect costs to an island community during an eradication campaign: biosecurity and monitoring costs after the eradication; loss of income from commercial activities during the eradication; costs of volunteering; and inconvenience from new practices (such as garbage disposal and animal husbandry) and regulations. In the United States and many other countries, if the proposed management involves a major federal action, an environmental assessment is required by law (see chapter 15). Indirect costs of management can sometimes be calculated from these government documents.

Compensation and abatement programs are a cost of management. In the United States, most states and related federal programs have **abatement programs** (i.e., consultation services, direct support from agency experts, or subsidies for abatement technologies), and about 25 states have compensation programs (Yoder 2002). For example, a farmer or rancher might be paid for verified wildlife damage to her or his crops, livestock, or property. Or he/she might receive educational and technical support from state or federal extension specialists or agents or financial assistance from state agencies to install deer fences or other forms of damage abatement. In Germany, landowners may be entitled to reimbursement for damage to forest species such as Norway spruce (*Picea abies*) by ungulates such as red deer, roe deer (*Capreolus capreolus*), European beavers (*Castor fiber*), hares (*Lepus europaeus*), and voles (*Muridae, Arvicolidae*). The compensation is usually paid by the shooting tenant (the individual or group who contracts with the landowner for hunting rights) of the land (Schaller 2002). In South Africa, farmers are compensated for relocating problem animals, e.g., cheetahs (*Acinonyx jubatus*), to approved areas (Cilliers 2003).

Increasingly, conservation organizations have supported compensation programs to encourage conservation of wildlife by local people. Whereas costs of damage lie with the ranchers or farmers surrounding the refugia, benefits of retaining wildlife are spread globally across the human community, which supports the presence of wildlife for existence, option, or bequest values. Compensation can be used to facilitate a more equitable distribution of costs and benefits.

Haney (2007) listed 48 wildlife species from around the globe for which compensation has been paid, including many "charismatic megafauna" such as African elephants, African lions, American black bears,

Eurasian lynx, gray wolves, and snow leopards. Compensation was paid by national, state, or regional governments and by nongovernmental organizations (NGOs) such as Mbirikani Predator Compensation Fund, Defenders of Wildlife, World Wildlife Fund, and Project Snow Leopard. Of interest is the fact that the presence of wilderness offered alternative prey for predators, tending to reduce the costs of compensation for livestock depredations, a form of **wilderness discount** (Haney et al. 2007). Compensation for losses was key to gaining the support of ranchers for the reintroduction of the wolf into Yellowstone National Park in the United States. Funds for the program were initially provided by Defenders of Wildlife.

Despite their popularity, compensation programs remain controversial. Some experts (e.g., Bodenchuk et al. 2002) argue that compensation programs are an efficient means of protecting damaging wildlife, whereas others (e.g., African Elephant Specialist Group 2002) argue that with an open market economy and inefficiencies in distribution of funds, compensation actually diminishes the value of the wildlife. Others have stated that compensation does not make economic sense, because the practice reinforces inappropriate husbandry practices, encourages economic dishonesty in both claims of damage and distribution of funds for compensation, and may even increase resentment of damaging wildlife and subsequent retaliation (e.g., Rondeau and Bulte 2007). For example, Fernandez et al. (2009) reviewed community-based conservation programs in Zambia that compensated villagers for wildlife damage. As wildlife populations increased, so did damage. Program policies, however, did not allow adjustment of compensation for increasing damage. Villagers became frustrated with uncompensated damage and looked to their own solutions, leading to additional human-wildlife conflicts. Montag (2003) argued that compensation programs were fraught with additional economic problems. She posited that the issues might be framed differently by ranchers or farmers who may see compensation programs as federal management of private rights, affecting equity, public grazing, and public or private land management.

Examples

BIRD-AIRCRAFT COLLISIONS. Bird collisions with aircraft occur worldwide, with airlines incurring costs for losses and airports incurring costs for control. Species weighing over 2 kg (e.g., geese in North America and lapwings, *Vanellus vanellus,* in the United Kingdom) tend to cause the most damage (Allan et al.

1999). An average bird strike costs about US$39,705 (about 11% for damage repair and 89% for delays and cancellations; Allan 2002). Based on air transport movement worldwide, Allan estimated worldwide costs of damage at US$1 billion to $1.5 billion per year, or about US$64.50 per flight. The estimates did not include helicopter traffic or costs to military aviation. This researcher estimated costs from bird damage to the U.S. Air Force at about US$33 million per year and to the Royal Air Force (RAF) of the UK at about US$23.3 million per year.

Thorpe (2003) reported the loss of 231 lives in civil aviation due to bird strikes between 1912 and 2002, while Richardson and West (2000) estimated about 141 deaths in bird collisions with military aircraft between 1959 and 1999 in western nations. Costs associated with loss of lives were not provided.

COSTS TO MANAGE DAMAGE AT AIRPORTS. Allan (2002) considered costs of minimizing habitat in and around an airport at about US$75,000 per annum in Western Europe. This estimate was based on costs for making and maintaining grass swards that discouraged the presence of species such as gulls, lapwings, golden plovers (*Pluvialis apricaria*), and starlings on airfields in the UK, an approach that might also favor the presence of small mammals that attract predatory birds (Barras et al. 2000). Costs were estimated at about US$65,000 to $130,000 per airport per annum as contracted by the RAF for patrolling, bird dispersal, and wildlife depredation, either during daylight or during all times, respectively. Similar harassment programs in the United States were estimated to range in cost between US$25,000 and $150,000 per airport. Thus, Allan (2002) estimated a cost of approximately US$200,000 per year to implement a full control program in the UK; less expensive were programs that might be cost effective at airports with minimal needs for bird control.

Dolbeer and Chipman (1999) estimated that airstrikes were reduced from about 170 to 50 per year during the period of time when shooting teams were employed to prevent birds from entering airspace at John F. Kennedy International Airport in New York. They estimated the airlines using the airport saved about US$4,764,600 per year, at an annual cost to the airport of about US$120,000 per year.

ANIMAL-VEHICLE COLLISIONS. Hardy et al. (2007) estimated costs of annual wildlife-vehicle collisions in the United States to be US$8.4 billion. This estimate is probably conservative. Many accidents involving damage to cars under $1,000 go unreported. And, although Hardy et al. based their analyses on collisions with wildlife, actual collisions do not always occur. Sometimes, for example, drivers lose control of a vehicle while attempting to avoid wildlife, causing human injury or death. And sometimes drivers stop along or on roadways to remove wildlife from a dangerous situation, and accidents ensue. Further, the report noted that travel delays, as well as secondary accidents to subsequent motorists if the vehicle or animal lies in the right of way, were not considered, nor were the costs related to emotional trauma of the motorists.

Williams and Wells (2004), in a U.S. study, found that when human injuries or fatalities occurred, they were costly, medical expenses averaging $2,702 per collision. Other costs averaged at $125 for towing and law enforcement services; $2,000 for the value of the deer; and $50 for carcass removal and disposal (Hardy et al. 2007). If a larger animal was involved, costs were greater—e.g., $3,000 for an elk and $4,000 for a moose. If the involved species was federally or state listed as threatened or endangered, costs were higher.

Excluding Russia, Bruinderink and Hazebroek (1996) estimated that about 507,000 collisions of vehicles with ungulates occurred annually in Europe. Ungulate-vehicle accidents resulted in about 300 fatalities and US$1 billion annually in property damage. In Australia, approximately 3,300 animal-vehicle collisions were reported to police between 1990 and 1994—reported because the vehicle needed to be towed or humans were injured (Kangaroo Advisory

Committee 1997). Cooper (1998) estimated that total costs for animal-vehicle collisions in Australia, including property damage and injuries to people, were about US$10 million annually. Collisions with animals, including camels, are also a problem in Asia and the Middle East, including Saudi Arabia (Bashir and Abu-Zidan 2006). Collisions with vehicles were also reported as a trouble in Japan (Bruinderink and Hazebroek 1996).

INVASIVE SPECIES. Pimentel et al. (2001) estimated that over 120,000 species of wildlife, both plants and animals, have invaded the United States, the United Kingdom, Australia, Africa, India, and Brazil. A few adapted to domestication—crops and animals that became livestock provide over 98% of the human food supply globally. Others cause economic damage to human interests, including the environment. Pimentel et al. (2001) estimated that annual costs for damage to the six nations discussed were over US $334 billion (table 13.1). This estimate does not include costs for damages such as extinction of wildlife species by cats and feral pigs. Costs were about US$240 per year when calculated as cost per capita for the six nations. If this amount represents an average cost worldwide, global damage by exotic invasive species exceeds US$1.4 trillion per annum, or about 5% of the world economy (US$31 trillion per annum). The estimate appears conservative in that estimates of losses to exotic invasive species will likely increase as additional information is quantified.

TABLE 13.1   *Economic losses of crops, pasture, and forests to invasive species, in US$ billion per year*

Introduced pest	United States	United Kingdom	Australia	South Africa	India	Brazil	Total
Losses to agriculture							
Weeds							
Crops	27.9	1.4	1.8	1.5	37.8	17.0	87.4
Pasture	6.0		0.6	—	0.92	—	7.52
Vertebrates							
Crops	1.0	1.2	0.2	—	—	—	2.4
Arthropods							
Crops	15.9	0.96	0.94	1.0	16.8	8.5	44.1
Forests	2.1	—	—	—	—	—	2.1
Plant path							
Crops	23.5	2.0	2.7	1.8	35.5	17.1	82.6
Forests	2.1	—	—	—	—	—	2.1
Subtotal	78.5	5.56	6.24	4.3	91.02	42.6	228.22
Losses to environment							
Subtotal	58.3	6.57	6.87	3.0	25.00	6.7	106.44
Total	136.8	12.13	13.11	7.3	116.02	49.3	334.66

Source: Adapted from Pimentel et al. 2001, with permission.

Vilà et al. (2010) provided a summary of findings on economic costs for management of invasive species for Europe, based on data provided through the DAISIE (Delivering Alien Invasive Species Inventories in Europe) project. Some species with the highest costs are toxic bloom of the marine alga *Chrysochoromulina polylepis* in Norway, at US$12.13 million; control/eradication of common water hyacinth in Spain, at US$4.97 million; and control of nutria in Italy, at US$4.22 million.

THE BENEFIT/COST ANALYSIS AND BENE-FITS OF MANAGEMENT

Statement

Assessment of potential benefits in relation to management costs helps practitioners decide how to judge trade-offs—e.g., whether management action is justified—and to choose among management options. Assessment of actual benefits helps practitioners determine potential values of adaptive improvements and whether a program should continue.

Explanation

Cost effectiveness is the economic bottom line of wildlife damage management. To assess net benefits, one must know the full range of benefits in relation to the costs of damage and costs of management (see the previous section in this chapter). In wildlife damage management, benefits are sometimes viewed as avoided costs. Sometimes benefits can be calculated by measuring the reduction in damage and associated losses or by measuring a concomitant increase in production. This direct approach applies most readily to wildlife damage to livestock, crops, forestry, and aquaculture production. If the damage is to property such as houses, automobiles, or aircraft, a reduced rate of loss over some reasonable period of time might allow an estimate of the benefits of management (e.g., 10% fewer wildlife-vehicle collisions) and a measure of the effectiveness of the control action. If there is a strong statistical relationship between damage and its actual cost or between activity of the population and cost of damage, then reduced damage or activity can be used to estimate benefits of the management.

Also important is to consider indirect or "spillover" benefits to secondary entities—e.g., increased livestock production from management of coyote predation to improve antelope production in Wyoming (Shwiff and Merrell 2004). If human health is improved as a consequence of management, as with malaria control programs or reduced incidences of ehrlichiosis or Lyme disease, indirect benefits accrue as fewer sick days and increased productivity as well as reduced costs for medical treatment and insurance payouts. There might be other induced or intangible benefits, hard to assess but real nevertheless (Shwiff 2004). For example, jobs might be created or additional moneys flow into local economies from improved production. Distribution of benefits, the environment, and resources may also be positively affected.

Tools available for economic assessments include the following: studies of economic feasibility (i.e., is cost of damage equal to or greater than cost of management?); economic efficiency (i.e., which of the available management options is least expensive?); cost-effectiveness analysis (CEA, i.e., which of the available management options offers the best return on investment?); and **benefit/cost analysis** (**BCA**, i.e., what is the actual benefit to be gained in relation to the cost of management?).

McCoy (2002) provides a generic example of an economic analysis, using a hypothetical multiyear goose management program. The analysis begins by assessing management costs, including initial investments, total (life-of-program) operating costs, and secondary costs, such as unintended loss of other wildlife. Total benefits—i.e., value of reduced goose damage—need to be assessed. Future expenditures and benefits can then be incorporated into present values using a future discount rate (i.e., roughly annual inflation), resulting in **net present values** (NPV). **Expected value** (**EV**) analyses can be used if there is uncertainty surrounding benefits. In his example, McCoy spends US$100,000 for goose management that includes habitat changes around an airport. There is a 10% chance that management action to reduce goose strikes to aircraft at an airport will produce no beneficial results, a 60% chance that the actions will reduce losses from US$1 million to US$650,000, a 25% chance that the actions will reduce losses to US$500,000, and a 5% chance that losses will be reduced to $0. An EV of US$315,000 is calculated from the probabilities, after subtracting the US$100,000 for cost of management (see McCoy 2002 for calculations). This information allows several economic analyses. For example, economic feasibility can be calculated. Here, any effort that yields an NPV greater than $0—i.e., a cost of US$315,000 or less—is considered economically feasible. Economic efficiency might be calculated by comparing potential benefits of habitat changes with other management strategies, such as hazing, repelling, or shooting geese, an approach McCoy (2002, p. 121) suggests can help the practitioner get the "biggest bang for the buck."

If only costs of management are known and benefits can be quantified on some scale (e.g., percent effectiveness of the method), a CEA can still be completed (McCoy 2002). With CEA, costs are compared to relative benefit. In the goose management example, McCoy provides the following management options: hazing with US$10,000 for 10% population reduction; repellents with US$20,000 for 10% population reduction; habitat modification with US$20,000 for 20% population reduction; habitat modification and hazing with US$25,000 for 25% population reduction; and habitat modification and repellent with US$35,000 for 30% population reduction. By comparing these figures, one can see that use of repellents alone costs relatively more than the other methods and that hazing alone can achieve the same results with a lower cost.

To complete a full-fledged BCA, all benefits and costs, direct and indirect, present and future (discounted to present value), must first be estimated. Total costs are then subtracted from total benefits to derive the **net benefit**. By comparing net benefits of different approaches, the practitioner can compare economic differences among the options (Portney undated). Shwiff and Sterner (2002) presented a general model that applies BCA to hypothetical proposed approaches for reducing wildlife hazards to aircraft. Four approaches were considered: bird harassment, habitat management, fencing, and doing nothing. This model is helpful in explaining the use of dependent variables, spreadsheets, decision trees, calculation of NPV, and sensitivity analyses (to assess impacts on outcomes when variables change)—all important components of BCAs—and can be applied to other wildlife damage situations.

BCA models are increasingly available to wildlife damage practitioners. The practicality of BCAs and their applications to wildlife damage issues have risen dramatically, along with the rising capacities and iterative calculating capabilities of computers. Many BCAs have specific applications; for example, Caudell et al. (2010) provide a BCA model to assess when Ovotrol G is more cost effective for controlling Canada geese than other methods, such as egg oiling or other forms of addling. Sterner and Tope (2002) provide models to determine when it is cost effective to use a commercial turf repellent (Rejex-It) to repel Canada geese from golf fairways and a commercial shrub/plant repellent (Deer I Repellent) to deter deer from

eating landscape shrubbery. VerCauteren et al. (2002) have developed models that generate BCA analyses for various levels of rodent control at confined swine facilities. Choquenot and Hone (2002) provide models to be used to evaluate BCAs of helicopter shooting and Compound 1080 poisoning to reduce feral pig predation on newborn lambs in Australia.

BCAs have also assisted conservation efforts. For example, Emerton (1999) provided a BCA for conservation at Lake Mburo National Park, Uganda. Her evaluations included costs of damage by wildlife to crops, from livestock kills, and from transmission of diseases to domestic herds. She used the analyses to suggest cost-effective approaches for managing damage.

Examples

BCAS COMPARING DEER FENCES. VerCauteren et al. (2006b) developed a model that provides BCAs for different fence designs to protect crops from deer damage. The model uses simulation software that can be adjusted for acreage (0–200 acres), field perimeter (254–11,390 meters), percent of damage (0.00–100.00), crop value per acre ($0.00–$4,500), fence efficacy (0.00–0.99), material costs per meter of fencing ($1–$15), labor costs per meter of fencing ($0–$15), and discount rate (0.00–0.15) for the life of the fence.

The user provides information specific to his/her situation. The model then calculates NPV (net present value). By using iterative calculations, practitioners can compare impacts of factors such as different fence designs and labor costs on net benefits. In their simulations, for example, fences designed to protect high-cash-value crops such as apple orchards tended to benefit from more expensive, long-term fences made from woven or welded wire. For lower-cash-value crops, such as soybeans and alfalfa, fences were only marginally cost beneficial, even with less expensive designs.

USE OF RODENTICIDES FOR ALFALFA DAMAGE. Sterner (2002) developed a model that calculates BCAs for the use of zinc phosphide to manage vole damage to alfalfa. The model uses the maximum crop value to assess potential savings from management and considers both personnel and materials as costs for application. This researcher assumed a single prebaiting, uniform distribution of steam-rolled baits at 10 pounds per acre, 2% zinc phosphide, and multiple cuttings of alfalfa. Sterner used production values and prices as well as costs, three sizes of areas, and six levels of crop losses to complete 1,260 combinations of calculations and provide a three-dimensional picture of potential

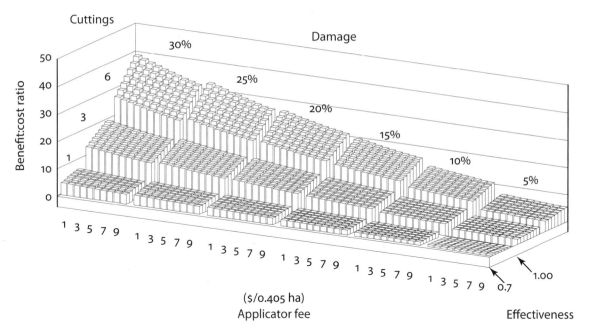

Figure 13.1 A three-dimensional graph that predicts benefit-to-cost ratios of zinc phosphide applications in relation to levels of rodent damage to alfalfa. Redrawn from Sterner 2002, with permission. *Illustration by Lamar Henderson, Wildhaven Creative LLC.*

benefits (fig. 13.1). The graph provides threshold values that a wildlife damage practitioner could use to determine when benefit exceeds costs and when treatment is likely to be economically feasible.

AUSTRALIAN MOUSE PLAGUE. Brown and Singleton (2002) estimated that an Australian mouse plague in 1993/1994 cost about US$60 million to crops, livestock industries, and rural communities. They evaluated economic factors associated with both outbreaks and control, including cost and effectiveness of use of rodenticides in bait stations, spraying grasses and weeds along fence lines, grazing stubble after harvest, cleaning up spilled grain, and cultivating after sowing to disguise seeds. Data came from wheat farmers in New South Wales. The information was simplified and then incorporated into a model they called the Decision Support System "Mouser." With it, a farmer would provide basic information on wheat production, mouse density, and anticipated rainfall; the model then predicts cost effectiveness of four mouse control options. Using established relationships between mouse abundance and crop loss, the program predicts monthly mouse abundance with and without the control (Anonymous undated).

ANIMAL-VEHICLE COLLISIONS. Huge costs are incurred by governments in attempts to prevent or reduce collisions. Hardy et al. (2007) identified 34 meth-

ods of alleviating this problem and compared BCAs for these techniques. Fencing, in combination with underpasses and overpasses, had the best economic return on investment, resulting in reduction in deer-vehicle collisions of over 80% and bringing net benefits of about $24,000 to $33,000 per km of a hypothetical highway per year. Standard signs and deer reflectors and mirrors were ineffective, costing more to install and use than was gained in benefits.

Summary

- Most assessments of wildlife damage are based on statistical sampling and surveys, using methods such as quadrat sampling or minimum distance sampling. When feasible, the best measure is damage itself. When this is impractical, a variable related to damage must be found. Damage can be direct or indirect.
- Cost of damage is based on numbers of goods damaged and the value of the goods. Goods can be property, crops, livestock, human health, or other wildlife. Value can be based on market value or on economic concepts such as willingness to pay. Costs can be direct or indirect.
- Costs of management can be direct or indirect. Direct costs can be obtained from fiscal budgets and records. Indirect costs may be harder to value. Costs are available for major areas of

wildlife damage such as animal-vehicle collisions or invasive species.

- Compensation and abatement programs are a cost of management. Abatement programs reduce damage by providing technical and material assistance for both preventing and managing damage. Over 48 wildlife species worldwide have been part of compensation programs, often underwritten by conservation organizations as a means of facilitating a more equitable distribution between benefits and costs to communities bearing the brunt of damage.

- To assess net benefits from managing wildlife, one must know all benefits in relation to all costs of damage and its management. Tools available for economic assessments include economic feasibility, economic efficiency, cost-effectiveness analysis, and benefit/cost analysis (BCA).

- Economic tools provide a means for assessing potential benefits in relation to management costs, to decide whether management action is justified and to choose management options.

- Many BCA models are focused on specific issues, such as various levels of rodent control at confined swine facilities, comparisons of economic differences among deer-fence designs in relation to situations, and economics of use of rodenticides in alfalfa fields to manage vole damage.

- Other BCA analyses focus on a particular area of wildlife damage, such as predator management or relative merits of animal-vehicle mitigation measures.

Review and Discussion Questions

1. Choose a primary publication that provides statistical information on wildlife damage. What information is provided on variability of the results around the means or medians? At what level of confidence? Choose a test comparing two or more means. Relate variability to significant difference (if there is one) or no significant difference (if there is not one). A diagram or two using normal bell curves (i.e., normal distributions) might help.

2. Compensation programs are sometimes used by conservation organizations to effect a more equitable economic distribution between those bearing the costs of damage and those benefiting from effective damage management. Does this approach improve the value of wildlife or not? Choose a specific damage situation, choose a particular position, and defend your position with references to literature.

3. We have seen an increase in economic computer models that allow the practitioner to assess relative economic merits of control options. What is the role of iterative calculations in these models? What is the importance of calculating a threshold value?

4. Federal predator management programs in the United States have been criticized as being noneconomical subsidy programs for the livestock industry, but Bodenchuk et al. (2002) calculate a 12:3 benefit-to-cost ratio for all livestock protection programs. Who is correct? Defend your view.

5. Are signs and deer reflectors an economically effective means of reducing deer collisions with vehicles in North America? Explain your answer, using literature.

14

Human Perceptions and Responses

In this chapter we explore societal, cultural, religious, and personal factors that influence wildlife damage and its management.

Statement

How one responds to wildlife damage is based largely on one's perceptions of its seriousness, especially the sense of personal risk and whether one is actually suffering damage. The responses also fall within the limits of one's values, beliefs, attitudes (VBAs), and mental state, such as fear or anger.

Explanation

VALUES. **Social values** are conceptions of what is good and bad, desirable and undesirable. Values form slowly during youth, change little later in life, and strongly influence attitudes and norms (group-held rules for acceptable behavior within social settings, or things one "ought" to do (Manfredo and Teel 2008). Social values therefore provide a broad frame within which we think and behave, and they serve to transmit social culture across generations, ensuring that individual behavior is in line with societal norms. These values underlie how people view wildlife damage management and their thoughts about which benefits are most desirable (Allen et al. 2009).

Wilson (1984) proposed an innately emotional connection between humans and other living things driven by a long co-evolutionary history, which he called **biophilia** (love for living things). He argued that natural selection favored individuals well tuned to nature and life, leading to a "cognitive archeology" with detailed attention to plants, animals, time, space, weather, water, and wayfinding (Kellert 2007). The result was biophilia, genetically based in modern humans. Biophilia serves as a paradigm, filtering what humans pay attention to and what responses they learn or resist learning (Wilson 1993).

Kellert (e.g., 1993) theorized that biophilia manifests in different people as different typologies (types of values) or biophilic values (table 14.1). He found typologies particularly useful in relating peoples' values to their behaviors. For example, Kellert (1980) found that hunters scored high on dominionistic values (other researchers use the term *doministic,*

TABLE 14.1   *A typology of biophilia values*

Term	Definition	Function
Utilitarian	Practical and material exploration of nature	Physical sustenance / security
Naturalistic	Satisfaction from direct experience/contact with nature	Curiosity, outdoor skills, mental/physical development
Ecological-Scientific	Systematic study of structure, function, and relationship in nature	Knowledge, understanding, observational skills
Aesthetic	Physical appeal and beauty of nature	Inspiration, harmony, peace, security
Symbolic	Use of nature for metaphorical expression, language, expressive thought	Communication, mental development
Humanistic	Strong affection, emotional attachment, "love" for nature	Group bonding, sharing, cooperation, companionship
Moralistic	Strong affinity, spiritual reverence, ethical concern for nature	Order and meaning in life, kinship and affiliational ties
Dominionistic	Mastery, physical control, dominance of nature	Mechanical skills, physical prowess, ability to subdue
Negativistic	Fear, aversion, alienation from nature	Security, protection, safety

Source: From *Biophilia Hypothesis* by Stephen R. Kellert and Edward O. Wilson. Copyright © 1993 by Island Press. Reproduced with permission of Island Press, Washington, D.C.

and these two terms are often used interchangeably, although they are not identical), whereas anti-hunters scored low on dominionistic values and high on humanistic and moralistic values.

Manfredo et al. (2009; see also Kluckhohn 1951) refer to values that characterize a culture as **value orientations**. Value orientations are a form of ideology. For example, the authors suggest that the Navajo were oriented toward harmony with nature, whereas Mormons were oriented toward mastery over nature. Two value orientations are particularly important for understanding human relations with wildlife in North America. One is a **doministic value orientation,** in which human well-being is seen as a higher priority than the well-being of wildlife. Doministic value orientation leads to acceptance of lethal or intrusive methods of control and utilitarian evaluation of methods. The other is a **mutualistic wildlife orientation,** in which wildlife is seen in trusting relationships with humans and is perceived as having rights similar to those of humans, as part of an extended family. Individuals with mutualistic orientations oppose lethal or intrusive control methods and tend to see wildlife from an anthropomorphizing view. Manfredo et al. (2009) thought doministic orientations were important during colonial times in North America, during the period of western expansion, and during the industrial revolution. The researchers believe that the mutualistic orientation is now, during the post-industrial era, regaining prominence in North America.

BELIEFS. **Beliefs** are "judgments about what is true or false" (Allen et al. 2009, p. 5). "Knowing" what is true

and false can come from many sources, such as scientific information, feelings, intuition, or cultural norms (Allen et al. 2009). Beliefs can be found in individuals or groups. For wildlife damage practitioners, "knowing" can be based on experience as well as peer-reviewed scientific information. An experienced feral pig trapper, for example, has an intimate personal knowledge and understanding of the habits and behaviors of these pigs. Allen et al. refer to this as "local knowledge" or "traditional ecological knowledge" if it passes from generation to generation.

Everyday beliefs can go wrong in at least three ways (Babbie 2001). First, the strength of one's commitment to a belief may be so strong that contrary evidence is ignored. Strong religious or political ideologies or family loyalties are examples. Second, one's analyses may be too casual, often relying on anecdotal evidence. Third, there may be no clear link between observations and explanations.

Beliefs, though they may not be correct, can nonetheless be culturally defined (Romney et al. 1986). For example, 62.3% of respondents from villages in the Tunduru, Liwale, Nachingwea, and Masasi districts of Tanzania believed that humans in their villages were attacked and killed by ghost lions, not real lions (Nyahongo and Røskaft 2011). The villagers believed the ghost lions were made by witch doctors for revenge. Casual observations about the lions and the circumstances surrounding their appearances supported their being ghosts.

Many societal values and beliefs are rooted in religion. All religions view animals as sentient and therefore foster humane treatment. Each religion teaches beliefs on how humans relate to other animals, communities, and ecosystems, and these in turn affect how wildlife damage is perceived and contribute to or constrain allowable management practices. Intertwined with religious beliefs are those values and beliefs provided by society as norms, ethics, and culture. **Ethics** are sets of moral values defining good and evil, right and wrong, by which people behave. **Culture** is the set of shared values, beliefs, and behaviors that characterize an institution, organization, or group. Understanding such collective values and beliefs can be important for wildlife damage practitioners. For example, in pantheistic cultures in which humans believe God exists in all living creatures and some nonliving things as well, a direct attack on damaging wildlife can also be a direct affront to God (box 14.1). Or, if animals are believed to possess souls that survive death, the animals might take revenge directly or through a living animal relative. Persons holding either belief are likely to choose nonlethal methods of management.

In the 1980s, Cornell University's Human Dimensions Research Unit developed a scale for assessing human attitudes toward wildlife (i.e., a wildlife attitudes and values scale, or WAVS). Applying the WAVS to a variety of human and wildlife situations in New York State and synthesizing the information, the research unit (Decker et al. 2002) related attitudes to basic beliefs. Although the beliefs fell into four basic categories, those relating to problem **tolerance** (i.e., **human tolerance**, beliefs surrounding acceptable risks associated with wildlife) seem particularly important to wildlife damage management. For example, Decker et al. (2002) pointed to a trend analysis by Butler et al. (2001) involving WAVS analyses of New York residents between 1984 and 1996. The analysis shows a declining tolerance for problems related to wildlife among both rural and nonrural residents.

ATTITUDES. **Attitudes** are "learned tendencies to react favorably or unfavorably to a situation, individual, object, or concept" (Allen et al. 2009, p. 5). Predicting support or opposition to wildlife damage actions, attitudes can be modified more rapidly than values or beliefs by such factors as education and experience.

In a survey presented to the U.S. Fish and Wildlife Service, Kellert (1979) showed many insights into the attitudes of the public regarding wildlife damage and its management. Only 27% of the public were aware of issues surrounding the killing of livestock by coyotes, whereas 52% were unaware. Regardless of awareness, the public opposed trapping and poison-

BOX 14.1 Karni Mata Temple

Karni Mata Temple is located in Deshnok, India, about 30 miles south of Bikaner. The following material summarizes information on the temple from several Internet sources. Karni Mata, who lived in the 1400s, was believed to be the incarnation of Durga, a highly revered Hindu goddess. According to believers, the following is true: Karni Mata lived for 151 years and performed many miracles throughout life, even as an infant. She disappeared into a divine light, altered the hand of an aunt, saved the lives of her father and others by rendering harmless otherwise toxic snake venoms, fed many soldiers from a single small pot, and provided water for multitudes on several occasions.

When a son-in-law drowned and the body was brought before her, Karni Mata remained in a closed cave with the body for three days. On the fourth day, a resurrected son-in-law left the cave. "Kabas" (ratlike creatures that are a reincarnation step between human forms) began emerging from the cave and subsequently lived and reproduced in the Karni Mata temple. When Karni Mata facilitated the rebirth of her son-in-law, she also changed the natural laws; all humans had to be reborn as kabas before being born again as humans. Harmless relatives of other humans, the kabas comingle with their human relatives under the auspices of the temple. The kabas never leave the temple, completing their life cycles within it, and the temple provides for their sustenance.

For the wildlife damage practitioner charged with managing rat problems in this part of India, it matters little whether the story of Karni Mata is true. The belief matters, and it must be respected and considered if any strategy to control rats (not the kabas) in the area is to succeed.

ing for coyote management but favored hunting offending individual coyotes. Livestock producers favored trapping and poisoning over hunting. The public and livestock producers opposed use of public funds to compensate producers. Respondents from concentrated populations over 1 million opposed (64%) shooting or trapping coyotes, whereas respondents from populations under 500 favored (56%) the methods. The general public disagreed with shooting (61%) or poisoning (about 90%) golden eagles if they were killing sheep, whereas livestock producers (81% for sheep producers, 72% for cattlemen) favored shooting the eagles. The public (about 65%) also opposed poisoning blackbirds that were affecting crops, assuming that some unintended (but not endangered) animals would also be poisoned. However, the general public supported (71% informed, 77% uninformed) killing rats to protect agriculture even if some unintended animals were killed. In general, Kellert found that persons holding utilitarian attitudes (later, *typologies*) supported more obtrusive management methods, including lethal control options, than persons holding humanistic attitudes. Persons with utilitarian attitudes were predominantly rural, had limited mobility and education, and were older males with lower incomes who had direct experience dealing with wildlife damage. While these findings were based on studies from North America, Treves (2008) points to similar findings from studies in other countries.

The society in which one lives and its concomitant culture—and through it the ethical standards and norms of the society—can strongly influence one's attitudes about wildlife and management of its damage. For example, Kellert (1991) found that Japanese appreciation for nature tended to be "narrow and idealized," focused on individual favored species, and devoid of ecological and ethical context. A high value was placed on mastery and control of nature. Given these societal perspectives on wildlife, one can anticipate a broad range of acceptable control options for the management of wildlife damage.

In addition to culture, society, and individual factors such as sex, age, income, and education, people who have experienced damage first-hand tend to take a stronger interest in reducing or eliminating that harm than those who are more removed. Thus, farmers adjacent to wildlife parks in Africa are often less tolerant of damage caused by elephants than those individuals less affected by the damage, just as persons experiencing damage from prairie dogs or geese are less tolerant of the animals than those who have not experienced it.

Attitudes are closely related to subsequent behavior (Manfredo and Bright 2008). Attitudes can predict behavior, but general attitudes do not predict specific behaviors. With wildlife damage management, for example, a generally positive attitude about use of traps would not necessarily predict support for trapping feral cats. Salient beliefs, those that come to mind in a given situation without prompting, are more likely to influence attitudes than those that are nonsalient. Strong attitudes—i.e., stable ones that

remain relatively unchanged over time—are more influential than weak ones in guiding behavior. Attitudes may be extreme, ambivalent, or carry some other level of conviction. For example, Manfredo and Bright state that two people may agree that shooting from a helicopter is good for controlling coyotes. One, however, may see the method as excellent, whereas the other may find it only acceptable. Here, the more predictable behavior would be by the person seeing the method as excellent (the one with the extreme attitude). Attitudes may be centrally embedded (or not) within a broader base of beliefs. Each of these factors influences the strength of the attitude, its likelihood of being changed, and its likelihood of affecting one's behavior.

Attitudes underlie behaviors important to wildlife damage, such as taking a management action or voicing support or opposition to such an action. Often the individual evaluates her/his own attitude in relation to appropriate group norms before taking any action (Allen et al. 2009). Attitudes can be influenced and changed by factors such as education and experience. Therefore, educational programs, including those provided by extension and natural resource agencies, play important roles in ensuring that accurate and objective information is available when individuals interested in issues of wildlife damage are receptive to the information, with subsequent changes in attitude. Values may also be influenced by education during early formative years. Project Wild (United States), Project Wet Foundation, Project Learning Tree, 4-H Youth Development Programs, the World Organization of the Scout Movement, and World Association of Girl Guides and Girl Scouts are examples of international environmentally oriented youth educational programs that already have ties with natural resource agencies and extension programs and could serve as conduits for furthering education of youth about wildlife damage and its management.

RISKS. Beliefs involving **risks** can be particularly important in forming human attitudes about wildlife damage. Here, the probability of occurrence of damage and severity of the consequences (measured as the level of the emotion dread—see below) is important (Decker et al. 2002). A cougar attack, for example, may be unlikely but carries a high level of dread. Deer damage to rural gardens may be likely but carries low levels of dread. Other insights relating risk to attitudes include the following: tolerance for risk decreases as probability of the occurrence increases; perceptions of risk rather than objective risk assessments often drive people's actions; people accept greater risk if they assume it voluntarily; perceptions of risk are elevated when consequences can be severe or are not distributed equitably; risks to children are less tolerated than those to adults; and perceptions of risk decrease when associated benefits are understood (Decker et al. 2002). As summarized by Madden (2004), public outcry is often not proportional to actual damage but rather to the perceived risk and lack of control of managing a solution.

Perception of high risk may not always drive one's attitude. For example, Zinn and Pierce (2002), in a survey of wildlife value orientations of Colorado metropolitan residents toward mountain lions, found that women perceived more risk than men from a mountain lion but were less willing than men to accept destroying the mountain lion. Storm et al. (2007) evaluated attitudes of exurbanites (people who live between suburbia and rural areas) around Carbondale, Illinois, toward white-tailed deer. Concerns for deer-vehicle collisions (low probability but high level of dread) were common among respondents (84%), but damage to ornamentals or fruit-producing plants (high probability but low level of dread), not fear of collisions, determined tolerance for deer.

EMOTIONS. As already indicated, understanding **emotions** is also necessary to comprehend how one might respond to situations involving wildlife damage or its management. Emotions, along with moods, constitute "affect," a general class of feelings that humans experience (Manfredo 2008, p. 51). Emotions relate to specific events and are short-lived and conscious, whereas moods are longer lasting and in the background of consciousness. Primary emotions include happiness, sadness, anger, surprise, disgust, frustration, and fear. Secondary emotions are combinations of primary ones. Emotions are probably rooted in the long evolutionary history of humans and are modified by culture.

In wildlife damage management, emotions are sometimes used in attempts to effect changes in attitudes and to raise funds. Most of us, for example, have seen photographs of animals caught in traps on fundraising mailers and brochures from organizations that oppose trapping. Do these appeals work? According to Manfredo (2008), research suggests such appeals can be effective, but it depends on the context. For example, Zinn and Manfredo (2000) found that emotional appeals, including pictures of animals in traps, were more memorable but no more effective than rational appeals in swaying people on banning trapping in Colorado.

Dread, or extreme fear, is another emotion important to wildlife damage management. Decker et al. (2008, p. 15) write, "practitioners in the area of human dimensions of large herbivore restoration and management should assess levels of fear of the animal in question to gain a deeper understanding of public attitudes regarding management efforts." In one study, for example, respondents in the Hochsuerlandkreis region had significantly more negative attitudes and less knowledge of bison than respondents in the Siegen-Wittgenstein region of Germany. The researchers used logistical regression analyses to show the role of fear in forming the public's attitudes. The differences led to a decision to restore bison in the region in which residents had more favorable stances.

Personal involvement with a wildlife problem tends to increase its importance. For example, ranchers near Yellowstone National Park expressed stronger concern about release of wolves than people more remote from the introduction. Likewise, people living around wildlife refugia in Africa, and who are likely to suffer crop losses or injury, tend to have strong emotional involvement with wildlife damage issues.

Personal moral decisions involve direct (i.e., "I am doing it") harm to another person or to wildlife. Impersonal moral decisions involve someone else actually doing the harm. Whereas personal moral decisions stimulate areas of the brain tied to cognition and emotion, impersonal moral decisions stimulate areas tied to in-depth cognitive processing. This difference may help to explain, for example, why some individuals will not trap and euthanize a squirrel from an attic but have no problem paying a management specialist to do the job. Here, Manfredo (2008, p. 61) sees great opportunity for examining human-wildlife relationships: "Are groups that differ on wildlife values or on wildlife issues more similar in their personal moral judgments than impersonal judgments? Are impersonal judgments more susceptible to cultural differences and culture shift? . . . Can we explore apparent inconsistencies in people's attitudes or wildlife value orientations based on this distinction?"

Memories are more easily made, as well as recalled, when tied to a strong emotional state. For example, decision making is facilitated by a positive emotional state wherein people tend toward more creativity and belief in success; decision making is opposed by a negative affect in which people tend to be more vigilant and analytic. In the context of wildlife management, Manfredo (2008, p. 60) asks, "What forms of conflict resolution and stakeholder-engagement might be devised that explore the utility of creating positive affect as a base for more effective and lasting compromise?"

In sum, Gigliotti et al. (2009) pointed to the changing culture of wildlife management. The authors argue that the visionary manager will see the shifting changes of society, embrace them rather than resist them, and use the new paradigms to strategic advantage. These researchers point to three such paradigm shifts: from management of individual species to that of whole ecosystems; from state to worldwide issues such as climate change and global warming; and from revenue based on hunting and fishing to more broadly based revenues that also include nonconsumptive stakeholders of wildlife.

We believe the advice provided by Gigliotti et al. (2009) applies directly to wildlife damage specialists, and we would add the shifting paradigm of wildlife damage management as the fourth to be considered and embraced by the whole wildlife profession. Wildlife damage management has broadened dramatically from the early days when its paradigm was that of a predator- or rodent-control program. Today's paradigm includes other aspects of wildlife management, such as invasive species, overpopulation and habitat impacts, wildlife diseases, and zoonoses. The practice of wildlife damage management has shifted from a pole position fully opposite the conservation of wildlife, including threatened and endangered species, to that of an adjoining ally able to provide key concepts and tools for wildlife conservation (Berryman 1992; Berryman 1994; Hawthorne 2004; Miller 2007; see chapter 18). Further, the paradigm has shifted to one of concern and compassion, with a heart open to humaneness and empathy for involved wildlife when lethal control is needed. Here also, the expertise and character of wildlife damage practitioners can contribute to the entire wildlife damage management profession.

Examples

Not convinced that values, beliefs, and attitudes are important to wildlife damage practitioners worldwide? Here are some excerpts from a newspaper article in the *China Daily*, dated November 3, 2003, and entitled "Monkeys terrorize India workers, tourists" (*China Daily* 2003):

In a capital city (New Delhi) where cows roam the streets and elephants plod along in the bus lanes, it is no surprise to find government buildings overrun with monkeys. . . . [Government officials have] . . . been bitten, robbed and otherwise tormented by monkeys that ransack files, bring down power lines, screech at visitors and bang on office windows.

A past initiative to scare off the army of Rhesus macaques with ultra-high-frequency loudspeakers didn't work. A plan to deport them to distant regions has stalled because local governments refused to have them. . . . There's an ape patrol of fierce-looking primates called langurs, led about on leashes by keepers. But whenever a langur looms, the pink-faced, two-foot-tall hooligans simply move elsewhere on government grounds. . . . A plea to not feed the monkeys on Raisina Hill, near the president's residence, Parliament, and Cabinet offices, didn't work because Hindus believe that the monkeys are manifestations of the monkey god, Hanuman, and worshippers come to Raisina Hill every Tuesday handing out bananas.

On October 24, 1992, a coyote research facility located in Millville, Utah, was attacked and 16 of 54 coyotes were released (Knowlton 1992). The facility was a field station of the federal National Wildlife Research Center (at that time the Denver Wildlife Research Center) and a part of Utah State University, Logan. The research building at Millville was burned to the ground, and much of the research data was lost. The office of the principal researcher (also a federal scientist) on the main Logan university campus was also attacked in a failed attempt at incineration. In all, over $100,000 in damage was sustained (Liddick, 2006). The Animal Liberation Front took responsibility for the attack, contacting a newspaper in nearby Salt Lake City and claiming that the research conducted there was inhumane.

To us, the attack points to the importance of emotions versus rational thinking in considering social attitudes and actions (behavior). In this case, the individual prosecuted for the attack claimed to have felt the pain of suffering wildlife and used suffering as the basis for acting against the facility. Unfortunately, most data lost in the attack were being used to find alternative methods for coyote control that emphasized humaneness.

And what about the pain suffered by the *released* coyotes? Fourteen of the sixteen were recaptured and returned to their pens after examination and treatment for bites by a veterinarian. One died from injuries due to fighting following release, and one was probably shot on an adjacent property. In a subsequent statement, the field station leader said: "Curiously, their (ALF) motivations and mine are not very different—namely trying to reduce human intervention in natural systems. Unfortunately, just as they use and abuse animals to attract attention to their views, their use of our facilities, animals, and research program to abet their cause temporarily reduced our ability to develop alternative techniques for resolving conflicts between wild animals and other human endeavors" (Knowlton 1992, p. 1).

Thornton and Quinn (2009) evaluated attitudes of residents in the urban fringe of Calgary, Alberta, Canada, toward cougars (*Puma concolor*). Interactions between humans and cougars have become more frequent as humans have moved into cougar habitat over the last several decades. The cougar population in the study area was high, about 4 per 100 km², and probably expanding, but predation on livestock was uncommon. In the study area, human growth was driven mostly by need for more rural residential subdivisions. The researchers observed that residents living in cougar habitat sometimes felt anxiety and fear from perceived risks to children, pets, or themselves (Riley and Decker 2000; Teel et al. 2002).

Using a Likert-like seven-point scale, the researchers mailed a survey with 37 closed-end questions to 1,508 randomly selected residents in a stratified design. The survey addressed four areas: attitudes and beliefs about wildlife, including hunting, wildlife rights, education, and government participation; attitudes about cougars, including knowledge and beliefs, measures of risk, and government participation; attitudes about cougar management, ranging from preventing problems to killing problem cougars; and demographic information. Twenty-nine percent of those receiving the surveys responded. The researchers assumed that cougars were not of particular interest to nonrespondents.

About 154 (36%) of the respondents owned cattle or horses, and about one-quarter of these respondents reported livestock losses due to cougars. Most (78%) indicated willingness to adjust husbandry to reduce predation. Overall, about 40% of respondents said they or family members had observed a cougar in the wild. About 43% said that the presence of cougars increased their quality of life, and 65% believed cougars were an acceptable threat to livestock (65%) and humans (54%). Most participants (72%) accepted hunting of cougars and other wildlife if it did not impact the sustainability of the targeted population, and most (87%) believed the presence of cougars did not affect opportunities for hunters. Most respondents (71%) wanted to be more involved in govern-

ment decision making, most believing that wildlife management lacked sufficient public participation. A majority of respondents (65%) did not feel at personal risk from cougars, and most felt that they could accept risks and co-exist with the cougars. Overall, male respondents tended to accept hunting, whereas females were more protective of cougars. However, females felt more at risk and expressed greater reluctance to go into wild areas known to have cougars. Relocation was the most common management action preferred by respondents when a cougar stalked a country skier, injured a hiker on a trail, charged and knocked down a person on a trail (but then left), or killed a person on the trail (but had cubs), wandered repeatedly around a neighborhood, or attacked and killed one or more pets. However, killing the offending cougar was preferred by respondents if the cougar attacked and killed a person in the neighborhood, killed a person on a trail and had a history of aggression, or killed a person on a trail (with no other information available).

PCI (potential for conflict index, which uses values between 0 and 1, wherein 1 represents greatest conflict and 0 least) analyses showed highest within-population agreement (0.022) for the statement that it is important to maintain cougar populations for future generations and greatest disagreement (0.52) for the statement "I believe that cougars would attack a human without being provoked." Relatively strong disagreement (0.42) was found for the statement "Cougars are an acceptable threat to humans."

The information provided in surveys can be helpful to the wildlife damage manager. Respondents indicated interest in public participation in the management process. The survey showed that respondents who had the least knowledge about cougars and their management—e.g., female respondents and respondents newly relocated from urban areas—also had the greatest fear. The researchers suggested that these groups be targeted for educational programs. Thornton and Quinn (2009) also suggested that the tendency toward relocation as the management method of choice might also indicate a "not in my backyard" mentality, with actual acceptance of cougars less than respondents implied.

In our view, the survey also indicated acceptance as long as cougar impacts were limited to livestock, pets, and other wildlife, at low frequency. We (as did Thornton and Quinn 2009) refer to the observations by Decker et al. (2002) that the level of perceived risk (measured as levels of dread) is a function of both the probability of its event and its severity. Here, cougar impacts are perceived as potentially severe but un-likely. Should the likelihood increase—e.g., as both human and cougar populations expand—we suggest that perceptions of risk (and dread) will probably also increase. Under that scenario, if the survey does in fact point to a not-in-my-backyard mentality with lower acceptance of cougars than indicated, we believe the wildlife damage manager can anticipate increased interest in, or demand for, lethal management actions.

Summary

- Social values, or conceptions of what is good and bad, have a strong genetic base, form early in life, and change little later in life. Values influence group norms and underlie how people view wildlife damage and its management.
- People with doministic value orientations see humans as being a higher priority than wildlife and accept lethal and other intrusive methods of control. People with mutualistic value orientations view wildlife in a trusting relationship with humans and oppose intrusive methods.
- Beliefs are judgments of what is true or false. Beliefs, whether based on facts or not, strongly influence how people view wildlife damage and its management.
- Values and beliefs may be rooted in religion. All religions view animals as sentient and foster humane treatment. Intertwined with religious values and beliefs are societal norms, ethics, and cultural or shared values.
- Attitudes are tendencies to react favorably or unfavorably to a situation, and they can predict support for, or opposition to, wildlife damage actions. Attitudes can be modified by education and experience.
- Levels of risk and emotions influence attitudes toward wildlife damage. Perceptions of risk, not necessarily actual risk, often drive people's actions regarding wildlife damage management.
- Emotions such as happiness, anger, or disgust relate to specific events and are short lived. They are sometimes used to persuade people about changes in attitudes or to raise funds.
- Personal involvement and frustration with a wildlife damage problem tends to increase its importance to a person. Those not suffering from or exposed to damage may have negative or nebulous attitudes toward wildlife damage management.
- Moral decisions, such as to take a personal management action rather than have someone else do it, stimulate the emotional parts of the brain.

Review and Discussion Questions

1. How might a person's genetic composition influence her/his attitudes toward wildlife damage and its management? Summarize scientific literature supporting your view, including references.

2. Relate Kellert's typologies to value orientations provided by Manfredo et al. (2009). Which typologies and value orientations seem particularly important to wildlife damage management? Explain, with examples.

3. Choose a recent publication on wildlife damage management that would have benefited by the addition of a study on the VBAs of stakeholders. Reference the study. Describe the information needed on VBAs and how the information might have helped the study.

4. What value orientations might lead a community such as Town and Country, Missouri, to recommend translocation of overabundant deer rather than bowhunting by sharpshooters (see Beringer et al. 2002)? How might one test for such value orientations? How might one attempt to influence the orientation?

5. Are people in North America shifting from a doministic value orientation to a mutualistic value orientation? State yes or no, and justify your response with examples and references. Explain how such a shift might influence the future of wildlife damage management.

15

Politics and Public Policy

In the world of politics, the needs of wildlife damage practitioners can be facilitated by empathetic individuals and groups or thwarted by those opposed to the methods; attention and resources can be quickly brought to bear on damage management issues from the local to international levels.

Statement

Wildlife damage practitioners are tied to the political worlds of the societies in which they live and work, and the final decisions about which wildlife damage problems are addressed and how they are addressed are often political ones.

Explanation

Dye (2008, p. 1) describes **public policy** as "whatever governments choose to do or not to do." In general, things governments choose to do are to regulate conflicts within society; organize conflicts with other societies; distribute rewards and material services to members of society; and extract money from society, mostly as taxes. Public policies are founded on constitutions, legislative acts, and judicial decisions and are embodied in actions, regulations, laws, and funding priorities.

Because wildlife is a public natural resource in many countries and is often managed under the stewardship of public organizations, the whole (including its private-sector aspects) of wildlife damage management is strongly influenced by public views and opinions.

Public policies are made or used when concerns are successfully raised about damage or how damage is being managed. For example, concerns about damage to ecosystems or to the commercial fisheries industry by zebra mussels may lead to policies or regulations on movement of ships into ports or through canals, regulations on how ballasts are decontaminated or released, or conditions under which recreational boats can be moved from one lake to another within a local area.

Concerns surrounding wildlife damage management are often based on one or more of these four factors: method effectiveness (i.e., is the method doing what is intended, with benefits that outweigh costs?); biological or ecological soundness (i.e., are the methods based on sound scientific principles and concepts?); animal welfare or **humaneness** (i.e., have efforts been made to minimize pain and suffering of impacted wildlife?); and animal rights (i.e., have legal and other rights of impacted wildlife been considered?). For instance, individuals or groups may believe the use of steel-jaw traps to capture coyotes is unnecessarily painful and oppose use of these traps. Or, individuals or groups may believe that the use of repellents to reduce coyote depredations on livestock is ineffective and want policies or regulations that allow use of more effective methods.

Successful efforts to deal with concerns result in new or amended public policies. The processes used to develop new public policies or amend existing policies vary by type of government (e.g., authoritarian, monarchic, democratic, theocratic, totalitarian) and legal system (i.e., civil, religious, or a combination of these). For example, the National Crop Protection Center in Laguna, Philippines, was established by Presidential Decree No. 936 in May 1976 (post–martial law era) as part of that country's efforts to become self-sufficient in rice production. The Center was established to offer scientific information on integrated pest management, including wildlife damage management.

Whereas individuals—e.g., elites such as high-ranking politicians or administrators—can sometimes influence overall public policies (see box), changes come mostly from efforts of the masses (although these may also be initiated by the elite—see below) organized as groups. Organized people can gain power and regulate the behaviors of others.

Groups, sometimes based on common VBAs (values, beliefs, attitudes; see chapter 14), can be formed at any level of society. For instance, some groups whose interests include wildlife damage management are organized to align with governmental levels and units (table 15.1). An illustration of this is the Columbia Audubon Society, aligned with the city of Columbia and one of 14 chapters of the National Audubon Society located in Missouri. Another is the Oktibbeha Audubon Society in Starkville, Mississippi. Other groups interested in wildlife damage are organized according to trade or profession. Examples in the United States include the National Sunflower Association and The Wildlife Society. Yet other groups, both national and international, are organized around advocacy for the sustainability of natural resources or wildlife, such as the National Fish and Wildlife Foundation and Teaming With Wildlife. Still others focus on particular spe-

Jack H. Berryman

Jack H. Berryman (1921–99) saw wildlife damage management as one of the most complex and challenging areas of wildlife management. When the federal Bureau of Sport Fisheries and Wildlife began undergoing major changes in 1964, Mr. Berryman was placed at the helm of its new Division of Wildlife Services, the division that managed wildlife damage. He was responsible for reorganizing and redirecting the predator program of the division and led the development of a national program for extension education for the U.S. Fish and Wildlife Service. He helped establish the North American Waterfowl Management Plan, resolve the lead/steel shot issue, and obtained new funding for conservation programs during times when federal budgets were cut.

Mr. Berryman began his wildlife career with the Utah Fish and Game Department. He later worked for the U.S. Bureau of Sport Fisheries and Wildlife (now USFWS), both in Utah and Minnesota. He was an associate professor and extension wildlife specialist in the College of Natural Resources at Utah State University, during which time he was president of The Wildlife Society (1964–65). In 1979 Mr. Berryman retired from the USFWS and became executive vice-president of the International Association of Fish and Wildlife Agencies (IAFWA), retiring in 1988. He subsequently served as counselor emeritus for the IAFWA, advisor to the Berryman Institute, and on numerous state, regional, and national committees, councils, and task forces dealing with critical issues in natural resources management.

During his career, Mr. Berryman received many honors, the Aldo Leopold Memorial Award from The Wildlife Society being among the most prestigious. The Berryman Institute at Utah State University, dedicated to education and professionalism in wildlife damage management, is named in his honor.

cies or taxa, such as Bat Conservation International and Partners in Flight.

Other groups are organized around specific issues such as animal welfare or animal rights. Advocates of **animal welfare** believe that wildlife damage should be managed in ways that minimize pain and suffering

TABLE 15.1   *Examples of groups that can serve as stakeholders in wildlife damage management*

CITY, STATE, OR NATIONAL, U.S.

Oktibbeha Audubon Society, Starkville, MS
Columbia Audubon Society, Columbia, MO
MI Audubon Society
MO Audubon Society
National Audubon Society

REGIONAL, U.S.

SE Association of Fish & Wildlife Agencies
Purple Martin Conservation Association

TRADE OR PROFESSIONAL, U.S.

American Fisheries Society
American Sheep Industry Association
American Society of Mammalogists
American Veterinary Association
National Cattlemen's Beef Association
National Cattlemen's Foundation
National Sunflower Association
The Wildlife Society
World Owl Trust

TRADE OR PROFESSIONAL, OTHER COUNTRIES

Australian Society for Fish Biology
Chinese Veterinary Medical Association
European Deer Farmers Association
Federation of Livestock Farmers Association of Malaysia
National Sunflower Association of Canada
Wildlife Conservation Society, Nepal

NATURAL RESOURCE OR WILDLIFE ADVOCACY, U.S.

National Audubon Society
National Fish & Wildlife Foundation
Natural Resources Defense Council
National Wildlife Federation
Sierra Club
Teaming with Wildlife
People for the Ethical Treatment of Animals

NATURAL RESOURCE OR WILDLIFE ADVOCACY, WORLDWIDE

African Wildlife Foundation
Conservation International
Convention on International Trade in Endangered
Species in Wild Fauna and Flora
Food and Agricultural Organization of the United Nations
International Union for Conservation of Nature
The World Organization for Animal Health (DIE)
World Conservation Network
World Wildlife Fund

TAXA, U.S. AND OTHER COUNTRIES

Bat Conservation International
Ducks (Quail, Trout) Unlimited
International Crane (Osprey) Foundation
International Association for Bear Research and Management
National Wild Turkey Federation
North American Wolf Association
Partners in Flight
Pheasants Forever
Raptor Research Foundation
Rocky Mountain Elk Foundation
Royal Society for Protection of Birds
The Wolverine Foundation
Timber Wolf Information Network
Whale and Dolphin Conservation Society
Whale Conservation Institute
Wolf Recovery Foundation
World Center for Birds of Prey

ISSUE—ANIMAL WELFARE, U.S.

American Society for Prevention of Cruelty to Animals
Animal Protection League
Animal Welfare Institute
Defenders of Wildlife
U.S. Humane Society

ISSUE—ANIMAL RIGHTS, U.S

Animal Liberation Front
Animal Political Action Committee
Animals in Politics
Culture and Animals Foundation
Friends of Animals
Greenpeace
Transpecies Unlimited

---

of individual animals. They support responsible use of animals to meet some human needs, from companionship to sports. An example is the Animal Welfare Institute. Advocates of **animal rights** believe that animals should have legal and other rights equal to those of humans. Extreme animal rightists believe that animals should never be used for meat, leather, or furs or kept as property, in zoos or circuses or as

pets, and they are opposed to any wildlife damage management. An organization representing this category in the United States is People for the Ethical Treatment of Animals (Wywialowski 1991).

Further, committees and subcommittees are often found within groups, and some may have particular interest in wildlife damage management. One case is the Wildlife Damage Management Working

Group of The Wildlife Society. At any given time, any number of groups at any levels of various organizations may have interest in a specific wildlife damage management issue, such as in a local community or park or regarding European lynx predation on sheep. Indeed, it is the wise wildlife damage practitioner who assumes such interest exists and adopts transparent policies (that is, open communication, including all interested individuals and groups from planning to action to evaluation of results) for the formulation and execution of any wildlife damage action. In many countries, such transparency is implicit in laws overseeing any governmental actions involving wildlife damage.

To exert social pressure, groups often form **coalitions** (agreement between two or more parties to advance common goals and secure common interests) that focus on specific wildlife damage issues. Such coalitions often fit models such as Group Theory formulated by political scientists (Dye 2008). One model that applies to wildlife damage management is the **iron triangle**. In the United States, points of power in the triangle are relevant positions and units (depending on the specific issue) of the executive branch (e.g., the secretary of agriculture, the deputy administrator of Wildlife Services [WS], who administers the WS program; and/or the director of the National Institute of Food and Agriculture, who administers the federal extension program); relevant positions and units in Congress (e.g., Agricultural Appropriations Committee members and many individual senators and representatives); and lobbyists (see table 15.1 for examples). According to the iron triangle concept, policies affecting wildlife damage management at the national level are debated, legislated, implemented, administered, and monitored within the triangle. Analogous triangles occur at all levels of government and in other countries, and legislation is often transferred to other levels or other countries.

Ogden (1971) expanded the iron triangle concept to a "web of power" that included more groups as players. He described the web as "power clusters" and said it exists in every area of public policy. With power clusters, groups act independently but come together on specific issues that affect their interests at local, state (province, etc.), and national levels. A power cluster has six elements: administrative agencies, legislative committees, special-interest groups, professionals, an attentive public, and a latent public that might be aroused to act on a given issue. Power clusters include behavioral patterns that help shape policymaking: close personal and institutional ties with key people,

active communication among cluster elements, internal conflicts among competing interests within clusters, internal cluster decision making, and a well-developed power structure. For example, Sims (1995) describes the efforts to make available the M-44 and the Livestock Protection Collar in Texas following the federal ban on use of predacides. This statesman mentions the involvement of another congressman, a Texas state representative, the Texas Department of Agriculture, an extension predator specialist, an inventor, Wildlife Services (then Animal Damage Control), and the Texas Department of Agriculture. Because Sims was not attempting an exhaustive listing, the power cluster may have included additional individuals and groups.

Other models in political science also apply to wildlife damage management. Dye (2008), for example, describes a theory of elites in which wildlife damage policies come from the interactions of individuals and groups organized as a pyramid. At the bottom of the pyramid is a generally apathetic (but latently powerful) public. Moving upward, there exist a public interested in the issues, a public both interested and active, officials and administrators who have power and responsibility to shape policy, and a few elites (heads of major corporations, associations, etc.) at the top of the pyramid who actually determine public policy. According to this theory, the elite decide public policy in advance and make it happen in a top-down manner by communicating and shaping the opinion of the general public. With this model, groups and the general public have little actual influence on the formation of public policy.

Our sense is that none of the models applies exactly but that each, particularly the power cluster model (we mostly use this model for the rest of the chapter) and the elite models, offers insights into how public policy and laws on wildlife damage management are formed in the United States and probably in other countries as well.

How do these groups, individually or as power clusters, affect changes in national public policies in the United States? Public policymaking in the United States can be seen as a process with the following steps (fig. 15.1):

- problem identification, wherein societal problems are publicized and action is demanded by mass media, interest groups, citizens' initiatives, and the public
- agenda setting, wherein issues that will be addressed by government are decided by elites, Congress, elective candidates, and mass media

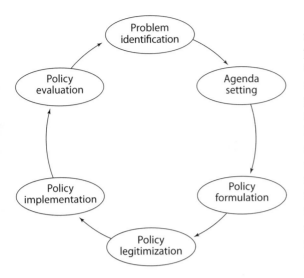

Figure 15.1 Steps in formulating federal public policy in the United States. *Illustrations by Lamar Henderson, Wildhaven Creative LLC.*

- policy formulation, wherein proposals to resolve issues are developed by think tanks, the president and executive offices, congressional committees, and interest groups
- policy legitimization, wherein a proposal is selected, political support is developed, the proposal is enacted into regulations and/or law, and its constitutionality is decided by interest groups, the president, Congress, and the courts
- policy implementation, wherein departments and agencies are organized, payments or services are provided, and/or taxes are levied by the president, White House staff, and executive departments and agencies
- policy evaluation, wherein government programs provide reports, policies are judged, and changes and reforms are proposed by executive departments and agencies, congressional oversight committees, the mass media, and think tanks

Although this description indicates an orderly, stepwise process, the actual process is usually disorganized, with steps occurring in various orders or simultaneously.

Individuals, or alliances of groups such as power clusters, initiate the process and may be involved in every step. An example is the move of Wildlife Services from the USFWS to the USDA in 1985 (see the Examples section at the end of the chapter). Groups within the power cluster were involved in the transfer of the unit from its inception, and the power cluster retains active involvement with the unit today.

In their guide to community-based deer management, Decker et al. (2004) provide case studies of power clusters formed at community levels to raise and resolve issues surrounding overabundance of deer. Individuals and groups play varying roles in the policymaking process, which often leads to consensus on a management action rather than a new regulation or law. In Bedford, Massachusetts, for instance, a deer manager specialist for the Division of the Fisheries and Wildlife served to both make the community aware it had a problem with deer and advocate for hunting as a solution. In Gettysburg National Military Park, the process, including the development of an environmental impact statement, led to a decision to reduce the number of deer in the park by shooting. The park had previously not allowed hunting within its borders.

Individuals, power clusters, and processes organize to impact wildlife damage management in other countries in ways particular to that country. For example, regulations and laws in countries such as Austria, Germany, Switzerland, and the Czech Republic allow individual owners to be compensated for damage to specific forest tree species caused by specific pests (Schaller 2007). These laws were developed through a power cluster that included landowners who hold hunting rights and lease hunting privileges, governmental units that regulate hunting (at the federal level) and enforce complaints law (at the local level), and hunter associations such as the Deutsche Jagdschutz-Vergand.

In England, the Department for Environment, Food and Rural Affairs (DEFRA 2010) described the framework within which public policy is made for resolving wildlife damage problems. The steps reflect those shown in fig. 15.1. DEFRA described the framework as a "process tree." The first step is problem identification (i.e., deciding whether government intervention is justified), followed by informing (gathering and presenting scientifically based information), deciding (choosing which action among options), communicating (informing **stakeholders**—i.e., persons who are affected by or can affect a problem or its management—of the decision and the reasons for it), implementing, and monitoring. DEFRA cited management of inland great cormorants (*Phalacrocorax carbo*) as an example. Great cormorants compete with anglers and fish farmers who then seek ways to reduce economic losses. DEFRA turned to data from Wetlands Bird Surveys for estimates of cormorant populations and used a model to relate different levels of licensing (i.e., numbers of kill permits) to effects on overall population sizes. By running thousands of iterations, uncertainty in projected future growth of cormorant populations was reduced. Then, by issuing licenses, monitoring population responses, and following an Adaptive Resource Management approach (see chapter 16), numbers of licenses can be adjusted annually to changing circumstances.

How effectively does the public policy process work? This is where the logic of the system can sometimes pose a challenge. Although the processes just described serve as the norm for developing most public policies, **deception** (wherein one is given only part of the whole story or the story is deliberately misconstrued, hence the term "spin") and **expediency** (wherein a politician may take the easiest route to re-election rather than the one in the public's best interest) are also part of political life, including both politically based groups and individual politicians.

Whereas most public agencies and politically based groups encourage transparent actions, political methodology sometimes favors sharing only as much information as is needed for a participant to do his/her part, without the participant knowing the full agenda. For example, the broad membership of an animal rights group may support objectives such as humane treatment of animals, including wildlife, but may not be told that an unstated goal of the leadership is to eliminate pets from houses and hamburgers from McDonald's. Further, information may not only be intentionally incomplete but also written with a spin that is intended to mislead. Even ballots or referenda are sometimes written so that the public is not sure exactly what a "yes" or "no" vote actually means.

Campaigns supporting referenda may also include photographs or videos that appeal to emotions rather than reason, intending to arouse the latent public and to persuade. For example, a campaign to ban traps as "cruel" may show photographs of animals caught in traps but fail to provide actual information on the humaneness of the targeted trap designs. The campaign might also fail to contrast the relative "cruelty" of the traps with methods left for practitioners if the traps are banned. The intent of the campaign is to get public support and votes for the ban, not necessarily to provide objective information. Groups may attempt to use deception and expediency to justify the importance of their causes (e.g., Vantassel 2008).

In these (as in all) situations, it is left to the voter to gather objective information and to make informed assessments and decisions regarding the proposal. The campaigners know that many voters are likely to respond emotionally or to see through the filters of existing ideologies and not bother with more rigorous and time-consuming analytical assessments.

In the United States, bills (i.e., proposed laws) unpopular to the public but important to lobbyists (and sometimes for the re-election of the involved officials) may be attached as **riders** to publicly popular but unrelated bills. An unpopular rider (which would not be passed on its own) can get passed on the wings of a popular bill, another form of deception. Thus a rider attached to a U.S. congressional federal budget bill in 2011 returned management of wolf populations to the states of Idaho and Montana, sidestepping checks and balances that are an intended part of the delisting process under the Endangered Species Act (IdahoStatesman.com 2011).

Political expediency can sometimes short circuit not only the political process but also the scientific and analytical evaluation of proposed solutions. For example, a recent decision to ban all congressional directives in the United States won votes and gained the support of some constituencies. The "one-size-fits-all" ban, however, eliminated scientific and professional programs in wildlife damage management that included those addressing problems with feral pigs, rodent research related to sugarcane, and academic educational programs. An analytical process that evaluated the need for and worth of individual congressional directives would have allowed the separation of "pork" from other contents of the barrel, but this approach was not politically expedient.

While discussion thus far has focused on the United States, the same strengths and problems are seen in other countries, although details become idiosyncratic. For example, a New Zealand farmer distributed rabbit hemorrhagic disease (RHD), short circuiting a process of federal evaluation, even as the government of New Zealand determined that the risk was too great for introduction.

Once made, public policies set the boundaries within which wildlife damage is managed. That is, wildlife damage is assessed and scientifically investigated and its management is conducted under the authority of, and within, public policies from local to federal levels in all countries where wildlife problems occur and are managed. For many countries, the laws, regulations, and policies include those governing the management of natural resources such as wildlife, air, and water; the activities of resource management agencies and the private sector; methods used in wildlife damage management; and redress for citizens who believe they have been injured by damage or its management. For example, public regulations and laws often determine the types of oral vaccinations that can be used to manage rabies in Europe and North America, the nature of delivery of the vaccinations (e.g., by aircraft or by hand), and the areas treated.

In the United States, wildlife damage management activities are governed at the national level by many specific laws and their amendments (table 15.2). These laws may be complex in how they are applied. The Federal Insecticide, Fungicide, and Rodenticide Act (FIFRA), for instance, provides for registration, classification, and regulation of use of all pesticides in the United States. Administration of FIFRA is the responsibility of the Environmental Protection Agency. The intent of the legislation is to ensure that benefits derived from use of pesticides outweigh costs to the environment. Pesticides, as interpreted by the EPA, include anything chemical or nonchemical (e.g., a mechanical device) that affects the behavior of an organism.

Registration of pesticides by the EPA is based on risk assessment by EPA scientists using data provided by the proposed registrant. Data are gathered under study plans (protocols) approved by the agency using guidelines called **Good Laboratory Practices** (**GLP**; see below).

TABLE 15.2 *Selected examples of public policies affecting wildlife damage management in the United States*

Act	Year	Responsible U.S. Agency	Impact
Airborne Hunting Act	1971	USFWS	Disallows shooting or harassing birds, fish, or other animals from aircraft except for certain reasons, including protection of wildlife, livestock, and human life
Animal Damage Control Act	1931, 1973, 1987, 1991	USDA	Allows operational & research activity; allows cooperative agreements and exchange of funds
Animal Welfare Act	1966	USDA	Regulates transport, sale, and handling of dogs, cats, non-human primates, guinea pigs, hamsters, and rabbits intended for research or other purposes; requires Animal Care and Use Committees, following USDA guidelines and procedures
Convention on International Trade in Endangered Species of Wild Fauna and Flora (CITES)	1973	USDA, USFWS	Regulates trade in endangered species of wildlife and their products
Federal Insecticide, Fungicide and Rodenticide Act (FIFRA)	1910, 1972, 1988	USEPA	Ensures benefits derived from use of pesticides outweigh environmental costs; decisions based on data provided by registrant, following GLP guidelines
Fish & Wildlife Coordination Act	1934	USFWS	Requires that all federal agencies coordinate efforts on environmental issues; authorizes Secretary of Interior to minimize damage from overabundant species and to control losses from diseases or other causes
Lacey Act	1900, 2008	USFWS, USDA	Regulates import, export of invasive species listed as "injurious to human beings, to the interests of agriculture, horticulture, forestry or wildlife resource"
Migratory Bird Treaty Act	1918	USFWS	Coordinates management of game and nongame migratory birds through permitting; exempts some offending blackbirds, grackles, crows, and magpies
National Defense Authorization Act	2008	USDOD	Section 314 requires brown tree snake report to congressional defense committees
National Environmental Policy Act	1969	Council on Environmental Quality	Requires an evaluation of environmental risk before any major federal action is taken; usually done progressively as an Initial Environmental Examination, then Environmental Assessment, then Environmental Impact Statement
Nutria Eradication and Control Act	2003	USDI	Authorizes funds for assistance to Maryland and Louisiana
Safe, Accountable, Flexible, Efficient Transportation Equity Act	2005	Natl. Highway System	Allows funds for control of terrestrial noxious weeds and aquatic weeds
U.S. Fish & Wildlife Conservation Act	1980	USFWS	Encourages all federal agencies to conserve and protect nongame wildlife and their habitats; authorizes some wildlife damage action to protect threatened or endangered wildlife
Water Resources Development Act	2007	Dept. of Army	Authorizes barrier demonstrations to prevent dispersal of Asian carp, Upper Mississippi

Let's assume an entity is requesting registration of a new chemical toxicant that is formulated as a bait. Data (assuming the pesticide is a chemical bait) are required on

- basic chemistry, including descriptions of color, melting point, freezing point, reactivity, stability in air/soil/water, and product shelf (storage) life
- product chemistry, including active and inactive ingredients (see chapter 11);
- toxicity to "target" and other animals, including LD_{50} (the lethal dose that kills 50% of the population), LC_{50} (the lethal concentration when fed as a bait that kills 50% of the population), and dermal toxicity to skin and eyes
- toxicity to other (nontarget) animals, including primary toxicity from feeding on the bait directly and secondary toxicity to prey or animals scavenging an exposed carcass
- field efficacy, possibly including both confined and open field studies

Laboratory and field studies must meet GLP standards in order to be accepted as data for consideration of registration. These requirements include a chain of custody documenting handling and storage of any materials used in the study; an archive of samples and all original data; documentation of qualification of laboratory and field participants; documentation of the conditions of storage of samples (e.g., that materials stored in a freezer were kept frozen without interruption); documentation of calibration of instruments used in the study; and establishment of approved protocols and standard operating procedures. Work conducted under GLP guidelines can be very costly. For example, Fagerstone et al. (2008) estimate that total costs for the 48–60 data requirements typically required for registering a new active ingredient can be up to US$2.7 million. A consequence has been a great reduction in the numbers of pesticides available for use, including those for wildlife damage management, and the formation of data-gathering coalitions, including both public and private-sector entities, to share costs of registration (Fagerstone and Schafer 1998).

Some laws affecting wildlife damage management may either be targeted at specific organisms or situations or embedded within laws that regulate broadly. Noxious weeds, nutria, and brown tree snakes are examples of organisms that have their own acts. The National Defense Authorization Act of 2008 includes Section 314, which requires that a report be provided on the status of the brown tree snake to congressio-

nal defense committees. The Convention on Trade in Endangered Species (CITES) influences the movement of invasive species between the United States and other countries (see table 15.2).

Other regulations, policies, and laws at state and local levels, including wildlife codes, may also apply. For example, purchase and use of pyrotechnics and firearms are regulated from federal to local levels. The wildlife damage practitioner must, of course, be aware of and compliant with all relevant policies and laws.

What are the legal responsibilities of governments and natural resource agencies when wildlife under their stewardship damage ecosystems or people's property or cause human, pet, or livestock disease, injury, or death? It depends partly on the country. In the United States, natural resource agencies often have little or no responsibility for the damage itself or for payment of compensation. This is because wildlife species are seen as *ferae naturae* (an ancient common-law doctrine stating that a wild animal cannot be owned by anyone). Legal interest in wildlife mostly lies with states or federal agencies managing lands, which act as a **sovereign** (representing the common interests of its citizens) rather than a **proprietor** (asserting ownership). Because the state has not asserted dominion, the state cannot be held accountable for damage caused by wildlife (Bader and Finstad 2001).

Tan (1990) points to steadfast federal and state courts that refuse awards for restitution of wildlife-inflicted damage when suits are directed at governmental entities such as public natural resource management agencies. In *Barrett v. State,* often used as a reference, a suit was brought against the State of New York in 1917 by a landowner who sustained "considerable commercial damage" from beavers on his land. The beavers were protected under a New York law stating that no one "shall molest or disturb any wild beaver on the dams, houses, homes or abiding places of same." The claimant argued that since the beavers were "owned" by the state, the state should pay for the damage. The court disagreed, citing ownership of wildlife in a sovereign capacity for the benefit of all people (Animal Legal and Historical Center undated). In another suit, farmers near Horseshoe Lake State Game Preserve in Illinois sued the state to recover damage to corn and soybeans destroyed in 1946 and 1947 by migratory birds, mostly Canada geese. Among other arguments, the claimants alleged that the federal government was responsible for damage because it had the geese in its possession and control, failed to protect the defendants, took actions in 1946 that stirred up the waterfowl and caused more damage, and is the owner of geese when they are in the United States. The court again struck down the arguments, stating that a claim would have to be based on negligence or wrongful act or omission of an employee of the government while acting within the scope of employment (Justia US Law undated).

While suits against federal or state agencies for wildlife damage may be denied, property and human injuries and deaths do occur as a consequence of wildlife damage. Costs can be high, and liability has been claimed under two circumstances of common law (Bader and Finstad 2001): first, the state has a duty to warn of known dangerous conditions, including those involving wildlife on state property; and second, the state may be required to compensate when it produced artificial conditions that led wildlife to cause harm. In those situations, negligence or mismanagement is presumed to allow wildlife damage that would not otherwise have occurred. Both tort (civil) and criminal lawsuits might be used, constituting another set of laws related to wildlife damage management. Claims may be against public agencies (less likely to be successful) or private entities (more likely to be successful). Successful examples include suits associated with predator attacks or animal-vehicle collisions at airports, where agencies and their staff were found to be negligent or guilty of mismanagement.

In one case, a family was awarded $1.9 million in a wrongful death suit involving a bear attack on an 11-year-old boy that occurred at a U.S. Forest Service campsite in 2007. The bear had attacked other campers 12 hours earlier, and the agencies were found negligent for not warning other campers or closing the campsite (Aiken & Jacobsen undated).

Dale (2009) summarized personal and corporate liability surrounding bird strikes. Because 74% of airstrikes happen at 500 feet or less above ground, airstrike prevention is often the responsibility of the airport and its management. A common liability is failure of the airport manager to take actions that are legally required to maintain a safe operating environment. Wildlife hazard assessments are legally required after these occurrences: an aircraft experiences a multiple wildlife strike; an aircraft experiences substantial damage from striking wildlife; an aircraft experiences an engine ingestion; and wildlife capable of causing damage is observed having access to airport flight patterns or aircraft movement areas (Cleary and Dolbeer 2005). Based on results of the assessment, the U.S. Federal Aviation Administration (USFAA) determines whether a wildlife hazard management plan is needed. If threatened or endangered species are involved, a biological assessment of impacts is also required. Management actions must follow all relevant public regulations and laws.

Failure to take appropriate action can be costly, both in damage following airstrikes and in subsequent litigations. As examples of tort claims, Dale (2009) included

- a November 12, 1975, aborted takeoff of a DC-10 that ingested gulls and caught fire, injuring 30 of 139 people on board and costing $15 million in suits by the airlines against the USFAA, the Port Authority of New York and New Jersey, New York City, and several aircraft companies
- an Air France Concorde that on June 3, 1995, ingested geese at 9 feet above ground upon landing, causing uncontained failure with over $7 million in damage to the Concorde and settlement out of court for $5.3 million between the French Aviation Authority and the Port Authority of New York and New Jersey
- an Air France A-320 hitting a flock of gulls on takeoff at Marseille Provence Airport, France; the impact destroyed the engine, and the airline was awarded $4 million in settlements because of negligence (failure to remove a hedgehog, hit earlier, that attracted the gulls) in operating the airfield

As examples of criminal charges, Dale (2009) points to

- a January 20, 1995, crash of a Falcon 20 that struck a flock of birds, killing all 10 people on board, with French authorities bringing charges of involuntary manslaughter against the Paris Airport Authority and 3 of its former officers
- the September 22, 1995, crash of a U.S. Air Force AWACS B-707 after ingesting geese at Elmendorf Airbase in Alaska, killing all 24 people on board; the senior airport controller and one other controller invoked their Fifth Amendment rights to avoid criminal prosecution

Lawsuits against wildlife damage management agencies in the United States are often based on faulty or inadequate address of laws applicable to the proposed action, such as failure to fully comply with the National Environmental Policy Act. Suits often involve wildlife species that the public sees as charismatic and that thus engender public interest. For example, a coalition of conservation and animal protection organizations unsuccessfully sued Wildlife Services over a program to kill black bears as part of integrated management action to reduce damage to forests in Oregon. The suit targeted the environmental assessment, claiming that it was inadequate (Umpqua Watersheds 2004). Also in Oregon, four conservation groups sued the WS program for its proposed role in killing two wolves at the behest of Oregon's Department of Fish and Wildlife (ODFW). The wolves had taken livestock, and nonlethal efforts by the ODFW had failed. Trying to prevent a pattern of killing livestock, the ODFW requested assistance from WS and set the terms of the permit so that the alpha female and her four pups were protected. The basis for the suit was that the required environmental analysis was not conducted (Oregon Wild 2010). This suit was also unsuccessful.

We have focused on policies in the United States, but public policies may differ among countries. Policies surrounding animal welfare serve as an example. The United Kingdom revised its pet laws in the Animal Welfare Act of 2006, providing for humane treatment of pets and other animals directly under the control of humans. India is proposing an Animal Welfare Act that will replace its Prevention of Cruelty to Animals Act of 1960. The proposed act is intended to be easy to enforce, provide a clear definition of animal abuse, and impose stiff penalties for violations.

One of the most ambitious programs in animal welfare is provided by the Australian government. In 2004, the Primary Industries Ministerial Council endorsed the Australian Animal Welfare Strategy, a plan that includes livestock and production animals, animals used for recreation or display, companion animals, wildlife, aquatic animals, and animals used in research. Development of the strategy involved a power cluster including animal welfare stakeholders, state and territory governments, livestock industries, and other nongovernmental organizations. This strategy, based on scientific as well as social and economic considerations, has been endorsed by all members of the power cluster (Harlock Jackson 2006).

Takahashi (2009) reviewed Japanese wildlife law and policy and found only a superficial resemblance to United States public policies. In Japan, there were no legal channels, such as lawsuits, for inputs by citizens and experts. Fishery laws were separate from other wildlife laws, and management of damage strongly emphasized culling. Furthermore, management of problem wildlife increasingly relied on an aging hunter population.

Examples

TRANSFER OF WS. The transfer of Wildlife Services (then Animal Damage Control, or ADC) from the USFWS back to the USDA in 1985 serves as an illustration of involvement of a power cluster in the public policy process. The transfer was the consequence of these factors: individual farmers and ranchers complaining that wildlife damage was inadequately addressed by the USFWS (problem identification); groups such as the National Cattlemen's Beef Association and the American Sheep Industry expressing concerns to congresspeople (problem identification); congressmen (particularly Jesse Helms, NC) and 19 western senators expressing concerns and making recommendations to President Reagan (problem identification, agenda setting, and policy formulation); the Department of Justice and the secretaries of agriculture and the interior finding mechanisms for transferring the program (policy formulation and legitimization); and the budgeting power of Congress putting the program's funds into the USDA rather than the USFWS in 1985–86 (policy implementation). Representatives of 21 members of the power cluster now meet annually as the National Wildlife Services Advisory Committee, which deliberates policies and recommends changes (policy evaluation) and thereby completes steps in the process.

INVASION OF ASIAN BLACK CARP. Simberloff (2005) summarized the politics surrounding risk assessments for biological invasions in the United States. One example was assessment of risks associated with introduction of the Asian black carp in hatcheries within the Mississippi Basin. The completed assessment included concerns regarding introductions of new diseases of fishes and mussels and concerns about foraging on native mollusks, causing competition with native species. The assessment recommended that only sterile triploid carp be permitted, solely in contained facilities away from open waters.

It soon became apparent, however, that black carp were being held near open water. A consortium of 28 state fish and wildlife agencies subsequently petitioned the USFWS to list black carp as "injurious" under the Lacey Act, and the USFWS began that process. In 2003, however, the USFWS made some procedural changes that left the fish unlisted, removed the USFWS fishery biologist from the listing consortium, and removed its funding from the consortium. Apparently, an influential state senator was responding to a different power cluster, one that included catfish farmers who saw the carp as valuable for controlling intermediate snail hosts of trematodes. In the words of Simberloff (2005, p. 220), addressing a more general context, "The problem that seems inadequately treated currently is that a substantial benefit that might accrue to a few has more political weight than a substantial cost that might be borne forever by all."

MANAGEMENT OF THE LEOPARD. Marker and Sivamani (2009) summarize the need for policy evaluations and changes in the management of leopard (*Panthera pardus*) attacks on humans in India. In 2006, at least 133 people were killed by leopards. Leopards also prey on dogs, cows, and goats. Leopards have adapted to environments occupied by humans, who continue to move into leopard habitat. People retaliate by injuring or killing the leopards. Indian federal law presently allows two options for managing a leopard that has killed a human: hold it in captivity for its natural life or translocate the animal. Holding the leopard captive is expensive, and translocation can exacerbate problems, because leopards are strongly territorial and attempt to return to their original home range; in doing so, they move through human habitations, resulting in further human encounters and attacks.

Based on recommendations from a workshop held in 2007, during which experts on predator management from India and other countries were invited to share information and ideas, the following policy changes were suggested:

- the formation of trained emergency response teams composed of a highly ranked forest conservator, a veterinarian, and at least five support members empowered to declare curfews, provide information to the public, alert police and order ambulances, and decide the fate of the leopard; that Section 11 of the Wildlife Protection Act be amended "to make them more amenable to conflict resolution"
- that laws, such as definition of a "problem animal," be clarified so that officials with responsibility for management actions know that their actions fall within legal bounds, thus avoiding personal liability

TNR OF FERAL CATS. Williams (2009) summarizes a visit to the University of Hawaii, Oahu, where a ten-year effort to manage feral cats with the Trap, Neuter, and Return (TNR) method is underway. The organizer of the TNR program at the university estimates

that about 80% of the 400 feral cats on campus are sterilized and states that TNR is working. There are 1,200 registered caregivers for feral cat colonies on the island. The Hawaiian Humane Society sterilized 2,573 feral cats for 461 people between July 1, 2007, and June 30, 2008. An estimated 100,000 feral cats exist on the island, and at least 71% of a colony needs to be sterilized before the program has any effect.

Wildlife agencies and their surrounding power clusters have mostly opposed TNR, instead supporting euthanasia of feral cats. This is partly because feral cats affect other wild species, such as birds and sea otters. The American Bird Conservancy estimates that nationwide, about 500 million birds are killed yearly by feral cats. Feral cats probably outnumber other wild predators in North America, and they transmit diseases such as toxoplasmosis, roundworm, and rabies. The evidence accrues that TNR is ineffective (e.g., Dauphiné and Cooper 2009; Longcore et al. 2009). Yet, the use of TNR continues to expand rapidly in communities throughout the country. Why?

Williams (2009) argues that growth of TNR is caused at least partly by the feral cat lobby being stronger than the wildlife power cluster. Alley Cat Allies alone has a staff of 25 and an operating budget of $4 million. Williams (2009) points to cases in which the two clusters have clashed. For example, the feral cat lobby prevented legislation from being enacted that was supported by the wildlife agencies and that would have removed invasive exotic species from National Wildlife Refuges. The lobby was concerned that feral cats might be taken.

Summary

- Public policy is what governments choose to do or not do, including regulating internal conflicts, organizing conflicts with other societies, distributing rewards and services, and extracting money from society, usually taxes.
- Public policies are made or used in wildlife damage management when citizens are concerned about damage or how it is managed. Concerns often focus on effectiveness of a method, its biological or ecological soundness, or animal welfare or rights.
- Individuals sharing common interests gain political strength by organizing into groups, and groups form power clusters (coalitions) around wildlife damage issues. Elite individuals can directly influence public policy.
- Most public policies involving wildlife damage management follow transparent processes, but

political deception and expediency also exist. Individuals, groups, and power clusters participate in any or all steps of the public policy process, from increasing awareness of a problem to evaluating effectiveness of new laws. Transfer of the federal Wildlife Services program from the USFWS to the USDA in 1985 is an example of public policy driven by public interest.

- Once made, public policies in the form of regulations and laws set the legal boundaries within which wildlife damage is assessed, studied, and managed. Because policies and laws occur at every governmental level, the wildlife damage practitioner must be aware of and compliant with them all.

- Public policy establishes the legal bounds for redress if a citizen believes she/he has been injured by wildlife damage or its management. Personnel of agencies and private individuals and entities can be held accountable for negligence or creating situations that are unnatural and lead to wildlife damage.

Review and Discussion Questions

1. Find a specific example of how a power cluster was formed around an issue of interest in wildlife damage management, such as the ban on trapping in Colorado. Who were the groups and individuals in the power cluster? Was it based on reason, emotion, or both? Was the general public well informed, or was deception or political expediency involved? How might the process have led to a different outcome?

2. Jacobson (2008) used institutional models to evaluate state wildlife agencies in an era of changing interests and needs. She found that state agencies that relied heavily on hunters for revenue tended to resist change, whereas agencies in states with stable funding from multiple sources were more proactive in seeking new stakeholders and programs, including wildlife damage mitigation. She used the notion of the iron triangle to explain resistance by some agencies. How might the power cluster model apply to this situation and be used to help state wildlife agencies become instruments and leaders of change? How important might wildlife damage management be—including its roles in addressing wildlife diseases, zoonoses, and invasive species—for the future of these agencies?

3. Feral cats are an increasing concern for conservationists and wildlife damage managers. The article by Williams (2009) argues that the feral cat lobby is stronger than that of professional wildlife management. Do you agree? Support your position with specific examples and references.

4. Feral pigs are also an increasing concern for conservationists and wildlife damage managers. Would you anticipate the same problems for management of feral pigs as Williams (2009) describes for feral cats? Why or why not? What issues would you anticipate in managing the feral pig? Defend your responses with specific examples and references.

5. How much publicity surrounded the use of Compound 1080 to remove arctic foxes from Kiska Island for the benefit of endangered geese? Do you agree that it is the wise wildlife damage practitioner who assumes that interest exists in any management action and adopts a policy of transparency? Does that not put the practitioner at a disadvantage with political groups and politicians who practice deceit? Defend your position.

PART VI • **STRATEGIES AND THE FUTURE**

In this section we consider overall processes and strategies to manage wildlife damage now and in the future and look particularly at the association between wildlife damage management and the conservation of wildlife.

16

Operational Procedures and Strategies

Here we explore the effective use of operational procedures and strategies in wildlife damage management.

Statement

Strategies emerge through planning the resolution of all aspects of a wildlife damage problem. An effective strategy can lead to the most viable solution to a wildlife damage problem, based on criteria such as safety, economy, effectiveness and/or humaneness, and stakeholder concurrence.

Explanation

A **strategy** is a broad plan of action to achieve a goal. A military concept, the term has been applied more recently in the sense of strategic management. **Holistic** (i.e., comprehensive) strategies are the consequence of careful consideration of every aspect of a wildlife damage problem (fig. 16.1). Such considerations include history and existing literature on the problem, applicable biological and ecological concepts, and human dimensions, and they take advantage of routine plans such as standard operating procedures and applicable decision support systems. Holistic strategies begin with fundamental considerations—for example, whether to use an autecological or synecological approach. Autecological approaches are chosen when an individual species—or an ecologically closely related species—is causing damage. Suppose, for instance, ring-billed gulls are nesting on air conditioning systems on a rooftop, fouling air that is being circulated into rooms and posing a fire hazard with flammable nesting material. An autecological approach would be most appropriate, because only one species is causing damage.

Synecological approaches are used in situations where several species might cause damage simultaneously, such as at airports or landfills. Here, a broad ecosystem or landscape approach may be needed, wherein the habitat is made unsuitable for many or all species, regardless of their ecological niches. Pools of water might be removed to reduce the attraction of the area to wildlife. Grass might be managed at a given height or removed entirely to deter geese, gulls, and other

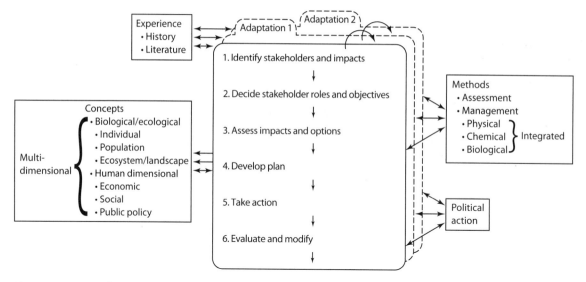

Figure 16.1   Steps in formulating strategies for solving wildlife damage problems. *Illustration by Lamar Henderson, Wildhaven Creative LLC.*

wildlife. Small mammals might be trapped or poisoned to reduce their presence and attractiveness to raptors. Fencing might be used to exclude wildlife from critical areas. Human garbage might be covered to reduce its attractiveness to birds, rodents, bears, or other wildlife.

Holistic strategies build on day-to-day operating procedures and tasks. **Standard operating procedures** (SOPs) are written, formalized directions to be followed during performance of tasks. As such, SOPs are themselves more like tactics or operational plans than strategies. An SOP ensures that a task, regardless of who does it, can be completed the same way each time it is performed, serving as a form of quality assurance. For this reason, SOPs are routinely included in environmental assessments of many programs required under the National Environmental Policy Act. They are widely used for both research and operational activities in the public and private sectors.

For example, steps for weighing a sample with an analytical balance may be described in an SOP at a research facility. Anyone weighing samples with the balance is expected to follow the detailed directions in the SOP. A written statement or signature may be required, ensuring that the SOP was followed. Steps used to calibrate the balance may be written as a separate SOP. Procedures used to freeze and store the weighed samples may have an SOP as well. The procedures for ensuring that the storage device remains below freezing may also be documented as an SOP. A

laboratory may therefore have hundreds of SOPs, especially if they are following GLP (Good Laboratory Practices) guidelines for registration of pesticides. In larger laboratories the SOPs are often maintained by a quality assurance officer who also looks after other quality factors, such as chain of custody of samples and archival records of laboratory studies. Collectively, these SOPs contribute to a broader strategy of quality control.

Sengl et al. (2008) described the SOP used by Wildlife Services, USDA, in Las Vegas, Nevada, when receiving phone calls from the public seeking help. The SOP describes each step, including

- gathering information on the caller (e.g., contact, location)
- identifying the conflict (e.g., species, type of damage)
- determining the responsible managing agency (e.g., USFWS, Nevada Department of Wildlife, or the USDA's APHIS / Veterinary Services)
- assessing which public policies may apply (e.g., conducting technical assistance or direct control)

Here the SOP also serves to ensure a high level of quality in response to public calls.

Universities use SOPs to describe standardized research, teaching, and extension activities related to wildlife damage management. As specific examples, the Institutional Animal Care and Use Committee at Mississippi State University follows an SOP designed for emergencies (Anonymous 2010), one for policy and procedures regarding possible noncompliance (IACUC-RVW-011) and one for training investigators on protocols (IACUC-TRN-001). SOPs are also used by private-sector wildlife damage specialists. We have not completed a count but venture that literally thousands of SOPs could be found globally that describe procedures to be followed for all aspects of assessing and managing wildlife damage.

Decision support systems (DSSs) do not make decisions, nor do they constitute stand-alone strategies. Rather, DSSs help to gather and store information and to analyze and present the information in ways that allow new insights and options for the practitioner. They are incorporated as parts of holistic strategies. Many DSSs are computer based, but they need not be. Hoare (2001), for example, provides a written DSS for managing elephant damage in Africa. Other DSSs are designed as apps (applications) for cell phones that can be downloaded and used in developing countries or remote areas where computers are unavailable (e.g., Eisen and Eisen 2011).

Computer-based DSSs have evolved from their inception in the 1960s to where many are interactive and provide real-time (virtually instantaneous) information. Most have a database (or knowledge base), a model, and a user interface. DSSs can be particularly useful when options, each involving complex iterative calculations, need to be considered as part of a broader strategy. An example is the DSS offered by Sterner (2002; see chapter 13) for controlling vole damage to alfalfa. The DSS does not decide on a specific strategy for the practitioner, such as when to treat with zinc phosphide; instead, it shows benefits in relation to costs for differing treatments and levels of damage.

DSSs are increasingly available for a wide range of uses. Ellis and Schneider (2000) describe a DSS that helps to evaluate the level of risk for the spread of zebra mussels in lakes and streams associated with boat use at infested sites in Illinois. Among other revelations, the DSS model predicted that quarantines of lakes with high boater populations could actually increase the risk of infection and that educational efforts could reduce the risk of infestation at larger lakes and their surrounding areas by 80%. Such information is helpful in designing strategies to reduce the movement of zebra mussels between bodies of water.

WeedSOFT was developed by the University of Nebraska as a DSS to help growers, consultants, and extension agents develop strategies for managing weeds (WeedSOFT undated). The software provides information according to specific field conditions and considers economic and environmental factors. Photographs help with weed identification, and mapping features allow consideration of ground-water contamination. Programs such as WeedSOFT have many useful applications in managing wildlife damage associated with exotic invasive plants.

Eisen and Eisen (2011) describe uses of geographic information systems combined with DSSs to predict outbreaks of vector-borne diseases such as Lyme disease, plague, dengue, malaria, trypanosomiasis, and West Nile virus, based on weather conditions and likely distribution of vectors. This research team points particularly to DSS packages that are freely available, can be downloaded on cell phones, provide easy data entry and visual display of analyses, and can be used in resource-limited areas where vector-borne diseases occur. These packages are helpful in formulating strategies to prevent zoonotic epidemics or spread of wildlife diseases.

Some DSS systems are designed to provide real-time warnings of imminent bird strikes. Examples include a U.S. bird avoidance model, a German Bird-strike risk forecast model, a Swiss/Dutch dynamic bird migration model, a U.S. avian hazard advisory system, the Dutch ROBIN system, and the German BIRDTRAM system (Ruhe 2005). Such systems can be incorporated into strategies to reduce or eliminate impacts of bird strikes. Huijser et al. (2009) created a DSS that predicts benefits and costs for various methods to mitigate deer-vehicle collisions over a 75-year period. This DSS model could be used to support decisions on mitigation measures for reducing collisions.

Whereas SOPs and DSSs can be viewed as tactics that contribute to broader strategies, other approaches can themselves serve as holistic strategies. An ideal holistic strategy can be characterized as **integrated** (involving a range of management methods that serve to minimize the use of pesticides), **multidimensional** (including human dimensions as well as biological and ecological concepts), or **adaptive** (wherein one proceeds without a full understanding of the problem and its solution, but the strategy includes sufficient monitoring and experimental data-gathering so that post hoc analyses lead to improvements in the strategy). Most holistic strategies have some combination of these attributes (fig. 16.1).

Many factors in an individual situation play into holistic strategies. For example, if the problem is deer damage to alfalfa, relevant aspects might include

- flavor preferences and foraging behavior of deer
- density of deer in the area

- ecological season
- surrounding crops and their closeness to water and shelter, such as forested areas
- extent of damage
- availability of suitable fencing, chemical repellents, or kill permits
- policies of the wildlife agency regarding issuing kill permits
- attitude of farmers toward hunters and the wildlife management agency
- attitude of hunters toward culling operations
- benefit-to-cost ratio for using fencing, repellents, sharpshooters, or hunters to cull

A holistic strategy would take into consideration each of these factors and guide the actions needed to resolve the damage to alfalfa effectively and acceptably. Holistic approaches include best management practices (BMPs), integrated pest management (IPM; box 16.1), integrated wildlife damage management (IWDM), adaptive management practices (AMPs), and adaptive impact management (AIM).

Best Practices, or **Best Management Practices**, are informal standards for techniques or methods that have proven successful over time. BMPs can serve as stand-alone strategies, such as using a trap to remove a raccoon from an attic. BMPs can also be utilized as approved SOPs for entities such as agencies or as part of a broader strategy—e.g., where trapping is part of a national strategy for management of raccoon-based rabies. By considering factors such as efficiency, selectivity, practicality, user safety, and animal welfare, BMPs ensure that every practitioner has the opportunity to follow the best-known technology for a management action, thereby ensuring the best possible outcomes (e.g., Association of Fish and Wildlife Agencies undated).

BMPs are widely used in wildlife damage management. For instance, as of 2008 the Ontario Ministry of Natural Resources includes use of BMPs as a guiding principle for its activities (Gatt undated), and Wittenberg and Cock (2001) provide BMPs as part of a toolkit for preventing and managing invasive exotic species. The Center for Human-Wildlife Conflict Resolution (Virginia Tech 2005) is serving as a repository for BMPs in Virginia and making these broadly available on the Internet.

Integrated wildlife damage management (IWDM), a form of IPM, is rapidly becoming a common strategy for wildlife damage practitioners. As with other IPM strategies, the idea is to reduce reliance on chemical toxicants and to emphasize a mix of methods that

---

BOX 16.1 Integrated Pest Management

Integrated pest management (IPM) can stand alone as a strategy or be part of a broader one. IPM was first conceived in the late 1960s by Raymond Smith and Perry Adkisson for managing insect pests in agriculture. The notion was to reduce use of pesticides by incorporating a range of nonchemical control options into a single management plan. The plan might call for use of pheromonal attractants along with mechanical traps, sterile male releases, natural enemy releases, and bred plant resistance to pests, thereby reducing dependence on chemical pesticides. IPM quickly succeeded in reducing farmer dependence on pesticides—e.g., by 50% in the United States (for which Smith and Adkisson received the World Food Prize in 1997).

IPM practices were also quickly applied to pests other than insects (e.g., Sterner 2008). For example, in the mid-1970s Dr. Smith helped to determine the physical, training, and equipment needs required to establish the National Crop Protection Center in Los Baños, Philippines. The center was designed to establish a research and training aspect of IPM that included damage from vertebrate species and weeds in addition to insects. The concept recognized that pest species occurred across taxa and that these species and their damage were often ecologically interrelated. IPM became a fundamental strategy underlying research and development of pest management practices by major international organizations such as the International Rice Research Institute. In growing rice, for example, bird damage was known to be related to weed management. Reducing weeds reduced perch sites and damage by birds, thereby reducing the need for bird toxicants or repellents. Similarly, some believed that reducing the number of insect larvae in rice stems would reduce damage by rats that seek the grub and might damage many stems to find one that is infested. The IPM approach and methodology has been utilized extensively in wildlife damage management programs to effectively address and manage problem wildlife species in a variety of situations.

work holistically to resolve the problem. The focus is primarily on reducing damage rather than reducing populations of offending wildlife, although reducing local populations is sometimes necessary to resolve a problem. The practitioner, following IWDM practices, first determines whether management is needed. If it is, the practitioner figures out the combination of methods that would work most effectively. SOPs and BMPs are often part of the mix. The approach takes into consideration the factors outlined in fig. 16.1. Rodent and bird species, with higher densities and predictable patterns of damage, seem particularly amenable to IWDM as an economically beneficial strategy (Sterner 2008).

IWDM is being used around the globe, as the following examples show. Nugent et al. (2008) used IWDM practices to manage a mix of gull species roosting nightly on Lake Auburn, an unfiltered municipal water source for Auburn, Maine. The gulls were contaminating the drinking water with fecal coliform bacteria. Cotton (2008) described IWDM of a threatened and recovering species, the Louisiana black bear (*Ursus americanus luteolus*), in Louisiana. Campbell and Long (2009) and West et al. (2009) provide excellent reviews of IWDM strategies for managing damage caused by feral pigs worldwide, emphasizing damage and management practices in the United States.

Ecologically based rodent control (chapter 4; Singleton et al. 1999, 2004) is a form of IWDM that reduces use of rodenticides by providing ecologically based, nonchemical methods such as the following: good hygiene that reduces food and cover for rodents; synchronized crop planting; and the trap barrier design system (where a crop, such as rice, is planted and fenced earlier than the rest of the crop in a small portion of a field; rodents, attracted to the older rice, move through holes placed in the fencing into the patch of older rice and are periodically removed; the method effectively gets rid of offending rodents without the need for rodenticides); see chapter 10. This strategy includes strong socioeconomic components, such as surveys of knowledge, attitudes, and practices of farmers—information that underlies the management plans. This approach has been used successfully in both developing and developed countries, for example, to develop management strategies for intensive organic piggeries and poultry farms in Europe (Singleton et al. 2004).

Use of Adaptive Management Practices underscores the belief that in a real and always-changing world, one is unlikely to get all the information needed to completely and assuredly solve a damage problem. At some point, the practitioner needs to act with the situation, information, and options on hand. This strategy embraces change and uncertainty. It is contrary to the **precautionary principle,** which states that no action should be taken until a problem and all consequences of proposed actions are fully understood, lest one be surprised with an unanticipated form of ecological backlash that causes damage to the environment or human health.

With AMP, the practitioner proceeds with a management action without a full understanding of the situation, monitors the outcomes, and adjusts future actions based on what was learned (see fig. 16.1). By gathering detailed information on an appropriate design while taking the management action, the practitioner learns more about the damage situation and how to improve future actions. In a sense, the manager becomes an amalgam of scientist and practitioner (or involves scientists). For the practitioner, in fact, a major challenge is achieving a balance between the resources required to acquire new information in a scientifically rigorous manner (Raffaelli and Moller 2000) and the resources needed to resolve the problem (e.g., Roy et al. 2009).

AMPs are increasingly used in wildlife damage management, often as part of IWDM (formerly called adaptive IWDM). For instance, Engeman et al. (2007) described use of track plots to adjust management of feral pig damage to wetlands in Florida. Bryce et al. (2010) described an adaptive approach for eradicating American mink (*Mustela vison*) from areas where they compete with native species in and around Cairngorms National Park, Scotland. The South Dakota Division of Wildlife uses an adaptive management strategy for managing wildlife, including wildlife damage problems (Anonymous 2006). Treves et al. (2009) suggested that mitigating wolf damage would be facilitated by including clauses in compensation agreements that permit AMPs from the outset. These researchers argued that the clauses should "articulate explicit goals for compensation programs lest the costs skyrocket without measurable success" (p. 16). Roy et al. (2009) stated that AMP could provide a means of filling data gaps for managing invasive species on islands. These scientists argued that some information, such as assessing risk, can be gathered only by carrying out eradications.

Wildlife damage management today is a truly multidimensional field, embracing not only its underlying biological and ecological concepts but also concepts related to its human dimensions, including economics, sociality, and public policy. Application of concepts such as **wildlife acceptance capacity** (the mixture of tolerances that can be found among the stakeholders in a wildlife damage problem) is as central to solving wildlife damage issues as the selection of appropriate management methods (Decker et al. 2002; see chapter 14). Decker et al. (2002) offer adaptive impact management (AIM) as a new twist on AMP. They characterize the approach as both the practitioner and the stakeholders agreeing that "we don't have all the answers needed for developing a management program that will fix this problem with certainty, but we'll apply what we know, use our best judgment in those things we are less certain about, and commit to learning from the experience of the specific strategy and tactics we employ" (p. 7).

We summarize by offering a generalized process for the wildlife damage practitioner (see fig. 16.1). The process is multidisciplinary, integrated, and adaptive. It applies to individual practitioners or those working for agencies and can probably be adapted to specific societies and public policies particular to a country. The first step is identifying impacts and stakeholders, following methods suggested by Decker et al. (2002). The extent of stakeholder involvement (in each of the steps) needs to be assessed in the second step, as do the specific objectives of any management actions. For example, an objective might be to measurably reduce concerns of homeowners of contracting Lyme disease. Damage surveys might be useful at this time, as might information on attitudes and acceptance capacities of the stakeholders (Decker et al. 2002).

The third step in the process is to assess the impacts and consider management options. Here, an overall strategy needs to be chosen (e.g., incorporating or choosing BMPs, IPM, IWDM, AMPs, and/or AIM), as do available DSSs and SOPs. Legal and public policy requirements and media involvement need to be determined. The assessment may take the form of an environmental assessment or an environmental impact statement (both evaluations of environmental risks and benefits, part of the National Environmental Policy Act). For example, effects of concern about Lyme disease might be fear of families to enjoy the outdoors, such as backyards or local parks. Management options might include education of families regarding use of repellents to prevent vector-borne zoonoses; periodic treatment of yards and parks for control of ticks; control of intermediary hosts such as habitat, to reduce the presence of field mice; use of fencing to reduce the presence of white-tailed deer; use of bowhunting or sharpshooters to reduce over-

abundant deer populations; or some combination of these options.

After all management options are compared and considered and one or more have been selected, a logistical plan of action needs to be developed (step 4). Often, this takes the form of an operational or management plan wherein the details of resources needed, actions, and timeframes are provided. Time management plans are sometimes used for complex management actions.

In step 5, the management plan is conducted. Consultations are completed and permits obtained. Equipment and supplies are ordered and personnel brought on board. The media are notified, if they are part of the plan. Management actions are taken. Records of accomplishment are kept in relation to plans in step 4. Sometimes an assessment is conducted before and after a specific action to measure its effectiveness. Additional data may also be collected as part of an adaptive management program. If, for example, education on use of tick repellents, tick control at local parks, and reduction of overabundant deer by sharpshooting were selected, permits, equipment, supplies, personnel, and notification of media would be done at this time. Perhaps additional information was needed on abundance of field mice for possible future adaptive actions, and indices of abundance for the mice would be put in place at this time.

Step 6 includes an evaluation of the overall effectiveness of the management action, based on the analysis of the information collected during step 5 and on adjustments of the management plan. Since the intent was to reduce the impact of concern for Lyme disease in the community, effectiveness of the management actions on this factor should be measured. At this point, adjustments in future action programs would be made.

During the process, information can be drawn from history and experience; from basic biological, ecological, and human-dimensional concepts; and from methods for assessment and management of damage (see fig. 16.1). Information gained from the process can be added to the existing base of knowledge.

We describe the process as one that follows a logical, stepwise sequence from identification of an impact to modification of future management actions. Individual practitioners may see it differently and adjust the content and order of some of the steps according to their own mindsets. We note also that the orderly sequence of events as described on paper is not always the sequence followed in the real world.

For example, political support of, or opposition to, the proposed management actions (steps 3 and 4) may necessitate adjustments in plan development, management actions, or deviations from time sequences for planned management actions (step 5, fig. 16.1; see chapter 15).

Examples

The Vermont Fish and Wildlife Department (Anonymous 2004) provides an example of BMPs for managing beaver damage. The introductory portion of the guide is organized to provide an overview of beaver management in Vermont, the biology and behavior of beavers, and descriptions of beaver problems. The BMPs are divided into damage-prevention techniques, culvert and dam obstructions less than two years old, and damage from older, more established beaver dams. The guide provides information needed to work within applicable federal and state laws. The BMP begins with a phone call to a local authority, who determines whether the issue can be resolved by preventive techniques. For example, beaver damage to ornamentals might be prevented by fencing until removal or lethal reduction can be completed. Concern for rabies or giardia might be mitigated with educational materials or with removal or lethal reduction of the population. Obstructed culverts and dams less than two years old might be mitigated by monitoring for potential beaver problems; arranging for removal or lethal reduction of the population; installing or maintaining water-control structures and, between June 1 and October 1 (to prevent effects on trout spawning, provided in appendices), notifying anyone who might be affected by a lowering of the water level downstream; installing fencing or a control device; and notifying a fish and wildlife warden or other identified agents. For well-established beaver dams, a site visit may be needed. Here, biologists would be consulted to provide a plan that reduces damage while preserving the value of the wetland. The manual includes diagrams and drawings of common wire and electric fencing methods, addresses of people who can help, information on diseases such as giardiasis, siphons and other control devices, dam removal methods, and information on relevant ordinances, statutes, and regulations, along with references and additional reading materials.

Witmer (2007) provides an overview of IWDM as applied to rodent pests on cropland. This researcher suggests that methods need to be selected and put into IWDM strategies based on the species causing problems, the physical environment, and the nature of the damage. Correct identification of the species causing damage is essential, and the amount and distribution of damage need to be known. He emphasizes the importance of knowing the distribution of the rodents within their habitat, how the rodents use the habitat, how the rodents relate to other species, and what human activities exacerbate damage problems. Witmer suggests that DSSs, such as the Mouser, can sometimes be helpful in determining when rodent control will be cost effective. He suggests that any potential strategy be evaluated not only in its ecological and biological context but also in the human-dimensional world of economics, societal concerns, and public policy and acceptability (see fig. 16.1). This scientist recommends that sufficient monitoring be put in place to allow treatment of the strategy as a large-scale experimental field trial, so that its effectiveness can be evaluated in sufficient detail to allow adaptive adjustments and improvements in the future.

Summary

- Holistic strategies are the consequence of careful consideration of every aspect of a wildlife damage problem, including its history and existing literature, applicable biological and ecological concepts, and human dimensions.
- Holistic strategies begin with fundamental considerations, such as whether to use an autecological approach or a synecological approach.
- Standard operating procedures and decision support systems are incorporated into holistic strategies.
- Ideal holistic strategies are integrated, in that they involve a wide range of management methods; multidimensional, in that they include human-dimensional as well as biological and ecological concepts; and adaptive, in that they include sufficient monitoring and data-gathering to allow post hoc improvements.
- Holistic strategies that can either stand alone or be parts of broader strategies include best management practices, integrated pest management, integrated wildlife damage management, adaptive management practices, and adaptive impact management.
- A process that helps the practitioner formulate holistic strategies has the following steps: identify impacts and stakeholders; determine extent of stakeholder involvement and specific management objectives; assess impacts and consider management options; develop a logistical plan of

action; conduct the plan; evaluate the effectiveness of the management plan; and adjust the plan as needed for improvements.

Review and Discussion Questions

1. How important are SOPs for quality assurance of responses by staff who routinely answer calls for assistance with wildlife damage—such as extension, Wildlife Services, or state wildlife office personnel? How important might SOPs be for registration of a new pesticide? How important might they be for defense of management actions in a courtroom?

2. We state that the precautionary principle is at odds with AMPs in wildlife damage management. Find a literature example where AMPs were used successfully. How about an example where AMPs led to ecological backlash? Which is better, to use AMPs or to follow the precautionary principle?

3. Find a published example of the use of AIMs in wildlife damage management, and provide its reference. What is meant by impact, and what impacts were involved? Show how the results from the initial management strategy were used to make improvements in subsequent strategies.

4. Choose a specific damage problem, such as muskrat damage to a privately owned pond, and apply to it the generalized process for wildlife damage management shown in fig. 16.1. Did the process help ensure a comprehensive consideration of the issue? If not, how might the process be modified to make it more useful?

5. Show how the generalized strategy for wildlife damage management, as seen in fig. 16.1, applies to management actions, such as reducing damage to cars and buildings and reducing health hazards posed by starlings in a city roost, of a public agency or a private-sector company.

17

Future Directions

This chapter probes the likely impacts of human population growth, demographics, and emerging technologies on the future of wildlife damage management.

HUMAN POPULATION SIZE AND DEMOGRAPHICS

Statement

Sheer numbers of people and their collective activity will intensify the need to manage wildlife damage over at least the next 40 years, with sharp differences in the nature of management between developed and developing regions of the world.

Explanation

Malthus (1798) was the first to predict exponential human growth (fig. 17.1). Total human population size reached 7.0 billion on or about October 31, 2011, just 12 years after it reached 6.0 billion. It will achieve 8.0 billion in 10 years or so. The United Nations (2009) projects a peak of 9.22 billion people in 2075, slightly above the level projected for 2050 of 8.92 billion. These projections are extrapolations from existing demographic data on individual countries, not from a single equation for the whole world population, as created by Malthus. The projections mean an increase of 29% or more during the 39-year period 2011–50, adding at least 2 billion to 2.5 billion more people to the planet. And while the human growth rate, at 1.1%, has declined from 2.20% in the 1960s, the population will continue to grow exponentially for at least the next 40 years.

The sheer increase in numbers of people, along with concomitant increases in activities such as travel and transport of goods and pets, will exacerbate wildlife damage. This argument assumes that present levels of lifestyle will continue in developed countries, that lifestyles in some developing countries such as China and India will become more similar to those of developed countries, and that energy production is able to meet these increasing demands for the next 40 years.

If these assumptions are met, increased human activity will continue to cause or exacerbate wildlife damage problems. For example,

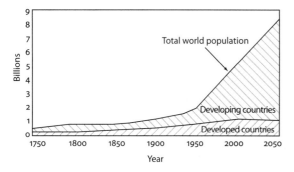

Figure 17.1 A general graph of human population growth in developing and developed countries. *Illustration by Lamar Henderson, Wildhaven Creative LLC.*

numbers of pets and agricultural livestock transported between different regions of the world will increase, creating more feral animals and transfer of parasites, insects, and diseases. Releases of pets or agricultural species into the wild—as has already occurred in Everglades National Park in Florida, Cairngorms National Park in Scotland, and broadly in Australia—will also continue or increase. Some releases will be successful, and the invasive species may take ecological roles as new mesopredators, as already observed with feral cats, dogs, pythons, and omnivorous feral pigs. Natural biodiversity of ecosystems will decline further, as with already-declining songbirds, leading to simplified systems and erratic population irruptions. Although the need for **ecosystem services** (services that ecosystems provide to humans, such as food or sewage cleanup) from natural ecosystems will dramatically increase, the simplified systems will have reduced capacity or will be completely unable to provide these services (Millennium Ecosystem Assessment 2005), leading to further wildlife damage.

Personal travel, including ecotourism, will also transfer insects, parasites, and diseases into new habitats, resulting in new episodes of wildlife diseases and zoonoses in natural ecosystems, agroecosystems, and urban environments, as already seen with crayfish plague and avian influenza H1N1. Wars, assuming they continue, will also facilitate wildlife damage associated with increased travel and drastically simplified and contaminated ecosystems. Diseases will spread more rapidly through simplified natural ecosystems—to livestock by intensified husbandry practices; to cropland by intensified monocultural agroecosystems; and as zoonoses to people, with their increased urban densities and fragmented rural land-

scapes. Numbers of ecotonal species will increase, some affronting people and their interests, as already seen with parasitic brown-headed cowbirds and deer.

In the future, roadways will intercept more natural habitat, affecting movement and distribution of some wildlife and exacerbating transport of exotic species and collisions with wildlife. Roads will be a particular problem in the United States, with 80% of the population already in metropolitan areas and growth expected in outlying suburbs (Dahl 2005).

Whether or not global warming is human made (IPCC 2007), it will continue to impact natural ecosystems by simplifying them, redistributing ranges and activities of wildlife, and changing the nature of wildlife interactions. Wildlife damage problems that characterize simplified ecosystems, such as pest irruptions and greater incidences of wildlife diseases, will intensify, as will opportunities for species to serve as exotic invasives and zoonotic diseases (e.g., Shope 1992).

The basic Malthusian equation for human population growth reveals a geometrically increasing human population without end. Malthus (1798) also argued, however, that increases in agricultural production were arithmetic, not geometric, so that human population would eventually outpace food production, exceed the earth's carrying capacity, and crash cataclysmically. Although we agree fundamentally with this prediction, we also have questions about how or when this might occur.

There is no doubt that the human population, unrestrained, would at some point exceed the earth's carrying capacity. Carrying capacity itself, however, does not consider lifestyle or quality of life. If the criterion is just providing enough food for everyone to eat, even if people are standing elbow to elbow and belly to back across the whole planet, then the earth's carrying capacity is considerably greater than if the criterion is a healthy and varied diet along with a defined minimum quality of life. In reality, quality of life becomes an issue long before biological carrying capacity.

Estimates of carrying capacity therefore vary widely, from less than a billion people (based on little environmental impact) to over 100 billion people (wherein all resources on the planet are used to provide just enough food, water, and shelter for people to survive). Most estimates for carrying capacity fall between 10 billion and 12 billion people (United Nations 2001).

Pimentel et al. (2010) estimated an optimal population of about 2 billion people for a sustainable economy and quality of life equivalent to present European standards, assuming that present use of fossil fuels is both unsustainable and undesirable. The World Wildlife Fund (2010) and the Global Footprint Network (2010) also set global human carrying capacity below the current population, basing their calculations on **ecological footprints**. If a worldwide plan were put into place that reduced the human population from its present 7 billion to 2 billion members over the next 50 or 100 years (e.g., Pimentel et al. 2010), we would anticipate gradually diminishing wildlife impacts, with a concomitant improved condition (i.e., increases in numbers and diversity) for nonhuman life on the planet. The role of wildlife damage management would increasingly focus on conservation efforts as interactions between wildlife and agriculture, urban areas, and frequency of travel and transport of goods gradually diminished. Even without such a global change, we anticipate the same change in emphasis on conservation-related issues in developed countries with human growth rates of less than two and age pyramids indicating stable or declining populations. Many European countries, as well as Australia and New Zealand, are examples. The United States is an exception in that it has a growth rate at replacement level, $r=2$, due mainly to its immigration policies.

Hopfenberg (2003) found a clear relationship between global food production and population growth. However, he noted that the relationship was based on localized food distribution and localized human carrying capacities, not on carrying capacity as a whole (see the example section below). These findings raise additional thoughts about future directions of wildlife damage management. First, populations at or near carrying capacity are often physiologically stressed (see chapter 5). This means increased vulnerability to diseases including zoonoses, leading us to predict increased emerging infectious diseases both locally and globally, emanating from locally stressed human populations. Second, damage management to prevent or control the zoonoses will almost certainly intensify in these localized areas. Third, localized environments will also become stressed by intense human activity, e.g., with movement of agriculture into marginally arable lands. Ecosystems will be further simplified, exacerbating wildlife damage effects, such as those due to rodent and bird population irruptions. Support for these arguments by Hopfenberg can be found in regions of Africa, Asia, and parts of South America, where localized populations already exceed localized carrying capacities, where diseases and starvation have already cost millions of human lives, and where conflicts between humans and wildlife are on the rise (e.g., Distefano 2005).

Furthermore, about 90% of the projected growth will be in developing countries, raising their share of the world population from 84% to 88% (see fig. 17.1). Again, we recognize that both birth and death rates are declining in developing countries. It is the demographic momentum of the youthful population, and its sheer size, that drives the upward growth. Highest growth rates are for people in Africa and Middle Eastern regions as well as people in some Asian and Southeast Asian countries. People in these regions already sustain increasing damage from wildlife. Wildlife-safe areas (such as refuges and park reserves) and changing agricultural practices (such as a shift to sugarcane in parts of India) have increased some wildlife populations. Increasing numbers of people and contact with wildlife will continue to exacerbate effects in these regions. Africa and Asia provide habitats for many of the world's favored wildlife. We see these charismatic megafauna increasingly stressed by human impacts, raising the need for conservation aspects of wildlife damage management in these areas (e.g., Distefano 2005). This need will also increasingly conflict with the wildlife damage management needs of growing urban populations in these areas.

Human population models predict an increasingly urban population. For example, the United Nations (2009) projects that urban areas will gain 2.9 billion people from 2009 to 2050 and go from 3.4 billion to 6.3 billion. This movement represents, in effect, the total anticipated population increase (2.3 billion) plus some additional movement (0.6 billion) from rural to city areas. Most growth will occur in urban areas of developing regions, including Asia, Africa, Latin America, and the Caribbean. By 2050, about 86% of the population in developed regions will be urban, and about 66% in less-developed regions will be urban, with about 69% urban for the world's population at that time.

The trend toward urbanization will further intensify the need for wildlife damage management and shift its focus. In developed regions such as Europe, Asia, North America, and Australia, the trend will probably lead to a broader base of wildlife damage interests and problems. Management will continue in relation to increased game production, such as managing predation on ducks in the prairie pothole regions of North America or predator control in Alaska to maintain an herbivorous game base. Enhanced protection of crops and livestock will be needed globally. Wildlife damage issues of particular interest to urban dwellers, including urban wildlife and protection of threatened and endangered wildlife, will also likely become increasingly important. Underlying such changes will be gradually shifting value orientations, driven by urban perspectives on wildlife, away from doministic ones toward more mutualistic ones (e.g., Manfredo et al. 2009). We think the trend will carry to countries such as China and India as middle classes emerge. We base this thinking partly on Maslow's (1954) hierarchy of needs, as people progress from physiological needs and safety to needs focused on love and belonging (mutualistic value orientation).

Urbanization in developing regions will probably have a drastically different effect on wildlife damage and its management, because the urbanization will occur in the poorest of the cities and within the slums (Dahl 2005). The focus there will not be on the love and belonging level of Maslow's hierarchy of needs but rather toward the most basic of physiological needs and human safety. We suspect that management of commensal pests, feral pests, and zoonotic diseases will be more heavily represented in the wildlife damage issues of the future. The outbreak of SARS in Hong Kong in 2002 and 2003 and its rapid spread throughout the globe point to both the increased prospects for emerging zoonoses in densely populated urban areas and to the global connectedness of such problems (e.g., Patel and Burke 2009).

Example

Hopfenberg (2003) used a sigmoidal equation for population growth that included carrying capacity as a time-dependent factor (i.e., not as a constant; fig. 17.2). He used the equation to relate population growth to food production estimates provided by the United Nations Food and Agriculture Organization (FAO). This researcher found that food production alone was sufficient to predict population growth. Estimates for food production for the year 2000 indicated an ability to sustain a human population of 23 billion, 3.79 times its actual size (fig. 17.2). Hopfenberg argued that human population growth is already limited by food production and probably has been since the advent of agriculture some 10,000 years ago. The limiting factor for population growth, however, is not global food production but the localized distribution of food and related factors, resulting in localized malnutrition, starvation, disease, and human mortality. Locally disparate food distribution therefore limits growth of the human population. But continued increases in food production allow continued growth of the global population.

The paradoxical relationship between food production and population suggested by Hopfenberg (2003) was described by Salmony (2006) as a positive accelerating feedback loop. In the words of Salmony, "in such circumstances, increasing food production for people who are starving is like tossing parachutes to people who have already fallen out of the airplane—the produced food arrives too late" (pp. A17–A18). Starving people need more food production, so production is increased. But starving people die before a food production cycle is completed, limiting the

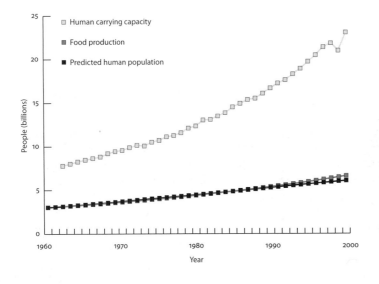

Figure 17.2 Relationships between human carrying capacity, food production, and human population growth. *From Hopfenberg, R. 2003. Human carrying capacity is determined by food availability. Popul. Environ. 25:109–117, fig. 2. Reproduced with kind permission from Springer Science + Business media. Illustration by Lamar Henderson, Wildhaven Creative LLC.*

world population. And the surplus food production allows for continued global growth.

TECHNOLOGICAL ADVANCES

Statement

Human technological advances will fuel unprecedented technical growth in methodology for wildlife damage assessment and management. These advances will allow dramatic improvements in methods as judged by such criteria as efficiency, cost effectiveness, and humaneness.

Explanation

A visit to the NOVA (2012) website can be an inspirational look at both the capabilities of humans and their technological future. Artificial life has been synthesized by human-made DNA; the first steps have been taken in tele-transportation; a machine will soon be built with the computing capabilities of the human brain; cellular biomass is being converted to biofuel by bacteria in a single step; a 1918 flu virus is being re-created to find ways to prevent flu epidemics; and molecules of graphene may store data and energy in sheets only one atom thick.

In the context of such dramatic technological advances, we see both enormous opportunities for future methods in preventing, assessing, and managing wildlife damage and a temptation to downplay practical technological advances that seem more mundane. However, we caution that some seemingly minor advances can in fact have far-reaching impacts.

PHYSICAL METHODS. Recent improvements in cable restraints are good examples. The addition of breakaway hooks, loop-stop ferrules, in-line swivels and anchor swivels, twisted link chain (Darrow et al. 2008), and insightful setting strategies, ideas contributed by many individuals over the last decades, have made cable restraints dramatically more reliable, selective, and humane (see chapter 10; Vantassel et al. 2008). A consequence has been less restrictive regulations in both water and land sets in the United States and their increased availability for use in wildlife damage management elsewhere. For example, Vantassel et al. found that nine states had reduced restrictions on cable traps during the past two decades; Muñoz-Igualada et al. (2010) found effective performance of Wisconsin cable restraints on trails for capturing foxes in Castilla–La Mancha, Spain; and Frank et al. (2003) found foot snares helpful in capturing lions that were wary of traditional capturing techniques such as cage traps and free darting.

Likewise, advanced designs, such as EGG traps (Tischaefer and Tischaefer undated) and similar traps for raccoons, have greatly improved the selectivity and humaneness of traps. Trapping systems are now available that send photographs to a cell phone or an e-mail address when wildlife such as feral pigs enter, and the door can be remotely closed (Wireless traps undated; Larkin et al. 2003). Remotely controlled drop and box traps have been made for capture of feral cats, birds, and many other types of wildlife (e.g., Plice and Balgooyen 1999).

Fall and Schneider (1968), describing structures in State College, Pennsylvania, were among the first to point to the importance of architectural design in wildlife damage management. Mechanical engineering is now underway to reduce damage caused by collisions with wildlife by modifying windshields, engines, and other parts of aircraft, and we believe that such modifications might also be made for automobiles. For example, lighting and sound might be used to make aircraft or automobiles more detectable by birds or other wildlife (e.g., Blackwell and Bernhardt 2004). We anticipate increased use of architectural and engineering designs to reduce wildlife damage problems in the future.

Crossings allow animals to get across human-made barriers such as roadways. They include underpass tunnels, viaducts, overpasses (e.g., green bridges and ecoducts), amphibian tunnels, fish ladders, tunnels, and culverts (fig. 17.3). Wildlife crossings and corridors already reduce vehicle collisions and allow safe passage for wildlife in parts of Europe (Bank et al. 2002) and Canada and are gaining more attention in the United States and other countries (Beckmann et al. 2010). For example, with about 25% of Dutch badgers (*Meles meles*) killed on highways each year, a special effort is being made in the Netherlands to use culvert systems that allow safe passage of the badgers. The U.S. federal Safe, Accountable, Flexible, Efficient Transportation Equity Act of 2005 (now expired) included a directive to design projects and processes to reduce road impacts on wildlife habitat and driver safety, and wildlife crossings and corridors are now being designed in many states, including Alaska, Arizona, Colorado, Florida, Montana, Vermont, and Washington.

One of the most extensive databases on wildlife crossings and fencing has been gathered from overpasses, underpasses, and fencing constructed across the Trans-Canada Highway for passage of wildlife in Banff National Park, Canada. Clevenger et al. (2001, 2009) collected information on more than 185,000 crossings in 12 years of monitoring, including

Figure 17.3 A Canadian underpass designed for small wildlife, with pathways for dry crossings and spillways for aquatic fauna, all fenced. *Photo by Cephas.*

crossings by grizzly bear, elk, red fox, striped skunk, and hoary marmot (*Marmota caligata*). Wolverine (*Gulo gulo*) and lynx have been observed using the passes. Given the number of crossings and the composition of species, the findings suggest that wildlife crossings could greatly reduce the number of collisions with vehicles as well as human and wildlife injuries.

CHEMICAL METHODS. Propiopromazine hydrochloride is an immobilizing agent being used experimentally to sedate wildlife, including coyotes and wolves, caught in leg-hold traps. The intent is to reduce self-inflicted injuries during restraint, sedating the animal without loss of consciousness (Savarie et al. 2004). The tranquilizer is placed in a rubber bladder on the restraint. Upon an animal's biting the bladder, the agent is released and the animal ingests it and is sedated. Diazepam, also a tranquilizer, is being tested for use with coyotes caught in neck snares (Pruss et al. 2002). The drugs are being evaluated for use in the United States, Australia, and New Zealand.

New pesticides, or new uses for old ones, will continue to be found that offer highly selective control and are humane and environmentally safe. Sodium nitrite for managing feral pigs is an example. Registered as HOG-GONE in Australia, sodium nitrite acts by blocking oxygen-binding pathways in the pig (Cowled et al. 2008). The mode of action is directly related to methemoglobin reductase activity in erythrocytes (similar to that of carbon monoxide, but reversible; Lapidge and Eason 2010). Pigs seem particularly sensitive to the effects of sodium nitrite and succumb within an hour or two after exposure. The animal is unconscious before death, and the method appears humane. Para-aminopropiophenone, which also functions through methemoglobin mechanisms, is being evaluated for use with foxes, wild dogs, and feral cats in Australia (Fleming et al. 2006) and with stoats (*Mustela erminea*), ferrets, and feral cats in New Zealand (Murphy et al. 2007).

New piscicides are also being sought. Studies in New Zealand included saponin-based piscicides such as teaseed cake, mahua oil, and a patented formulation from teaseed cake called SWIMTOP (Clearwater et al. 2008). Saponin-based piscicides are often toxins found in plants. Other compounds being considered include those from at least 15 species of plants used by tribal people in India to stupefy fish (Jadhav

2010); 2 new potential glycosidic piscicides isolated from fresh blue iris (*Iris spuria*), an ornamental cultivated in Egyptian gardens (Farag et al. 2009); extracts from poinsettia (*Euphorbia pulcherrima*; Singh and Singh 2009); a substance called *carpine* in the seeds of papaya (*Carica papaya*; Ayotunde et al. 2011); and a new oleanane glycoside from seeds of a fish poison tree (*Barringtonia asiatica*) used in Polynesian cultures for fishing (Burton et al. 2003).

Wildlife contraception will play an increasingly important role in wildlife damage management. Contraception has long been of interest for pet cats and dogs and for managing overabundance of some zoo animals, and it has attracted increasing public interest for species such as feral cats. Fagerstone et al. (2010) pointed to DiazaCon, a synthetic cholesterol used as Ornitrol in the past, for possible use to manage invasive monk parakeets (Avery et al. 2006) and black-tailed prairie dogs (*Cynomys ludovicianus*; Nash et al. 2007). Sperm-head glycoproteins that bind to zona pellucida have been identified and are being tested for contraception of red fox, European rabbit, and tammar wallaby (*Macropus eugenii*) in Australia. The DNA of laboratory rodents may provide the bases for sperm-directed contraceptive vaccines. A single-dose contraceptive could result from binding gonadotropin-releasing hormone analogs with cytotoxins, and preliminary studies of this approach have suppressed luteinizing hormone production for six months in female mule deer (Fagerstone et al. 2010).

Immunological technologies might have uses other than contraception in wildlife damage management. For instance, such technologies might be used to either block breakdown or accelerate production of self-made powerful pain killers such as endorphins, providing a humane form of anesthesia or euthanasia for wildlife.

BIOCONTROL. Biocontrol could open new vistas for managing wildlife damage. An example is the testing of a bacterial parasite (*Wolbachia* spp.) that halves the lifespan of mosquitoes (*Aedes aegypti*) and limits the spread of dengue fever in northern Australia (Hoffmann et al. 2011). Another is the release of a species of tsetse fly (*Glossina moristans*) in an area of Tanzania where a primary species, *G. swynnertoni*, was endogenous and killed cattle; by overwhelming the *G. swynnerton* with high numbers of releases of *G. moristans*, the former species eventually disappeared. Because *G. moristans* was poorly adapted to dry habitat, it too died out, leaving the area available for livestock production (Gould 2007).

Genetic pest management (**GPM**, wherein genes of pest populations are altered to either decrease pest densities or to render pests harmless), a form of biocontrol, could also contribute new solutions for wildlife damage problems (Gould 2007). New techniques in molecular biology, wherein genes such as "selfish" ones (those that tend to replicate in higher frequency in organisms even when their presence may be detrimental to the phenotype) or Meade genes (genes that kill all embryos that lack a trait carried by the mother) open new applications for wildlife damage management (Gould 2007). For example, this researcher describes modifying genes to manipulate pest populations or attaching codes that sterilize the pests, make them incapable of vectoring a disease, or render them otherwise harmless. Although insects have been the main focus of such technology, Gould suggests its use for species such as Scotch broom (*Cytisus scoparius*) or commensal rodents, where much is already known about chromosomal and population genetics. While Scotch broom itself is not well studied, the weed is a member of the family Fabacae, which includes soybean, alfalfa, peanut, and chickpea. Genetic composition of the whole family has been intensely studied for agriculture. Commensal rodents are laboratory animals, and their genetics have also been well studied.

Some researchers also see genetic tools as means of answering questions about the nature and distribution of damaging populations. Rollins et al. (2006), for example, used genetic evaluations to mark details of the spread of the European starling into western Australia.

Attenuated viruses and bacteria have been used to immunize humans for many years, but their applications to wildlife damage management are new GPM technologies. Attenuated viruses could self-disseminate (spread quickly through a targeted population) and carry with them sterilants, toxins, or other compounds or induce other effects, such as viral-vectored immunocontraception for rodents (Singleton 2009) and other species. Research continues, to manage rabbits in Australia and the brushtail possum in New Zealand, with scientists exploring the use of such self-disseminating biological agents (e.g., Ji 2009). This scientist suggests that nonself-disseminating GPM approaches might be more socially acceptable and includes bacterial ghosts (gram-negative bacterial cell envelopes without cytoplasm) and genetically modified plants (e.g., carrots genetically modified to deliver immunocontraceptive vaccines) as possibilities.

We believe that facilitation of native species interactions, as with lure crops, is another underexploited area of biocontrol that could yield future methods. For instance, King et al. (2009) described an initial trial using a beehive fence to deter elephant raids on crops in Laikipia, Kenya. The presence of the hives reduced raids on the farm, allowing the farmer to harvest maize, sorghum, beans, and potatoes. Sales of honey helped to pay for the costs of fencing. Subsequent trials confirmed the effectiveness of the method (King et al. 2011), and a manual for construction of beehive fences is now available (King 2011).

Examples

Long et al. (2010) evaluated two feral-pig feeder systems. Each system required an upward push by the nose of a pig, a form of species specificity, to gain access to the self-feeder. The systems included a non-target exclusion feeder system (NEFS) developed in the United States and a Boar-Operated-System (BOS) developed in the United Kingdom. Prototypes were constructed from local materials and tested at sites in Texas with feral pigs, raccoons, white-tailed deer, and collared peccaries (*Pecari tajacu*). The sites were at least a kilometer apart from each other, in areas of known high feral pig activity. Camera traps were used to monitor activity at each site for four weeks, and feeder systems were serviced daily.

Analyses were based on over 400,000 digital images. During a period of pre-activation, bait was removed by raccoons (54%), feral swine (32%), deer (10%), and collared peccaries (4%). Once activated, feral swine removed 100% of bait from the BOS feeders, whereas raccoons removed 70%, feral swine 20%, and collared peccaries 10% of the bait from the NEFS. The BOS system was the first in five years of trials to offer feral-pig-specific bait delivery.

Pallister et al. (2011) attempted to disrupt the life cycle of the cane toad. The researchers chose Bohle iridovirus (BIV), a ranavirus originally isolated in one region of Australia. The researchers attempted to interfere with metamorphosis by injecting adult globin in the tadpoles. The tests were unsuccessful in that no changes were detected in globin, in the life cycles of injected toads, or in the survival of the metamorphs. However, the researchers were able to show that a globin "switch" occurs at metamorphosis in the cane toad, a discovery that might be used in future studies with RNA interference. In addition, the study is important as a first attempt to modify an infectious agent so that it no longer causes disease but rather disrupts the life cycle of the toad.

INFORMATION MANAGEMENT

Statement

Information on wildlife damage management, including instantaneous analyses through systems such as GIS, GPS, and DSSs, will dramatically improve bases for real-time decision making by wildlife damage practitioners.

Explanation

Information systems have advanced dramatically and are already being used to advantage in wildlife damage management. For example, damage information is available on the Internet, including such sites as the Internet Center for Wildlife Damage Management (see chapter 3) and the Global Invasive Species Database (see chapter 7). Databases for specific applications could offer ways of improving effectiveness or efficiency of some management practices. For example, such databases could be used to predict when and where risks are highest for wildlife collisions with aircraft and where efforts would be most useful in reducing potential collisions (e.g., Dolbeer and Wright 2009). Development of uniform databases might also help to unify regulations for managing species such as bears. The databases could also provide insights into where human encounters with bears are most likely, allowing managers to focus limited resources in areas with the highest clusters of potential conflicts (e.g., Baruch-Mordo et al. 2008).

GIS and GPS systems are increasingly used in wildlife damage management and will take on greater importance, because they can spatially relate different forms of information, including biological, ecological, and human demographic data, in real time.

Example

Fischer and Dunlevy (2010) described the importance of GIS and GPS technologies in the eradication of rats from Lehua Island, an uninhabited island in Hawaii. Polynesian rats have been on the island since at least the 1930s and have significantly affected the ecosystem, including at least 17 species of seabirds known to frequent the island. An attempt to exterminate the rats was begun in January 1999, using aerial broadcasts of 3,900 pounds of pellets containing diphacinone. Even distribution of the pellets was essential, both for maximum impact on the rats and to ensure compliance with the USEPA label for use of the pesticide. The problem was complicated because the topography of Lehua Island is steep, making even distribution a three-dimensional rather than a two-dimensional problem.

The researchers used an AgGPS Trimlight 3 System with in-cockpit display that recorded island boundaries and flight paths for each application and saved the data on compact flash cards. While reloading baits, the researchers analyzed data from previous applications using ArcGIS ModelBuilder so that adjustments in flight paths could be made. ArcInfo was used to construct a three-dimensional image of bait distribution. The eradication itself is in progress (Pitt et al. 2011) at this writing. The report illustrates the present and future importance of GIS and GPS systems as tools for data assessment and management of wildlife damage.

Summary

- Sheer numbers of people will exacerbate wildlife damage problems. Increased travel, trade, and communications will facilitate transfers of exotic invasive wildlife as well as new wildlife diseases and zoonoses; roadways will intercept more natural habitat, affecting patterns of movement of wildlife and increasing opportunities for animal-vehicle collisions; wars will simplify and contaminate ecosystems; and climate change will redistribute ranges and activities of wildlife, contributing to more problems with wildlife.
- Human populations will continue to be limited by distribution of foods in localized areas; stress will increase vulnerability to diseases, including zoonoses, that will have both localized and global impacts; ecosystems stressed by activity of people will become simpler, exacerbating wildlife damage impacts such as bird or rodent population irruptions.
- These impacts will be particularly focused on developing regions, where most growth will occur, and where urbanization will increase needs for management of commensal pests, feral pests, and zoonotic diseases.
- Impacts of urbanization in developed regions will probably mean a broadening of wildlife damage to include activities associated with mutualistic value orientations, such as conservation of threatened or endangered species.
- Human technological advances will fuel unprecedented technical growth in methodology for wildlife damage management.
- Physical methods will spur improvements in restraining devices as well as architectural and structural improvements in buildings, aircraft, and automobiles to resist damage by wildlife; crossing designs will allow wildlife to move across or around human-made barriers such as roadways and dams.
- Chemical methods will include new immobilizing agents, highly selective and humane toxicants such as sodium nitrite for feral pigs, saponin-based piscicides or ones yet to be discovered from known plants, and wildlife contraceptives.
- Biocontrol will be increasingly used to manipulate genes in damaging populations, using concepts such as selfish genes and Meade genes. Attenuated viruses and bacteria or their ghosts will be used to move sterilants, toxins, or other compounds through damaging populations. Interactions between native species will be exploited to manage pests.
- Advances in information technology will make wildlife damage information increasingly available throughout the world, using vehicles such as the Internet. Systems that integrate geographic with natural resources information, such as GIS and GPS, will increasingly find use in day-to-day wildlife damage management operational activities.

Review and Discussion Questions

1. Hopfenberg (2003) argued that food production could be used to predict human population growth globally; at the same time, he argued that the human population in 2000 was at least fourfold under its carrying capacity. Explain this apparent discrepancy.
2. Would wildlife damage management exist in a world where the human population was less than two billion? If your answer is no, defend your view with references. If your answer is yes, describe its nature with references.
3. List the top five technologies being advanced by humankind. Give a specific example of how each might be applied to wildlife damage.
4. From a global perspective, how important is the Internet to wildlife damage management? Explain your answer, with particular reference to management problems in developing countries.
5. Thinking as a wildlife damage practitioner, what advances would you like to see in technology over the next decade? In information management? In applications of human dimensions?

18

Wildlife Conservation

This final chapter explores wildlife damage management applications to the conservation of endangered or threatened species, particularly on islands, as well as charismatic megafauna and game species.

CONSERVATION OF ISLAND SPECIES

Statement
Wildlife damage specialists have joined forces with conservationists to eradicate offending species from islands.

Explanation
Since Darwin's accounts of his discoveries on the Galápagos Archipelago, islands have been seen as safe havens for some of the world's most diverse and unique wildlife. Isolated from the unceasing evolutionary pressures of mainland areas, species finding sweepstakes routes (that is, extremely uncommon routes, as unlikely as winning a sweepstakes) onto these natural refuges were free to evolve new and unique life forms at leisurely paces counted in eons. Birds, sea reptiles, and other oceanic species found islands and used them as safe areas for reproduction and other activities. Because defenses against common diseases, herbivory, or predation were mostly unnecessary, island species invested their energies in more unique evolutionary paths.

Humans changed all of that. Sometimes accidentally, sometimes intentionally, humans brought herbivores, omnivores, predators, and diseases to many of these natural refuges (see chapter 2). Many island-adapted species were "sitting ducks," defenseless against onslaughts. Feral pigs, goats, cats and dogs, or other pets efficiently removed unprepared wildlife from the Galápagos Islands and thousands of other islands. Introduced foxes preyed on endangered endemic species and nesting seabirds. With human help, three species of rat colonized over 80% of the world's archipelagoes, a process that continues today. Rats and mice attacked seabirds sitting on nests, sometimes consuming the birds live, and because of their omnivorous habits, the rodents also broadly affected whole ecosystems. Many threatened and endangered species saw the edge of extinction, and some succumbed. Three species of rats accounted for the extinction of at least 50 species on 40 different islands (Towns et al. 2006; Jones et al. 2008). Because of their broadly based herbivory, feral goats were identified in the 1970s as threatening insular plant species (Lucas and Synge 1978). Here, however, seed banks and restricted accessibility of localized plants made extinctions unlikely. Instead, broad-scale impacts on vegetation were common, with cascading effects on other wildlife.

The house mosquito was carried by ship to the Hawaiian Islands, where it now vectors wildlife diseases such as avian malaria, avian pox, and West Nile fever, raising havoc with honeycreepers and other endangered species. The same type of mosquito now regularly hitches rides on aircraft and tourist boats from the South American mainland to and among the Galápagos Islands, leading Bataille et al. (2009) to assert that serious wildlife diseases on these islands are disasters waiting to happen.

Thanks to damage management methods and collaboration of wildlife damage specialists, and to the experience of conservationists who practice eradication, there have now been over 780 successful eradications of invasive vertebrates from islands (Rauzon 2007; Keit et al. 2011). Notable successes have included cats, goats, pigs, and rats (Carrion et al. 2011).

Howald et al. (2007) reported at least 332 successful rodent eradications worldwide from at least 284 islands. Rodenticides, particularly anticoagulants such as brodifacoum, were used in almost all of the eradications. Bait stations were most often utulized, always where secondary hazards were a concern, followed by hand broadcasting and aerial broadcasting (which made up 76% of the area treated). Helicopters provided most aerial deliveries, and GPS and GIS systems pinpointed distribution of broadcast baits. Campaigns using bait stations often lasted up to a year, whereas baiting by aircraft was accomplished in one or two treatments over weeks. When practical, baiting was timed to maximize effects of additive mortality—e.g., when rodent populations were lowest. Failures tended to involve house-mouse eradications, perhaps because the small home ranges of mice required more intense broadcast rates than are presently used (Howald et al. 2007).

Campbell and Donlan (2005) reported at least 120 feral goat eradications from islands, mostly islands under 20,000 hectares. The goats were trapped, hunted, poisoned, biologically controlled (e.g., by burning habitat or releasing dogs or dingoes), or some combination of those (Veitch and Clout 2002). Hunting was most common and included shooting from helicop-

ters, hunting with trained dogs, and Judas goats (goats equipped with devices that reveal locations of other goats on returning to a herd). GPS and GIS systems helped to locate goats. Over 10,000 goats were removed from Raoul Island, New Zealand; 29,000 from San Clemente Island, United States; and 66,000 in less than 3 years from Santiago Island, Galápagos.

As opportunistic predators on islands, cats have consumed mammals, reptiles, birds, and insects (Nogales et al. 2004). Mammals involved included rodents, such as hutias (*Geocapromys* spp.), unique to islands in the Caribbean, and endemic rodents (*Nesoryzomys* spp. and *Oryzomys* spp.) from the Galápagos, and over ten taxa on the islands of Baja California, Mexico, reached extinction or near extinction. Cats have also reduced numbers and altered distributions of reptiles such as the tuatara (*Sphenondon punctatus*) on New Zealand, iguanas (*Brachylophus* spp.) and skinks (*Emoia* spp.) on the Fiji Islands, and iguanas (*Cyclura* spp.) on Caribbean islands. Feral cats have caused the extinction of at least 33 bird species, including the Stephen Island wren (*Traversia lyalli*) in New Zealand, a dove (*Zenaida graysoni*) on Socorro Island, and the Guadalupe storm petrel (*Oceanodroma macrodactyla*) on Guadalupe Island, both in Mexico (Nogales et al. 2004).

Nogales et al. (2004) reported cat eradications from 48 islands, most (16) in Baja California, Mexico, but also in New Zealand, Australia, Mauritius, and the Caribbean. Most of the islands were small, under 5 km^2. Practitioners used combinations of conibear traps or cage traps (less frequent), hunting with dogs, shooting, and poisoning with Compound 1080. A feline disease was used on one island. Documentation of recovery of island wildlife has received limited attention. A 90% reduction in mortality was seen with black-vented shearwaters (*Puffinus opisthomelas*) on Natividad Island, Mexico. Recolonization of the Coronados Islands, Mexico, by Cassin's auklets (*Ptychoramphys aleuticus*) occurred within four years. And hole-nesting petrels were seen recovering on Marion Island (Nogales et al. 2004).

Eradication of other species has also occurred. Soria et al. (2002), for example, reported removal of 3 of the over 600 invasive weeds—a vine, a scrambler, and a timber tree—from the Galápagos Archipelago. Brown and Sherley (2002) reported eradication of brushtail possums from Kapiti Island, a nature reserve off the coast of New Zealand.

Rauzon (2007) summarized the removal of arctic foxes from islands in the Alaska Maritime National Wildlife Refuge. In a program ongoing since the 1950s, foxes have been removed from all but 5 of the

450 islands with established populations. Several federal agencies, including the USDA's Wildlife Services, and private-sector associations collaborated in the eradications. Aleutian Canada goose populations have subsequently increased from under 1,000 birds in 1975 to over 100,000.

Kessler (2002) reported that Sarigan Island, of the Mariana Islands, was free of feral ungulates after 68 pigs and 904 goats were removed by aerial and ground shooting, trapping, and tracking with dogs. Vegetation was already improving at the time of the report. Feral pigs were also eradicated from Santiago Island, Galápagos Archipelago, after 30 years of effort, and over 18,000 pigs were removed by both ground hunting and poisoning with Compound 1080 (Cruz et al. 2005). Parkes et al. (2010) described the removal of 5,036 feral pigs from 25,000-hectare Santa Cruz Island, California, in 411 days.

Some responses to eradication had unexpected consequences, pointing to the importance of adaptive management practices. For example, mice irrupted on Buck Island Reef National Monument in the Caribbean Sea following the extermination of roof rats (Witmer et al. 2007; see chapter 6). European rabbits irrupted on sub-Antarctic Macquarie Island following removal of feral cats, causing major changes in the island ecosystems. This irruption was anticipated, but Myxoma virus, already in place, failed to control the rabbits (Bergstrom et al. 2009).

As eradication campaigns attract more public attention, and as efforts move closer to mainlands and islands inhabited by humans, concerns will approximate those faced routinely by mainland wildlife damage practitioners (e.g., Howald et al. 2007; see chapter 16). Such concerns include the following: economic benefits of the eradication to the community; hazards of pesticides to nontarget animals, particularly livestock, pets, and game species; legal status of wildlife on the islands; and confidence in the individual or agencies doing the work (e.g., Oppel et al. 2010).

Brooke et al. (2007) have examined cost as related to conservation gain (i.e., benefit) as a way to prioritize island eradications of rats, cats, and goats to conserve birds. Costs related directly to the size of islands, with rodent eradications costing more than the eradication of larger animals. Conservation gain included factors such as nearness to extinction of the bird species and intensity of impacts by invasive species. The greatest conservation gain to threatened birds per unit cost occurred on smaller islands. Economically

based prioritization has also been suggested by others, including Capizzi et al. (2010) regarding protection of seabirds by eradication of black rats on Italian Islands and Ratcliffe et al. (2009) for the United Kingdom, Channel Islands, and Isle of Man.

Oppel et al. (2010) noted that chances for successful eradication drop dramatically if even one or a few community members oppose it. If the targeted species are already seen as problems by community members, eradication is easier. If they are not viewed as problems, an outreach campaign may be needed. When the targeted species is valued by a segment of the community, eradication may not be possible even if it offers great ecological benefits. This research team pointed to the successful removal of rats from Lord Howe Island, Australia, where residents suffered rat damage to crops and wanted the island to be rodent-free. An attempt to eradicate feral goats on Lord Howe Island initially failed, however, because community members felt the effort was inhumane.

Even with the best-made plans, social or political factors can sometimes be difficult or impossible to predict. For example, Carrion et al. (2011) described a politically based threat by local fishermen aimed at the Galápagos Marine Reserve, managed by the Galápagos National Park. The fishermen threatened to restock feral goats on islands from which they had been successfully eradicated unless numbers of fishing permits were increased (fig. 18.1).

A common conservation goal of eradications is to return island wildlife to pre-invasion conditions. Here, there is a difference between **recovery**, wherein

Figure 18.1 Threats from fishermen in the Galápagos to restock goats on goat-free islands if demands for increasing fishing licenses are not met. *Photo by C. J. Donlan, with permission.*

changes occur by ecological succession over time without further human intervention, and **restoration**, wherein humans attempt to speed up natural ecological processes (Atkinson 1988). For example, some seabirds, such as sooty terns (*Onychoprion fuscatus*), possess population and behavioral characteristics that allow rapid re-establishment on islands, whereas others, such as hole-nesting procellarids, have attributes that make re-establishment slow and difficult. In addition, there is some question as to whether full recovery will occur on its own or whether island ecosystems could get stuck in some intermediate ecological sere (Beisner et al. 2003). In the latter case, human assistance may be needed (see the examples section below).

Examples

Jones (2010) conducted a series of experiments that sheds light on biota and seabird recovery and restoration following rodent eradications on islands. To simulate the ecological effects of guano from a fully recovered colony of seabirds (fig. 18.2), this researcher spread guano-like fertilizer in plots on an island that had no history of rats or seabirds (Maud Island, Cook Strait, New Zealand). She also dug holes in varying densities to simulate burrows. She then measured the movement of nitrogen, its decomposition, and recovery of flora and fauna, comparing these factors as functions of simulated seabird densities. Measurements included those that would detect fast responses, such as net primary plant productivity and arthropod (i.e., spider) abundance, and those that would measure slower responses, such as nutrient flow.

On islands in Cook Strait, Jones (2010) also conducted a natural observational experiment in which she paired similar islands into four categories: never invaded by rodents; invaded by rodents not eradicated; rodents eradicated; and rodents absent with seabird restoration. For these categories, she used transects to sample soil, vegetation, active seabird burrows, and spider abundance. She measured nitrogen-15 levels as indicators of portions that came from oceanic sources such as seabirds. She measured carbon-to-nitrogen ratios in soil, plants, and spiders from samples along the transects as well as nitrate and ammonium concentrations as indicators of ecosystem function (these are forms available after nitrification of nitrogen in guano).

The fertilization experiment indicated that after invasive predators are removed, island ecosystems may fail to recover on their own in a "contemporary

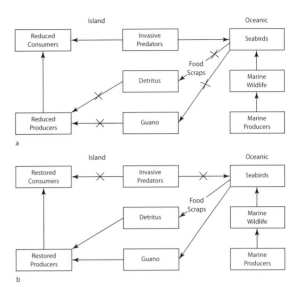

Figure 18.2 Eradication of invasive predators can help restore island ecosystems based on seabird guano. Nutrient flow with (*a*) and without (*b*) invasive predators. *Illustration by Lamar Henderson, Wildhaven Creative LLC.*

timescale." Densities of fertilizer that emulated full-density seabird populations were needed to initiate the net primary productivity, arthropod levels, and decomposition rates that allowed full ecosystem recovery. At these levels, recovery was rapid.

The results of that natural observational experiment also indicated a slow recovery of ecosystem functions, even though some of the islands were rodent-free for 8 to 13 years. The islands had not recovered their former seabird densities (6.33 burrows/m² on Stephen's Island versus 0.67 and 2.33 on Mana and Maud Islands, respectively), which were needed to allow full ecosystem recovery. Instead, levels of productivity, arthropods, soil nitrates, ammonium, and nitrogen-15 decomposition were intermediate, indicating that an intermediate ecological sere may have been achieved. Jones concluded that restoration actions, such as chick translocations, use of decoys, vocalization playbacks, and mirrors may be needed to supplement passive recovery.

CONSERVATION OF CHARISMATIC MEGAFAUNA

Statement

Charismatic megafauna often represent flagship species for the conservation movement. The same methods, including lethal ones, used in the past to depress numbers of these megafauna are now being used to conserve them.

Explanation

We greatly support the importance of wildlife damage management in today's efforts to conserve **charismatic megafauna** (i.e., large animals with widespread popular appeal, such as elephants, tigers, lions, leopards, bears, hawks, and eagles, often used by activists to achieve conservation goals). We recognize this as a worldwide issue.

Unguided wildlife damage management—or misguided or shortsighted management—can push targeted species toward extinction, just as today's forward-looking wildlife damage management can be essential to the survival and recovery of these same species. Woodroffe et al. (2005a), for example, cited lethal methods for the control of prairie dogs and the consequent near extinction of the black-footed ferret; the extinction of the Guadalupe caracara (*Polyborus lutosus*), the thylacine, and the Carolina parakeet (*Conuropis carolinensis*); and the lethal control of many apex predatory species globally. Impacts on charismatic megafauna, even when the species was not brought to extinction, have included greatly reduced ranges (i.e., range collapses), suppressed populations, and indirect effects on populations, ecosystems, and behavior (Woodroffe et al. 2005a).

Aside from the more specialized situation of conservation of species on uninhabited islands and the situation of common species such as commensal rodents, coyotes, and red-billed quelea, which seem resistant to almost anything humans do to them, the need is clear for wildlife damage management strategies that allow both the protection of people and their livelihoods and the conservation of damaging species and their environments. This is evidenced in part by increased public attention to the issue and by the rising amount of available resources. Sillero-Zubiri and Switzer (2004) and Witmer et al. (2005), for instance, provide summaries of research and experience on how to reduce wildlife damage by canids while at the same time conserving them. And Distefano (2005) used case studies to provide a global look at causes and management of damage by mostly charismatic megafauna.

We offer some thoughts. First, most charismatic megafauna live along with humans in multiple-use areas, not in wildlife-safe areas. For example, over 80% of African and Asian elephants range outside protected areas (Woodroffe et al. 2005b). Amur tigers (*Panther tigris*), wolves, bears, wolverine, and lynx also range near people. Conservation of these species, then, depends on their survival in areas where they cause damage and where the damage is managed. The goals and methods of damage management must support those of conservation if the species are to survive. Or, in the words of Woodroffe et al. (2005a, p. 389), "if these species cannot be conserved in multiple-use landscapes, there is a very real probability that they cannot be conserved at all." Second, we anticipate increasing geospatial overlap as human and recovering wildlife populations expand, with greater need for managing wildlife damage (see chapter 17).

Third, the methods already described for wildlife damage management, including lethal ones (Woodroffe et al. 2005b), remain the tools available. Osborn and Hill (2005) reviewed methods used to protect crops, from mostly elephants, primates, wild pigs (*Potamochoerus* spp.), and hippopotamus (*Hippopotamus amphibius*) in Africa. The methods included the following: lethal control by hunting, trapping, or poisoning; guarding and scaring; fences and barriers; repellents; and translocation, methods already described in chapters 10–12. The difference in these situations is not the methods. Rather, the methods need to be used by specialists and practitioners who understand the importance of conservation to managing wildlife damage, under public policies made by forward-looking and conservation-oriented legislators.

Fourth, each method has both limits to its suitability for conservation efforts as well as special aspects that could make it more applicable. For example, electric fencing is potentially effective for managing damage by a broad range of species, even those as large as elephants (Graham et al. 2010). Electric fences have limitations, however (Thouless and Sakwa 1995). For example, some primates learn to negotiate electrified fences. Material and construction costs can be prohibitively high, limiting use in developing countries to smallholdings, organizations and individuals that can obtain government subsidies, and national or international developmental or conservation organizations. Once constructed and working effectively, parts of electric fences may be stolen or left in disrepair. Theft of fencing can contribute to the demise of some species in an additional way—for instance, stolen fence wire can be used to construct snares and illegally poach wildlife (Hille et al. 2002).

Fifth, some historic methods should be revisited. Woodroffe et al. (2005b), for example, suggested reconsideration of shepherding and husbandry practices. Shepherding is mostly a lost art in much of North America and Europe, where livestock are left to roam unattended over wide areas mostly free of predators.

Predator populations are recovering, however—witness the reinvasion of coyotes in the high plateau of Texas, wolves on North America's mainland, and populations of lynx, bear, and wolverine throughout much of Europe. Guarding practices need to be reconsidered. Guard animals are presently being used in livestock protection with some success, and improvements in methods such as fladry are being investigated.

Sixth, wildlife damage management for conservation will increasingly depend on understanding the human dimensions. Economic, legal, and community developmental aspects are immediate concerns (Woodroffe et al. 2005b). For some wildlife, funds may be available, but uncertainty exists about its most effective use in furthering conservation goals. Compensation schemes, for instance, are now seen as having real-world limitations.

Compensation programs, performance payments, and consumptive and nonconsumptive uses of wildlife can offer economic incentives. However, these programs are subject to corruption and other unintended consequences, forms of "human-dimensional backlash" that can be difficult to predict and manage. For example, the relationship between a compensatory benefit and an action supporting a conservation activity may be unclear to a farmer. In a case described by Woodroffe et al. (2005b), the link may be unclear between not spearing a lion that just killed a cow and the availability of a health clinic, paid for indirectly by tourists who came to see the lion; the farmer may be able to both kill the lion and use the clinic. And, we add, should the person make the connection and not kill the lion (and use the clinic), the lion is now available not just to entertain another paying tourist but also to kill another cow.

Legal protection also has limits. Whether the species is jaguars in Brazil or prairie dogs in the United States, people resent being at the mercy of damaging wildlife without having legal recourse. Here, less stringent laws can engender public support, as with designating wolves as experimental rather than endangered outside of Yellowstone Park (Woodroffe et al. 2005b).

Examples

Linnell et al. (2005) described zoning practices as they relate to carnivores, especially large ones with extensive home ranges, but the basic concepts apply to other species targeted for conservation. As with zoning laws in cities where some areas are set aside for residents, other areas for commerce, and some for both, **zoning** in conservation sets aside some areas for human activity, others for wildlife, and some for both. Under the zoning concept, for example, a wildlife refuge would constitute one zone, with others designated for just human activity or for combined activities of wildlife and humans (such as buffer zones around refuges; Linnell et al. 2005).

With zoning, incompatible human or wildlife activities are prevented by excluding people or wildlife from some zones, and compatible activities allow both humans and wildlife in other zones. One benefit of zoning is that it extends areas suitable for wildlife beyond safe areas such as refuges into those inhabited by humans. This aspect is particularly important for carnivores, because their large home ranges often take them beyond the boundaries of refuges. Wildlife densities, mitigation of damage, and human activity can be managed in each zone to achieve its designated function. For instance, targeted wildlife might be removed by lethal control from a zone where damage is not tolerated and humans compensated should such damage occur anyway. In contrast, most human activity could be made illegal and laws strictly enforced in zones designated for conservation of the carnivores. Examples include fencing dingoes from sheep-farming areas of Australia; introduction of wolves into Yellowstone National Park, with their concomitant designation as experimental rather than endangered outside of the park boundaries; management of bears near the border of Croatia in Slovenia; and use of a region-specific quota system for cougar management in New Mexico, thereby enhancing their value as a game species while conserving them and keeping them from killing bighorn sheep (*Ovis canadensis*; Linnell et al. 2005).

CONSERVATION OF GAME SPECIES

Statement

Some goals of the wildlife damage practitioner overlap those of the game manager; other goals are uniquely those of the wildlife damage practitioner and may be at odds with those of the game manager.

Explanation

The concerted efforts of state fish and wildlife agencies throughout North America, in partnership with federal agencies and American hunters, anglers, private landowners, managers, and trappers—along with shifting land-use patterns and associated ecological seral changes—led to the unprecedented recovery of wildlife and fisheries species such as white-tailed

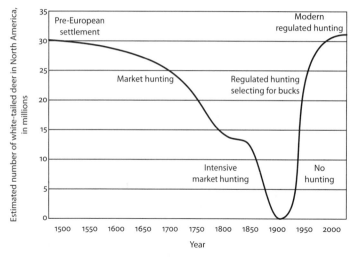

Figure 18.3 Human influence on white-tailed deer populations in North America since pre–European settlement. *Redrawn from VerCauteren 2003, with permission of Pope and Young Club.*

deer, wild turkey, geese, and many other game and nongame species throughout the continent (VerCauteren 2003; fig. 18.3). These continuing efforts now give us river otters and trout in our streams, turkeys and bears in our forests, and elk and wolves on our plains.

Both of us personally witnessed the remarkable restoration programs that initiated the recovery of many game species throughout North America, paid for mostly by conservation funds provided by sportsmen—meaning fees from licenses, stamps, and permits and excise taxes on firearms, ammunition, and archery equipment. The presence of game has added richness to the lives of both consumptive and nonconsumptive users. Both of us moved into our respective professions to be part of the conservation effort and focused on wildlife damage management because this was clearly an area of growing need. Indeed, the North American Model of Wildlife Conservation, based on the principle that wildlife belongs to all citizens and should be managed to be sustained forever, has served as a basis for future conservation of all wildlife in North America and serves as a model for conservation programs developing in other countries as well.

We view these successes with both a sense of pride in what humans can accomplish and a feeling of warmth about the ideal of what future conservation efforts might bring to the generations to come. But the story is also one of some adaptable species becoming overabundant, with concomitant damages to property, personal human injury and zoonotic diseases, and impacts on ecosystems and other species, all calling for a different management focus.

We do not feel a need to dwell here on the critical importance of wildlife damage management to the future conservation of game species in North America, in Europe, or across the globe. However, we feel it is important in this chapter to briefly emphasize, or re-emphasize, a few thoughts that tie conservation of game species together with wildlife damage management.

One is that the designation of a species as game versus nongame is a human determination, based strictly on human interest. In North America, we may eat venison burger provided by deer but generally not armadillo burger. The same ecological, biological, and sociological concepts and principles that underlie wildlife damage management of nongame species are therefore in play for the management of game species. In fact, the same biological, ecological, and sociological rules that led to successes in managing the growth of game populations are pertinent to managing their overabundance.

Second, not just the principles and concepts but also the goals of game conservationists and wildlife damage specialists sometimes overlap. The overlap can be absolute when wildlife damage is also a limiting factor in game production—for example, with predation and duck production in refuges or wolf predation and elk or moose production in parts of the northwestern United States, such as the Yellowstone ecosystem and some state game lands in Alaska. Here the methods and strategies may be the same, and even human-dimensional aspects, including the political and sociological ones, may be shared.

Third, within instances of overlap, there are differences. Some tools, assessment methods, management methods, and strategies are uniquely used by the wildlife damage practitioner, many for management of nongame species and for use in urban areas. Whereas the goals of the refuge manager and the wildlife damage specialist may overlap in reducing predators of migratory birds within a refuge where the manager is the stakeholder, the goal may be vastly different for the wildlife damage specialist working just outside the refuge; there the farmer is now also a stakeholder and crop losses may become paramount, conflicting with the need of the refuge manager to provide food for migratory birds.

Differences are also found in how the concepts and principles are applied. For instance, early in this work we pointed to the use of additive mortality rather than compensatory mortality when population size needs to be reduced rather than sustained. Another example is management of populations for optimal harvest rather than for expansion, as may be desired for conservation purposes, or for reduction, as may be needed for damage management. In that case, removal of predators might help maintain an optimal harvestable population size by reducing mortality without enhancing the breeding capacity of a population (Woodroffe et al. 2005b). Reducing breeding capacity may require entirely different approaches, such as addling or contraception.

Even when methodologies overlap, there may be differences in purpose. For example, both the game conservationist and the wildlife damage specialist may be using data from hunter harvests of snow geese. One, however, may be using the data to model future population growth and allowable takes to maintain the population as a game species. The other may be using the data to search for methods to reduce damage to habitat and nesting success of other ground-nesting species that may entail reducing the snow goose population significantly.

Example

Heather moorlands are dominated by the dwarf shrub heather and are largely restricted to the United Kingdom. This habitat is home to a mixture of species, but moorlands carry conservation significance primarily because they support golden eagles and peregrine falcons (*Falco peregrinus*) as well as golden plovers and curlews (*Numenius arquata*). Hen harriers (*Circus cyaneus*) are predators, as are red foxes, stoats, and carrion crows (*Corvus corone*), inhabiting the moorlands (Thirgood and Redpath 2005).

About half of the upland moorlands are privately owned and managed for red grouse (*Lagopus lagopus*) production. This management is generally acknowledged to provide ecological, social, and economic benefits to the area. Usually performed by professional gamekeepers, this management is complex and includes burning to provide a mix of heather and grasses that are optimal for grouse production and management of predators and parasites (Thirgood and Redpath 2008).

Hen harriers prey on adult red grouse and chicks in the summer, and hunters believe harriers limit grouse harvests, although limited scientific studies support this notion. Despite legal protection of the harrier in the United Kingdom since 1952, enforcement is limited and the risk of being caught for illegal kills is low. Harriers are illegally removed, limiting their populations, so there are essentially no successfully breeding harrier hens in game lands managed for grouse (Redpath and Thirgood 2009).

Conservationists want existing laws protecting harriers to be enforced, but logistics make this unlikely. Biologists have suggested that low levels of harriers could be supported on managed game lands without impact on the grouse. The low levels would have to be maintained by lethal management, however, and conservationists oppose such management. Hunters therefore refuse to recognize illegal killing of harriers as an unacceptable practice, and conservationists refuse to consider possible solutions involving lethal management that would limit harrier densities (Thirgood and Redpath 2008).

Several options have been considered to resolve the conflict. Compensating hunters for grouse losses to harriers would be prohibitively expensive, and there is uncertainty about who would pay. Habitat management, use of golden eagles to manage harriers, and zoning concepts have all been considered, but none have been fully explored. And according to Thirgood and Redpath (2008), both sides favor the status quo. Hunters continue to kill harriers without serious legal risk. Conservation groups demand increased enforcement, knowing it is unlikely, but they gain publicity from illegal kills.

Diversionary feeding and translocation of harriers appear to offer technical solutions. With diversionary feeding, carrion is provided to the harriers. The practice has had no effect on mortality of adult grouse, but it greatly reduces predation on chicks. Long-term effects on harrier populations are being studied. Strategic translocation of harriers, so they are distributed thinly over grouse-managed moorlands, would allow their presence, with little impact on the grouse. The expanded habitat could contribute to a total increase from 600 to about 1,600 female harriers. Excessive densities would need to be managed by shooting. Recent progress includes field trials on the merits of feeding and translocation and new forums for discussion and mitigation by the groups (Redpath and Thirgood 2009).

We see the issue, and its attempted resolution, as a microcosmic example of the role of wildlife damage management in the conservation of game and other wildlife.

Summary

- Wildlife damage specialists have joined forces with conservationists to conserve threatened or endangered species on islands and to conserve charismatic megafauna and other game and nongame wildlife worldwide.
- Human introductions of herbivores, omnivores, predators, and diseases have resulted in the extinction or endangerment of many unique and exotic island species.
- Through collaborative efforts of wildlife damage specialists and conservationists, there have been over 780 successful eradications of invasive vertebrates from islands, including feral cats, goats, pigs, and rodents.
- Recovery and restoration of native wildlife are underway on many of these islands.
- Unguided, misguided, or shortsighted wildlife damage management can push targeted species to extinction. Today's forward-looking wildlife damage management can be essential to the survival and recovery of endangered species and other desired wildlife species.
- Conservation-oriented wildlife damage management is needed because most charismatic megafauna live in multiple-use areas where management of wildlife damage is practiced, and the areas of overlap between wildlife and humans are increasing.
- Methods already described, including lethal ones, remain the tools available for managing damage and conserving wildlife. Each of these methods has limitations and adaptations particularly suited to conservation. Some historic management methods need to be rediscovered if predator populations are to continue recovering in parts of North America and Europe.

- As more conservation issues on mainlands or on islands populated by humans receive the attention of wildlife damage practitioners, successful management will increasingly depend on understanding and managing the human dimensions of the problem.
- Success in restoring game species has been a major accomplishment in North America but has brought with it problems of overabundance for some particularly adaptable species.
- Goals of wildlife damage practitioners may be in concert or at odds with those of game managers.

Review and Discussion Questions

1. For many years, eradication had been viewed as either unattainable or undesirable in wildlife damage management, and the term was largely dropped, except for management of diseases. How does eradication relate to extinction? How important is it as a conservation goal? Are there situations where eradications have been used to support specific conservation goals on a mainland? Cite two specific examples.

2. Is the use of Compound 1080 more socially acceptable when it involves the conservation of an endangered species such as the Canadian blue goose or when it is used for the protection of livestock such as sheep from coyotes? Is it more politically acceptable? Is it more ethical? Explain your responses, using literature references.

3. Is lethal control important for conservation of charismatic megafauna? If not, why have some conservationists advocated its continued use? If it is important, why have some conservationists opposed its use?

4. Cite two specific examples where the goals, strategies, and methods of wildlife damage management are in concert with those of (a) the conservationist and (b) the game manager.

5. Cite two specific examples where the goals, strategies, and methods of wildlife damage management differ from those of (a) the conservationist and (b) the game manager. How might these differences be resolved?

Glossary

Abatement programs—local, state, or federal public programs that provide consultation services, direct support from agency experts, or subsidies for technologies for abating wildlife damage

Abiotic—nonliving

Active ingredients—biologically or pharmacologically active ingredients of a pesticide

Acute rodenticides—rodenticides that are effective in a single dose, such as zinc phosphide

Adaptive management—management that proceeds without a full understanding of the problem and its solution but includes sufficient monitoring and experimental data gathering so that post hoc analyses lead to improvements in the strategy

Additive mortality—the notion that removal of population members by trapping, hunting, or poisoning might be timed so that it adds to other forms of mortality, thereby facilitating a decreased overall population size

Addling—"loss of development," the destruction of eggs by any physical or chemical means, by puncturing, freezing, or coating with vegetable oil

Agents—the causes of diseases, often biotic but sometimes abiotic, such as radiation

Animal rights—the belief that animals should have legal and other rights similar to those of humans

Animal welfare—the belief that wildlife should be managed in ways that minimize the pain and suffering of individual animals

Anthropomorphic—described or thought of as having human attributes.

Antibiotics—pesticides used on bacteria

Anticoagulants—compounds that reduce the ability of blood to coagulate, used variously as medicine and poison

Apex predator—predator at the top of its food chain, often also a keystone species

Attenuation—decrease in intensity of a behavioral response to a repeated stimulus

Attitudes—"learned tendencies to react favorably or unfavorably to a situation, individual, object or concept" (Allen et al. 2009, p. 5)

Attractant—in baits, the inactive ingredient that selectively attracts intended animals

Autecological—ecology of individual organisms or species

Avicide—pesticide used on birds

Baits—mixes of liquids, solids, or powders, parts of organisms, or whole organisms that attract targeted animals; the animals may be lured into traps, or the baits may deliver a management compound such as a pesticide, a repellent, or a contraceptive

Bait shyness—avoidance of a bait following consumption of a sublethal amount, based on flavor aversion learning

Basic reproduction number (R_0)—used to predict likelihood of a disease outbreak from a species jump, it is the ratio of secondary cases (i.e., number of individuals in the new population that are infected from primary cases) to primary cases (i.e., number of individuals infected by the jump)

Beliefs—judgments about what is true or false

Benefit/cost analysis (BCA)—an economic analysis that considers all benefits and costs, direct and indirect, present and future (discounted to present value), and wherein total costs are then subtracted from total benefits to derive a net benefit

Bequest value—the value of knowing that something, such as wildlife, will be present for future generations

Best Management Practices (BMPs)—informal standards for techniques or methods that have proven successful over time

Binder (sticker)—in baits, the inactive ingredient used to ensure that the active ingredient adheres to the carrier

Biocontrol (biological control)—using interactions among organisms to limit damage or presence of damaging species, e.g., using lure crops to attract pests away from crops of higher cash value

Biogeography—the study of the geographic distribution of organisms in time and space

Biological clocks—intrinsic abilities of organisms to time certain activities according to the rotation of the earth, the movement of the moon around the earth, the movement of the earth around the sun, or the juxtaposition of celestial bodies

Biological control (biocontrol)—using interactions among organisms to limit damage or presence of damaging species, e.g., using lure crops to attract pests away from crops of higher cash value

Biomes—as defined by Raven and Johnson (1992, p. 518), "climatically delineated assemblages of organisms that have a characteristic appearance and that are distributed over a wide land area"

Biopesticide—pesticide made from living organisms or their genetic parts

Biophilia—love for living things, an innate emotional connection between humans and other living things driven by a long co-evolutionary history

Biosphere—the largest biological community, composed of all life on the shell of the earth

Biotic—living

Bud caps—paper or plastic netting placed over the tops of transplants and held in place with one or more staples

Cable device (snare)—a steel cable with a loop on one end for capturing animals

Camera traps—wherein individuals are recognized from photographs taken at stations rather than actually captured, marked, released, and recaptured

Carnivore—organism that eats animals

Carrier—in baits, the matrix in which all of the ingredients are mixed

Carrying capacity (*K*)—the upper limit of population growth; i.e., the maximum number of population members that an area of habitat can support in a sustained manner

Character displacement—morphological or behavioral changes that occur over time within a

population in response to competition with other species

Charismatic megafauna—large animals with widespread popular appeal often used by activists to achieve conservation goals, such as elephants, tigers, lions, leopards, bears, hawks, and eagles

Chronic rodenticides—rodenticides, mostly anticoagulants, that require repeated feedings

Chronobiology (ecophysiology)—study of cues and rhythms

Circadian rhythms—biological clocks associated with the daily rotation of the earth

Circannual rhythms—biological clocks associated with the movement of the earth around the sun, including both seasonal and annual activities

Classical method of biocontrol—where the control agent is released one or more times so that the agent becomes established and gradually manages the pest

Climatic climax community—mature seres for a major biome or region, often described by the dominant plant species for that biome or sere

Climax community—stable final stage of ecological succession where energy used equals energy input; climatic climax communities are final stages for a particular climate, often described as biomes; edaphic climax communities are the most stable communities achieved in areas that are prone to periodic climatic disasters such as flooding or fire

Clumped distribution—the most common distribution of organisms, wherein the distances between neighbors are minimized; clumped distribution probably occurs because of gathering around limited resources, family or clan grouping, defensive gatherings such as herds, or limitations in movement

Coalitions—agreements between two or more parties to advance common goals and secure common interests

Commensals—uninvited guests that have "come to the table" to eat food intended for humans, including wildlife such as rats and mice

Common chemical sense—stimulation of the pain receptors of the trigeminal system

Community—all the groups of organisms living and interacting in an area at a given time

Community physiognomy—the physical structure of a community or ecosystem, including both its vertical and horizontal dimensions

Compensatory mortality—the notion that some members of a population can be removed by hunting or trapping without overall long-term impacts on population size because an equivalent number of these members would have died anyway from other causes, such as disease or starvation, had they not been hunted or trapped

Competition—when organisms of different species interact and both species are affected negatively by reduced environmental range or roles, either directly or indirectly, by reducing the supply of a needed resource

Competitor release effect—explosive increases in population size when a competitor is removed

Conscience—the sense of right and wrong with which humans are endowed

Conservation—the wise use of wildlife resources

Conspecifics—individuals of the same species

Convergent evolution—when species that occupy similar ecological niches evolve similar characteristics in response to similar selective pressures, thereby becoming more and more similar in appearance, behavior, and physiological attributes

Coprophagy—consuming feces

Corridors—thin runs of habitat similar to, and connecting, that in patches

Crepuscular—active at dawn or dusk

Critical minimum—the minimum population size, below which the population is no longer viable and dies

Crowding—population density becomes too high, affecting the health of population members when they become stressed

Culture—a set of shared values, beliefs, and behaviors that characterize an institution, organization, or group

Deception—a political method wherein one is given only part of the whole story or the story is deliberately misconstrued to achieve political "spin"

Decoy (lure) crops—crops provided as alternatives, to attract pests away from crops having higher cash values

Definitive host—final targeted organism of a disease or parasite

Density dependent—in populations, used to describe factors that affect the size of the population in a manner that is related to the number of members per unit habitat, e.g., stress due to crowding, which causes poorer care and survival of young, as the density increases with some species

Density independent—in populations, used to describe factors that affect the size of the population in the same manner regardless of the number of members per unit habitat, e.g., flooding of burrows in some rodent populations

Depletion (removal) trapping—when members of a population are intensely trapped and permanently removed; data gathered can sometimes be used to estimate population size

Detritus—the remains of a formerly living organism

Detritus food circuit—the flow of energy through a food web that begins with consumption of dead organisms and feces

Detrivore—an organism that eats the remains of a dead organism

Diethylstilbestrol—a reproductive inhibitor used to block egg production in some bird species

Digastric—having a two-stomach digestive system

Diseases—abnormal conditions that impair the functions of organisms

Distress—stress at levels that cause harm

Diurnal—active during the day

Diversionary feeding—providing food to a pest to protect something of greater value, e.g., feeding bears to keep them from damaging forest trees

Domestication—a special case of symbiosis in which humans artificially select characteristics of plants or animals for qualities that they desire, and the plants and animals benefit through resources and protection afforded by humans as well as managed harvesting with seed collection for the next season

Dominance—social "pecking order" wherein one or more members of a population claim more resources than are made available to others; or, the relative importance of one or more species in an ecosystem

Doministic value orientation—a value orientation in which well-being of humans is seen as a higher priority than that of wildlife

Ebola hemorrhagic fever—a viral disease with malaria-like symptoms, often fatal, first recognized in the Democratic Republic of the Congo in 1976, with bats or other wildlife as reservoirs

Ecological backlash—when the negative ecological consequences, often unanticipated, of a management action, such as the introduction of an exotic for biological control, outweigh the positive gains of the action

Ecological density—the number of individuals per unit habitat

Ecological equivalents—organisms or species that occupy the same or similar ecological niches in different geographic locations; the species may evolve convergently

Ecological footprint—the amount of land required to provide raw material required to sustain something, including processing of the wastes it produces

Ecological niche—the range of all factors within which an organism must live in a kind of multifunctional array, along with the functions, or roles, of the organism within that environment

Ecology—the study of plants or animals "at home," that is, the study of plants or animals interacting with both the living (biotic) and nonliving (abiotic) parts of their environment

Ecophysiology (chronobiology)—study of cues and rhythms

Ecosphere—the biosphere as a system with energy flowing through it and interacting with its components, both living and nonliving

Ecosystem—a biological community interacting with its abiotic components

Ecosystem services—services that ecosystems provide to humans, such as food or sewage cleanup

Ecotones—edges between communities or ecosystems, having some species characterizing each community as well as some unique species of their own

Edaphic climax community—seres maintained in a stable fashion for prolonged periods of time by recurrent factors such as floods or fires

Emerging zoonosis—"a pathogen that is newly recognized or newly evolved, or that has occurred previously but shows an increase in incidence or expansion in geographical, host, or vector range" (WHO undated b)

Emigration rate (*e*)—the number of members leaving a population per unit time, often expressed as the number of members leaving per 1,000 population members per unit time

Emotions—along with moods, constitute affect, a general class of feelings that humans experience; emotions relate to specific events and are short-lived and conscious, whereas moods are longer-lasting and in the background of consciousness

Endoparasites—parasites that live inside infected organisms

Energy subsidies—in ecology, energy sources that reduce the cost of self-maintenance of an ecosystem, thereby making more energy available for production

Enhance—to increase the intensity of a behavioral response or the useful lifespan of behaviorally based methods

Enhancer—in baits, the inactive ingredient that encourages consumption of sufficient amounts of the formulation

Enmeshing—tangling fish or other aquatic organisms to catch them; less selective than gill nets

Environmental resistance—collectively, all factors, such as starvation, disease, storms, and flooding, that exert pressure against growth of a population

Enzootic (sylvatic enzootic) cycle—movement of a disease between wildlife as hosts and vectors; the natural portion of its biological life cycle

Epidemic—an infectious disease spreading rapidly among large numbers of people

Ethics—sets of moral values defining good and evil, right and wrong, by which people behave

Eury—a prefix used to define a broad range or function; e.g., eurythermal means that an organism can live in a broad range of temperatures

Euryhalic—able to tolerate a broad range of salt concentrations

Euryphagic—able to eat a broad range of foods

Eurythermal—able to live in a broad range of temperatures

Eustress—low levels of stress that might be beneficial

Even distribution (evenness)—evenly spaced distribution of organisms, such as fields of corn, occasionally found in nature, e.g., creosote bushes; the distribution can be a consequence of competition for resources that are uniformly distributed over an area, such as reproductive territories for penguins

Exclusion—a method of reducing or eliminating damage by separating, physically or otherwise, wildlife from the items that could be damaged

Existence value—value ascribed to something such as wildlife just because it is there, e.g., the value of simply knowing that snow leopards exist in Asia

Exotic—a species appearing in a new habitat

Expected value (EV)—a form of economic analysis that can be used when there is uncertainty surrounding benefits from a management program

Expediency—in politics, taking the easiest route to re-election rather than the one in the public's best interest

Exponential growth curve—geometric growth of an unrestricted population, often resembling a J shape when graphed with numbers of population members as the ordinate (y) and time as the abscissa (x)

Feeding—eating

Ferae naturae—an ancient common law doctrine stating that a wild animal cannot be owned by anyone

Feral—domesticated wildlife that are released or escape back into the wild and that live to reproduce and become established as a population, often causing damage to humans, their property, or other wildlife

First law of thermodynamics—energy can be neither created nor destroyed but can be converted from one form to another, e.g., binding the photonic energy from the sun into a chemically bound form in adensine triphosphate

Fladry—flagging placed in lines on fences to direct the movement of wildlife; used, e.g., as an ancient method to direct wolves into traps or to keep them away from sheep

Flavor aversion learning—learned association of the flavor of a food with postingestional illness and subsequent avoidance of the flavor

Fleas—wingless insects of the order *Siphonoptera* with mouthparts specialized for sucking

Food chain concentration (biomagnification)—concentration of some compounds as they move up a food chain, often by an order of magnitude at each level of transfer, as occurs with, e.g., lipid-soluble pesticides

Food hoarding—storing food

Foraging—searching for food

Foregut fermentation (rumination)—elaborate and time-consuming but efficient digestion, using a two-stomach digestive system; used by artiodactyls, kangaroos, sloths, and colobus monkeys; microorganisms in the rumen detoxify poisons important for wildlife damage management

Forgone opportunity cost—costs accrued when an activity is not conducted in an area, e.g., crops or livestock are purposely not grown, because of concerns for wildlife damage management

Foundation species—a dominant primary producer in an ecosystem, such as kelp in an ecosystem with sea otter as the apex predator and keystone species

Fundamental ecological niche—the theoretically greatest potential habitat and ecological roles of an organism; a limiting factor keeps the organism or population within this niche

Fundamental equation for growth—$r = (b + i) - (d + e)$, where r is the population growth rate, b is natality, i is immigration, d is mortality, and e is emigration

Fungicide—pesticide used on fungi

Fyke nets—specialized fishing nets with wings and internal cones to direct fish into a collecting area

Gause's law—if two species occupy the same niche in the same habitat at the same time, one will out-compete the other until it disappears, or one or both species will evolve in divergent manners until their niches are sufficiently separated to reduce or eliminate the competition

Genetic pest management (GPM)—when genes of pest populations are altered to either decrease pest densities or render pests harmless

Genetic pollution—mixing genes, thereby causing hybridization or introgression and affecting survival of the species

Genetic resistance—survival by members of a population that are genetically predisposed to resist treatment by a management compound so that the whole population eventually resists treatment, e.g., genetic resistance of some populations of Norway rats to the anticoagulant rodenticide Warfarin

Geophagia—feeding drive to eat dirt, soil, earth, or starchy things

Giardia—disease caused by parasitic protozoa; sometimes carried by beaver

Gill nets—nets designed so fish or reptiles of a particular range of sizes are caught when they try to pass through the net

Global climate change—any change in climate over time due to either natural variability or human activity

Global warming—increase in the earth's air and oceanic temperatures

Good Laboratory Practices (GLP)—standardized practices that need to be followed in the conduct of studies used for registration of pesticides and drugs and that will stand up to both scientific and legal challenges

Granivorous—feeding on grains

Grazing circuits—the major path of energy flow in systems where sunlight is the direct source of energy, such as a grassland or savannah

Growth rate—change in numbers of population members over time, i.e., the sum of birth (or natality) rate b and immigration rate i minus the sum of mortality (death) rate d and emigration rate e; $r = (b + i) - (d + e)$

Gustation—sense of taste

Habitat—the place one goes to find an organism, i.e., the environment of the organism, including both living and nonliving components

Habitat fragmentation—when a habitat needed for wildlife is continually made smaller and smaller or is isolated from previously interconnected habitats

Herbicide—pesticide used on plants

Herbivory—eating plants

Heterogrooming—when one animal grooms another

Hindgut fermentation (monogastric)—the digestive system used by perissodactyls, elephants, lagomorphs, and rodents; this is a monogastric system involving one stomach, fast but inefficient; toxins ingested with the food are absorbed into the blood, and the liver detoxifies them

Holistic—used to describe comprehensive strategies that are the consequence of careful consideration of every aspect of a wildlife damage problem

Home range—the area within which an animal conducts its daily activities

Hormone—substance produced by an organ of an animal and released into the bloodstream that affects target organs, stimulating them to either produce more or less of another substance

Hosts—organisms that harbor pathogens

Humaneness—often measured as the extent of effort to minimize pain and suffering of animals, as with a management method such as trapping

Human tolerance—beliefs surrounding acceptable risks associated with wildlife

Immigration rate (i)—the number of new members recruited into a population, often expressed as the number of new members per 1,000 population members per unit time

Inactive ingredients—ingredients in a bait that do not have biological or pharmacological activity but serve other functions, such as binding, attracting, enhancing consumption, masking flavors, or preserving

Instantaneous growth rate—dN/dt, or the instantaneous change in the rate of growth dN for a population with size N at the instantaneous change in time dt.

Instinct—unlearned response to an environmental stimulus consisting of encoded stereotyped behaviors, often observed with insects, amphibians, reptiles, and birds

Integrated pest management (IPM)—management that is effective, involves a variety of methods, and minimizes the use of pesticides

Intermediate (secondary) hosts—vectors that carry parasitic pathogens for prolonged periods of time

Interspecific interactions—interactions between an organism and other species

Inundative method of biocontrol—when large numbers of agents are released periodically throughout a single season so that sheer numbers overwhelm the problem population

Invader species—the first organisms moving into an area devoid of life

Invasive species—nonnative species that adversely affect the habitat they invade

Iron triangle—a model formulated by political scientists that ascribes three foci of political power held together in an irrevocably bound system; In the United States, the foci are relevant positions and units of the executive branch, Congress, and lobbyists

Isocline—a curve through points at which the function's slope will always be the same, e.g., in Lotka-Volterra models, the sloped line where $dN_1/dt = 0$

Jesses—cables with loops, usually made of monofilament plastic, attached to cages to catch raptors

J-shaped growth curve. *See* **Exponential growth curve**

K/2—the inflection point of a sigmoidal population growth curve wherein overall natality is high and density-dependent control factors are still minimal; the point is sometimes viewed as the point of maximum sustainable yield for the population and targeted in game management plans

Kevlar cord—a para-aramid synthetic fiber string often used in grid systems to repel wildlife such as geese, ducks, or gulls from bodies of water

Keystone species—a species that has a much greater effect on its environment than would be indicated by its abundance; examples include jaguars in Central and South America and grizzly bears, beaver, and prairie dogs in North America

K-growth species—species that characterize mature seres of ecosystems, tend to have relatively fewer offspring, and invest parental care in the survival of the offspring

Landscape ecology—geographic description of communities and ecosystems, often based today on remote sensing, imagery, and layering provided by geographic information systems

Learning—complex behavioral responses that are modified by experience

Lice—wingless obligate parasites, of which there are over 3,000 species

Liebig's law of the minimum—the factor in the environment that occurs in the least quantity in relation to need will limit the growth of an organism or population

Life table, cohort or **horizontal**—a table that provides information on factors such as natality, mortality, or survivorship rates as functions of age and that is based on all members of a population that were born within the same period of time

Life table, static or **vertical**—a table that provides information on factors such as natality, mortality, or survivorship rates as functions of age and that is based on individuals of different ages, all sampled within one period of time

Lincoln-Petersen Index—a method for estimating population size based on capturing and marking population members and then recapturing a portion of the marked members

Lingering—said of a wildlife disease that has been known since ancient times

Livestock protection dogs (LPDs)—dogs used to protect livestock from predators

Lotka-Volterra models—a series of mathematical equations that describes the theoretical interactions of competing species, including those between predator and prey

Lunar rhythms—biological clocks tied to the rotation of the moon around the earth

Lure (decoy) crops—crops provided as alternatives to attract pests away from crops having higher cash values

Lyme disease—borreliosis, a disease caused by bacteria of the genus *Borrelia,* transmitted by ticks of the genus *Ixodes,* and causing flu-like symptoms in humans, eventually affecting the joints, heart, and nervous system if left untreated

Macroparasites—parasites that can be seen with the naked eye, e.g., protozoa and helminths

Malthusian growth curve. *See* **Exponential growth curve**

Management—getting something or someone to do what you want them to do

Marginal value theory—foraging models that are based on energy intake versus energy used to forage, such as where food available in one patch is diminished to the point where an animal must decide whether to keep foraging at greater and greater energetic costs or to invest the time and energy to locate a new patch with more food

Market value—what a good sells for in the local market

Masking agent—in baits, the inactive ingredient that overshadows or otherwise hides flavors in baits that might induce avoidance

Matrix—a broader ecosystem or habitat surrounding a smaller one

Maximum Sustainable Yield—$K/2$, the population size where maximum growth occurs

Median—the middle value of all values when ranked from lowest to highest, e.g., median damage of 5.7% in corn fields with damage ranging from 0.3% to 15.7%

Mesopredator release—dramatic increase in numbers of middle-level predators as a consequence of reduction in numbers of apex predators

Metapopulations—spatially separated populations of the same species that have some genetic interaction; often considered management units for species existing within fragmented landscapes

Microparasites—microscopic parasites that can complete life cycles within one organism, e.g., viruses, fungi, obligate intracellular bacteria (e.g., *Rickettsia*)

Midges—tiny "no-see-ums," or sand fleas

Minimum distance sampling—when a point is randomly selected within a field and a distance is measured between that point and the nearest damage, then the distance measured between that damage and the nearest second point of damage, then between the second and third points, and so on

Mites—along with ticks, small arthropods of the order *Acarina*

Molluscicide—pesticide used on snails

Monocultural agroecosystem—an agricultural ecosystem simplified to produce a single crop

Monogastric (hindgut) fermentation—the digestive system used by perissodactyls, elephants, lagomorphs, and rodents; this system involves one stomach and is fast but inefficient; toxins ingested with food are absorbed into the blood, and the liver detoxifies them

Mortality rate (*d*)—death rate, often expressed as the number of deaths per 1,000 population members per unit time

Multidimensional management—management that includes human dimensions as well as biological and ecological concepts

Mutualistic value orientation—a value orientation in which wildlife is seen in trusting relationships with humans and is perceived as having rights similar to those of humans

Mylar tape—a red and silver plastic tape used to repel birds

Myxomatosis—a disease of the European rabbit that causes skin lumps, puffiness, and sometimes conjunctivitis and blindness and reduces resistance to bacterial infections such as pneumonia; used successfully in the past to manage European rabbits in Australia

Natality rate (b)—birth rate, often expressed as number of births per 1,000 population members per unit time, or number of births per 1,000 reproductive-aged females per unit time

Nectivorous—feeding on nectar

Neophobia—fear of something novel

Net benefit—the benefits derived by considering all benefits and costs, direct and indirect, present and future (discounted to present value), and then subtracting total costs from total benefits

Net present values—present and future year expenditures versus benefits for the life of a management program, expressed as a present net, where future years are incorporated into present values by using a future discount rate (roughly annual inflation)

Nocturnal—active at night

Omnivore—eating a broad range of foods, both plant and animal

Operational programs—when professional experts conduct wildlife damage management directly on behalf of landowners or natural resources agencies

Optimal foraging theory—a model of foraging that theorizes that organisms maximize their energy intake per unit time; the model is usually based on three categories of factors—type of food or prey; currency used, such as time or energy; and constraints on time, such as risk of predation

Option value—the value of knowing that one will have the opportunity to use or enjoy something, such as wildlife, at a future time of one's choice

Oral rabies vaccinations—vaccinations that are delivered to targeted wildlife in massive baiting campaigns and that have successfully reduced the incidence of rabies in wildlife species, first in Europe and subsequently in North America

Outbreaks—when infectious diseases occur at higher frequencies than expected

Overabundant—used to describe populations that threaten human life or livelihood, affect densities of favored wildlife, are too numerous for their own good, or affect ecosystems and cause their dysfunction

Overshadowing—when two or more conditioned stimuli are presented along with a single unconditional stimulus, an organism associates the unconditional stimulus only with the most salient conditioned stimulus

Pandemic—when the spread of disease is larger than epidemic and potentially global

Parasitism—when one organism negatively affects another organism of a different species wherein the parasite, usually the smaller of the two species, benefits by consuming the food, nutrient, or a physical part of some of the other species, the host

Patches—areas of relatively homogeneous habitat surrounded by a broader but different habitat

Pathogens—biological agents that cause illness or disease

Peer review—critical evaluation of a piece of work by well-qualified individuals; information is published in most scientific journals only after passing through peer review

Pest—the individual, population, or species causing damage

Pesticide—any mix of substances (or devices) that affects the behavior of pest plants or animals

Phagic—eating

Pheromones—substances, usually highly volatile ones, secreted into the environment by an individual, that have some behavioral or physiological effects on other individuals of the same species

Phoresis—seed or young dispersal with the help of environmental factors such as wind or other organisms

Photoperiod—length of daylight time; changes in photoperiod are detected by many organisms and used as a cue to time certain activities

Photosynthesis—the process of converting carbon dioxide into organic compounds, often sugars, using energy from sunlight

Physiognomy—landscape or ecosystem structure

Piscicide—pesticide used on fish

Plague—a disease caused by the bacterium *Yersinia pestis* and manifested as bubonic, septicemic, or pneumonic

Population density—the number of population members per unit area of habitat

Population growth rate (r)—the change in numbers of population members as a function of time,

i.e., the sum of birth (or natality) rate b and immigration rate i minus the sum of mortality (death) rate d and emigration rate e;
$$r = (b + i) - (d + e)$$

Population index—estimates of relative changes in population size, often using changes in activity such as movement to and from a station

Populations—groups of actually or potentially interbreeding organisms living in the same place at the same time

Population size (N)—the number of population members at a given moment in time

Prebaiting—providing a bait or its flavor without the active ingredient so that the animal becomes familiar with the flavor and neophobia, and sometimes primary flavor aversion, is extinguished

Precautionary principle—states that no action should be taken until a problem and all consequences of proposed actions are fully understood

Predacide—pesticide used on predators

Predation—attacking and eating another animal, wherein the predator, usually the larger of the two species, benefits by killing and consuming some of the other species, the prey

Preservation—the wildlife management philosophy underlain by the notion that humans should not interfere with nature; to let wildlife take its course

Preservative—in baits, the inactive ingredient that prevents degradation of the bait and protects it from the weather

Primary flavor aversion—unlearned avoidance of certain flavors, often bitter

Primary scientific literature—original reports of studies that use the scientific method, related observational or experimental designs, or both, reviewed by peers prior to publication

Primary succession—succession that occurs when organisms enter an area devoid of life, such as after a volcanic eruption or a prolonged major flood

Prions—abnormally folded protein bodies, possibly the cause of bovine spongiform encephalopathy, or "mad cow disease" in cattle, chronic wasting disease in cervids, and scrapie in sheep

Propagule pressure—the likelihood of a successful invasion, based partly on the number of individuals emigrating and partly on the frequency of invasions

Proprietor—someone acting as an owner

Public policy—"whatever governments choose to do or not to do" (Dye 2008, p. 1)

Pyrotechnics—use of explosives; this might involve a broad range of activities, from use of "whistlers" and "bangers" (i.e., fireworks) to scare wildlife to plastic or other explosives to blow up beaver dams

Quadrat sampling—where a grid of specific dimensions, e.g., a 10-by-10 grid of cornstalks, is used to sample damage, starting with a randomly determined coordinate and continuing until a pre-established number of quadrats are sampled

Rabbit hemorrhagic disease virus (RHDV)—causes hemorrhagic disease, a highly contagious disease specific to both domestic and wild rabbits; used to manage rabbits in Australia and New Zealand

Rabies—a viral neuroinvasive disease that causes acute encephalitis in warm-blooded animals

Random distribution—the least common type of distribution of organisms, it shows no clear pattern, and the position of one organism is independent of the position of another; e.g., distribution of tropical fig trees that are pollinated by winds

Range of tolerance—range within which any organism must live for each of the physical components of its environment; organisms cannot survive in environments where any of these ranges are exceeded

Range of tolerance (Shelford's law of tolerance)—the factor in the environment that occurs in the most excess in relation to need will limit the growth of the organism or its population

Realized ecological niche—the actual niche occupied by an organism or population; it is less than the fundamental ecological niche due to interactions between the organism or population and other biota

Reasoning—behavioral responses based on rational thought and strategy

Recovery—in conservation, allowing ecosystem changes to occur by ecological succession over time without further human intervention

Recruitment—the combination of birth and immigration rates

Reflex—the response of an organism or part of an organism to an environmental stimulus; more sophisticated than a taxis and modifiable by learning

Removal (depletion) trapping—when members of a population are intensely trapped and permanently removed; data gathered can sometimes be used to estimate population size

Reptilicide—pesticide used on reptiles

Reservoir (nidus)—long-term host for a disease

Restoration—in conservation, when humans attempt to speed up recovery of an ecosystem by assisting natural ecological processes

R-growth species—species characterizing early seral stages of ecosystems; they tend to produce large numbers of offspring but make little parental investment in their survival

Richness—number of species in relation to the total number of individuals in an area

Rickettsia—obligate intracellular bacterial microparasites

Riders—in politics, a largely unsupported bill attached to a popular one

Rinderpest—cattle plague, a major infectious viral disease of cattle, domestic buffalo, and some other ungulates, declared eradicated by the World Organisation for Animal Health in 2011

Risk—the probability of occurrence of damage and severity of the consequences

Rodenticide—pesticide used on rodents and other small mammals

Rumination (foregut fermentation)—elaborate and time-consuming but efficient digestion, using a two-stomach system; used by artiodactyls, kangaroos, sloths and colobus monkeys; microorganisms in the rumen detoxify poisons important for wildlife damage management

Rut—fall mating season for ungulates

Salient—in flavor aversion learning, associated with an illness

Salt drive—urge to eat salt

Sand flies—blood-sucking Dipterans

Sanguinivorous—eating blood

Scientific method—a method that attempts to minimize opportunities for personal or other biases that can lead to misinterpretation of results and to maximize the odds that results observed are actually due to the factors being controlled and studied; usually a series of steps followed sequentially (in theory, often not in practice): gather background information and current literature; form a hypothesis and its negative form, the "null hypothesis"; develop an experimental or observational design to test the hypotheses; run the experiment, gathering data; analyze the data following appropriate statistical methods; interpret the results; communicate the findings to the rest of the world

Secondary (intermediate) hosts—vectors that carry parasitic pathogens for prolonged periods of time

Secondary literature—accumulations and summaries of primary scientific literature, including also other experiential and sometimes anecdotal information, that often leads to recommendations for specific management actions; in wildlife damage management, this is often where the art as well as the science can be found

Secondary succession—succession that occurs when an ecosystem is pushed back to an earlier successional stage by a force such as fire, herbicides, or intensive grazing

Second law of thermodynamics—some energy is lost as heat whenever energy is converted from one form to another

Seines—nets to capture fish and other aquatic organisms

Semiochemical—odor of another species, often a predator

Sere / seral stage—each successional community within an ecosystem; over time, seral stages replace each other in an orderly and predictable manner; i.e., the presence of one community alters the ecosystem in such a manner that it is replaced by the next seral stage until a mature, climax community is achieved

Shelford's law of tolerance (range of tolerance)—the factor in the environment that occurs in the most excess in relation to need will limit the growth of the organism or its population

Sigmoidal growth curve—growth of a population under conditions where resistance to growth increases with increasing population density; the shape of the curve often resembles an "S" when graphed with numbers of population members as the ordinate (y) and time as the abscissa (x)

Snare (cable device)—a steel cable with a loop on one end for capturing animals

Social facilitation—when the presence of one animal improves the performance of another

Social values—conceptions of what is good and bad, desirable and undesirable

Sovereign—used to describe a state or federal agency representing the common interests of citizens

Species jump—when a pathogen overcomes a species barrier and infects a new type of organism

Specific hunger—eating preference for a particular item, such as soil or salt

S-shaped growth curve. *See* **Sigmoidal growth curve**

Stakeholders—persons who are affected by or can affect a problem or its management

Standard operating procedures (SOPs)—written, formalized directions to be followed during performance of tasks

Steno—able to live in only a narrow range or function; e.g., stenothermal means that an organism can survive in only a narrow range of temperatures

Stenohalic—able to survive in only a narrow range of salt concentrations

Stenophagic—having a diet restricted to few foods or food types

Stenothermal—able to survive in only narrow ranges of temperatures

Stereotyped—used to describe a carefully orchestrated and rigidly set sequence of behavioral and physiological events

Strategy—a broad plan of action to achieve a goal

Succession—orderly replacement of a community of organisms with another, starting with invader organisms if the area has not had prior living communities, and ending with a stable climax community where energy used by the system balances energy input; each distinctive community is called a seral stage, or sere

Sylvatic (enzootic) cycle—movement of a disease between wildlife as hosts and vectors; the natural portion of its biological life cycle

Sympatric—used to describe species that occupy the same or similar ecological niches in the same geographic location at the same time

Synecological—a holistic approach to ecology, involving groups of organisms or landscapes

Taxis—tropism, but in a decidedly directed manner

Territorial behavior—when one member, usually a male, establishes a geographical area and defends it against other conspecifics

Theoretical ecological niche—the theoretically greatest potential habitat and ecological roles of an organism; a limiting factor keeps the organism or population within this niche

Tolerance. *See* **Human tolerance**

Toxoplasmosis—a parasitic infection caused by the protozoan *Toxoplasma gondii*; felids, including domestic and wild cats, are definitive hosts, while many warm-blooded animals serve as intermediate hosts

Trammel nets—nets designed for species not easily caught in gill nets, like flatfish or sturgeon

Transect—a path along which one counts and records occurrences, such as bird sightings or calls

Trawl net—a long bag or sock-shaped net that is pulled through water

Trophic level—refers to an organism's position in the food chain

Tropism—the general attraction to or avoidance of some environmental stimulus, such as temperature or light

Tubing—tubes made of paper or plastic that are used to protect seedlings from herbivores such as gophers, mice, rabbits, mountain beavers, beavers, and deer

Uniform distribution. *See* **Even distribution**

Value orientations—values that characterize cultures; a form of ideology

Vectors—carriers of diseases

Vertebrate pest control—the aspect of wildlife damage management that deals with vertebrates, usually mostly birds and mammals, but sometimes including reptiles, amphibians, and fishes

Warfarin—a first-generation anticoagulant rodenticide named after the Wisconsin Alumni Research Foundation

West Nile virus—a virus of the family Flaviviridae that infects humans, pets, and wildlife; often transmitted with the bite of a mosquito

Whirling disease—a protozoan disease that infects salmon and trout species

Wilderness discount—the portion of damage that is prevented or managed by a natural ecosystem, e.g., wilderness offering alternative prey for predators, thereby reducing livestock losses

Wildlife—any nondomesticated plant or animal

Wildlife acceptance capacity—the mixture of tolerances for a wildlife damage problem that can be found among its stakeholders

Wildlife conservation—a philosophy that advocates the wise use and stewardship of wildlife resources

Wildlife damage—a real or perceived threat to humans, their property, or something that they value (e.g., other wildlife)

Wildlife damage management—the science and art of diminishing the negative aspects of wildlife while maintaining or enhancing their positive aspects

Wildlife diseases—abnormal conditions that impair bodily functions, are associated with specific symptoms and signs, and are carried by nondomesticated organisms

Wildlife preservation—the philosophy that humans should not interfere with nature, i.e., let wildlife take its course

Willingness to pay—estimate of the value of something based on how much a person is willing to pay for the activity, such as a hunting, fishing, or hiking trip

Zeitgebers—German for "time giver"; internal cues and rhythms adjusted by external cues

Zoning—in conservation, separating areas according to use by humans, wildlife, or both

Zoonoses—diseases that are transmitted between wildlife and humans or their livestock or pets

References

Abbott, K. 2005. Supercolonies of the invasive yellow crazy ant, *Anoplolepis gracilipes*, on an oceanic island: forager activity patterns, density and biomass. Insect Soc. 52(3):266–273.

Advice to the Minister for the Environment and Heritage from the Threatened Species Scientific Committee (TSSC) on amendments to the List of Key Threatening Processes under the Environment Protection and Biodiversity Conservation Act 1999 (EPBC Act). Undated. www.environment.gov.au/biodiversity /threatened/ktp/christmas-island-crazy-ants.html. Accessed 1/13/11. 27 pp.

AFPMB. Undated. Literature retrieval system. www.afpmb .org/content/welcome-literature-retrieval-system. Accessed 10/13/2011.

African Elephant Specialist Group. 2002. Review of compensation schemes for agricultural and other damage caused by elephants. Tech. brief, Human-Elephant Conflict Work. Group, Afr. Elephant Spec. Group. IUCN, Gland, Switz.

Agaba, P., T. Asiimwe, T. Moorehouse, et al. 2009. Biological control of water hyacinth in the Kagera River Headwaters of Rwanda: a review through 2001. www.readbag.com /cleanlake-images-rwanda-bio-paper. Accessed 1/7/2013. 9 pp.

Aguirre, W. and S. Poss. 2000. *Cyprinus carpio* Linnaeus 1758. Non-indigenous species in the Gulf of Mexico ecosystem. Gulf States Marine Fisheries Commission, Ocean Springs, Miss.

Ahmed, M. and L. Fiedler. 2002. A comparison of four rodent control methods in Philippine experimental rice fields. Int. Biodeterior. Biodegrad. 49:125–132.

Aiken & Jacobsen. Undated. Family wins wrongful death lawsuit after son is killed by a bear. www.personalinju ryblogsacramento.com/2011/05/family-wins-wrongful -death-lawsuit-after-son-is-killed-by-a-bear.shtml. Accessed 7/10/11.

Akenson, J., M. Nowak, M. Henjum, et al. 2003. Characteristics of mountain lion bed, cache, and kill sites in northeastern Oregon. Pp. 111–118 in S. Becker, D. Bjornlie, G. Linzey, et al., eds. Proc. 7th Mt. Lion Workshop. Wyo. Game Fish Dep., Lander, Wyo.

Alderman, D. 1996. Geographical spread of bacterial and fungal diseases of crustaceans. Rev. Sci. Tech. Off. Int. Epiz. 15 (2):603–632.

Ali, S., N. Kamal, A. Hasanuzzaman, et al. 2008. Scientific assessment report on bamboo flowering, rodent outbreaks and food security: rodent ecology, pest management, and socio-economic impact in the Chittagong Hill Tracts, Bangladesh. United Nations Development Programme, Bangladesh. 50 pp.

Allan, J. 2002. The costs of bird strikes and bird strike prevention. Pp. 147–153 in Clark, L., J. Hone, J. Shivik, et al., eds. Human conflicts with wildlife: economic considerations. Proc. Third NWRC Spec. Symp., Natl. Wildl. Res. Cent. Fort Collins, Colo.

Allan, J., J. Bell, and V. Jackson. 1999. An assessment of the world-wide risk to aircraft from large flocking birds. Bird Strike Committee Proceedings, 1999 Bird Strike Committee—USA/Canada, First Joint Annual Meeting, Vancouver, B.C.

Allen, L., R. Engeman, and H. Krupa. 1996. Evaluation of three relative abundance indices for assessing dingo populations. Wildl. Res. 23:197–206.

Allen, S., D. Wickwar, F. Clark, et al. 2009. Values, beliefs, and attitudes—technical guide for forest land and resource management, planning, and decision making. USDA For. Serv., Pac. Northwest Res. Station, Gen. Tech. Rep. PNW-GTR-788. 113 pp.

Allin, C. Undated. Mute swan: an invasive species and its management in Rhode Island. www.dem.ri.gov/programs /bnatres/fishwild/pdf/muteswan.pdf. Accessed 4/10/11.

Alroy, J. 2001. A multispecies overkill simulation of the end-Pleistocene megafaunal mass extinction. Science 292:1893–1896.

Amano, T., K. Ushiyama, G. Fujita, et al. 2004. Alleviating grazing damage by white-fronted geese: an optimal foraging approach. J. Appl. Ecol. 41:675–688.

Amstrup, S., G. York, T. McDonald, et al. 2004. Detecting denning polar bear with forward-looking infrared (FLIR) imagery. BioScience 54:337–344.

Andelt, W., K. Burnham, and D. Baker. 1994. Effectiveness of capsaicin and bitrex repellents for deterring browsing by captive mule deer. J. Wildl. Manage. 58:330–334.

Anderson, G., E. Delfosse, N. Spencer, et al. 2000. Biological control of leafy spurge: an emerging success story. Pp. 15–25 in N. Spencer, ed. Proc. 10th Int. Biol. Control Weeds Symp., Bozeman, Mont.

Animal Legal and Historical Center. Undated. New York Supreme Court, *Barrett v. State*. www.animallaw.info/cases /causny220ny423.htm. Accessed 7/16/11.

Anonymous. 2002. The oral vaccination of foxes against rabies. Eur. Comm. Health Consum. Prot. Dir.-Gen., Rep. Sci. Com. Anim. Health Anim. Welfare. 55 pp.

Anonymous. 2004. Best management practices for resolving human-beaver conflicts in Vermont. Vt. Fish Wildl. Dep., Vt. Dep. Environ. Conserv. 28 pp.

Anonymous. 2005. Living beyond our means: natural assets and human well-being. Statement Board, Millennium Ecosyst. Assess. 28 pp. www.millenniumassessment.org /documents/document.429.aspx.pdf. Accessed 5/30/11.

Anonymous. 2006. Draft adaptive management system, strategic planning framework. S. D. Wildl. Div. Version 06-2 http://gfp.sd.gov/wildlife/docs/strategic-plan -framework.pdf. Accessed 7/30/11.

Anonymous. 2010. Unanticipated adverse events. Miss. State Univ. IACOC Stand. Oper. Proc. IACUC-RVW-026. 2 pp.

Anonymous. Undated. Mouser Version 1.0. www.cbit.uq.edu .au/RDActivities/OtherActivities/ProjectArchive /MouserVersion10.aspx. Accessed 5/27/11.

Anthony, R., J. Evans, and G. Lindsey. 1986. Strychnine-salt blocks for controlling porcupines in pine forests: Efficacy and hazards. Proc. Vertebr. Pest Conf. 12:191–195.

APHIS. 2003. Snake Repellents. APHIS Wildl. Serv. Tech. Note. www.aphis.usda.gov/lpa/pubs/tn_wmsnakere pellent.pdf. Accessed 4/23/11.

APHIS. 2008. Sodium lauryl sulfate: European starling and blackbird wetting agent. www.aphis.usda.gov/wildlife _damage/nwrc/registration/content/WS_tech_note_so dium.pdf. Accessed 4/26/11.

Aramini, J., C. Stephen, and J. Dubey. 1998. *Toxoplasma gondii* in Vancouver Island cougars (*Felis concolor vancouverensis*): serology and oocyst shedding. J. Parasitol. 84:438–440.

Arjo, W. 2003. Mountain beaver: the little rodent with a large appetite. West. For. July/August, 10–11.

Arjo, W. 2006. Efficacy of chlorophacinone for mountain beaver control. Unpubl. Rep. QA 1135. Natl. Wildl. Res. Cent., Fort Collins, Colo.

Arlinghaus, R. and T. Mehner. 2003. Socio-economic characterisation of specialised common carp (*Cyprinus carpio* L.) anglers in Germany, and implications for inland fisheries management and eutrophication control. Fish. Res. 61:19–33.

Association of Fish and Wildlife Agencies. Undated. Best management practices for trapping in the United States: introduction. 13 pp. www.fishwildlife.org/files /Introduction_BMPs.pdf. Accessed 3/28/11.

Atkinson, C., K. Woods, R. Dusek, et al. 1995. Wildlife disease and conservation in Hawaii: pathogenicity of avian malaria (*Plasmodium relictum*) in experimentally infected liwi (*Vestiaria coccinea*). Parasitol 111, Suppl.: S59–S69.

Atkinson, I. 1988. Presidential address: opportunities for ecological restoration. N. Z. J. Ecol. 11:1–12.

Atzert, S. 1971. A review of sodium monofluoroacetate (Compound 1080), its properties, toxicology, and use in predator and rodent control. U. S. Dep. Inter. Fish Wildl. Serv. Spec. Sci. Rep.—Wildl. No 146.

Avery, M. 1992. Evaluation of methyl anthranilate as a bird repellent in fruit crops. Proc. Vertebr. Pest Conf. 15:130–133.

Avery, M. 1996. Food avoidance by adult house finches, *Carpodocus mexicanus*, affects seed preference of offspring. Anim. Behav. 48:1279–1283.

Avery, M., J. Humphrey, and D. Decker. 1997. Feeding deterrence of anthraquinone, anthracene, and anthrone to rice-eating birds. J. Wildl. Manage. 61:1359–1365.

Avery, M., J. Lindsay, J. Newman, et al. 2006. Reducing monk parakeet impacts to electric utility facilities in south Florida. Pp. 125–136 in C. Feare and D. Cowan, eds. Advances in vertebrate pest management. Vol. IV. Filander Verlag. Furth, Fed. Repub. Ger.

Avery, M., M. Pavelka, D. Bergman, et al. 1995. Aversive conditioning to reduce raven predation on California least tern eggs. Colon. Waterbirds 18:131–138.

Ayotunde, E., B. Offem, and A. Bekeh. 2011. Toxicity of *Carica papaya* Linn: haematological and piscicidal effect on adult catfish (*Clarias gariepinus*). J. Fish. Aquat. Sci. 6(3):291–308.

Babbie, E. 2001. The practice of social research. 9th ed. Wadsworth Thomson Learning, Belmont, Calif. 498 pp. plus append.

Bader, H. and G. Finstad. 2001. Conflicts between livestock and wildlife: an analysis of legal liabilities arising from reindeer and caribou competition on the Seward Peninsula of western Alaska. Environ. Law 31:549–579.

Bailey, R. 1939. Kudzu for erosion control in the Southeast. U.S. Dep. Agric. 32 pp.

Baker, H. 1965. Characteristics and modes of origin of weeds. Pp. 147–169 in H. Baker and G. Stebbins, eds. The genetics of colonizing species. Academic Press, New York. 588 pp.

Balciunas, J. 2000. Code of best practices for classical biological control of weeds. P. 435 in N. Spencer, ed. Proc. X. Int. Symp. Biol. Control Weeds. 4–14 July 1999, Mont. State Univ., Bozeman, Mont.

Bangs, E. and J. Shivik. 2001. Managing wolf conflict with livestock in the northwestern United States. Carnivore Damage Prev. News 3:2–5.

Bank, F., C. Irwin, G. Evink, et al. 2002. Wildlife connectivity across European highways. U. S. Dep. Transp., Fed. Highw. Admin. HPIP/08-02(7M)EW Publ. No. FHWA-PL-02-011. 45 pp.

Barnacle, B. 1997. Protect pine tree seedlings from deer browsing with paper bud caps. Silviculture Field Tip No. 6. Minn. Dep. Nat. Resour., Div. For., 1 p.

Barnes, E. 2005. Diseases and human evolution. Univ. N. M. Press, Albuquerque. 484 pp.

Barras, S. 2004. Double-crested cormorants in Alabama. Ala. Wildl. Fed. 68:28–29.

Barras, S., R. Dolbeer, R. Chipman, et al. 2000. Bird and small mammal use of mowed and unmowed vegetation at John

F. Kennedy International Airport, New York, 1998 to 1999. Proc. Vertebr. Pest Conf. 19:31–36.

Barras, S. and S. Wright. 2002. Civil aircraft collisions with birds and other wildlife in Ohio, 1990–1999. Ohio J. Sci. 102:2–7.

Barras, S., S. Wright, and T. Seamans. 2003. Blackbird and starling strikes to civil aircraft, 1990–2001. Pp. 91–96 in G. Linz, ed. Management of North American blackbirds. Natl. Wildl. Res. Cent., Fort Collins, Colo.

Bartelt, R., J. Kyhl, A. Ambourn, et al. 2004. Male-produced aggregation pheromone of *Carpophilus sayi*, a nitidulid vector of oak wilt disease, and pheromonal comparison with *Carpophilus lugubris*. Agric. For. Entomol. 6:39–46.

Baruch-Mordo, S., S. Breck, K. Wilson, et al. 2008. Spatio-temporal distribution of black bear–human conflicts in Colorado, USA. J. Wildl. Manage. 72(8):1853–1862.

Bashir, M. and F. Abu-Zidan. 2006. Motor vehicle collisions with large animals. Saudi Med. J. 27:1116–1120.

Bataille, A., A. Cunningham, V. Cedeño, et al. 2009. Evidence for regular ongoing introductions of mosquito disease vectors into the Galápagos Islands. Proc. R. Soc. B 276(1674):3769–3775.

Baxter, A. 2007. Laser dispersal of gulls from reservoirs near airports. Ninth Annu. Meet., Bird Strike Comm. USA/Canada, Kingston, Ont. 10 pp.

Beasley, J. and O. Rhodes. 2008. Relationship between raccoon abundance and crop damage. Human-Wildl. Conflicts 2:248–259.

Beckmann, J., A. Clevenger, and M. Huijser. 2010. Safe passages: highways, wildlife, and habitat connectivity. Island Press, Washington, D.C. 424 pp.

Behnke, R. 1992. Native trout of western North America. Am. Fish. Soc. Monogr. 6. Bethesda, Mary. 275 pp.

Beisner, B., D. Haydon, and K. Cuddington. 2003. Alternative stable states in ecology. Front. Ecol. Environ. 1:376–382.

Belant, J. 1997. Gulls in urban environments: landscape-level management to reduce conflict. Landscape Urban Plan. 38:245–258.

Belant, J. and S. Ickes. 1997. Mylar flags as gull deterrents. Proc. Great Plains Wildl. Damage Control Workshop 13:73–80.

Belay, E., R. Maddox, E. Williams, et al. 2004. Chronic wasting disease and potential transmission to humans. Emerg Infect Dis June. www.cdc.gov/ncidod/EID/vol10no6/03-1082.htm. Accessed 10/21/2011.

Belshe, R. 2009. Implications for the emergence of a novel H1 influenza virus. New Engl. J. Med. 360:2667–2668.

Bennett, D. 1995. A little book of monitor lizards. Viper Press, Aberdeen, U.K.

Bergstedt, R. and M. Twohey. 2007. Research to support sterile-male-release and genetic alterations techniques for sea lamprey control. J. Great Lakes Res. 33 (Spec. Issue 2):48–69.

Bergstrom, D., A. Lucieer, K. Kiefer, et al. 2009. Indirect effects of invasive species removal devastate World Heritage island. J. Appl. Ecol. 46:73–81.

Beringer, J., L. Hansen, J. Demand, et al. 2002. Efficacy of translocation to control urban deer in Missouri: costs, efficiency, and outcome. Wildl. Soc. Bull. 30 (3): 767–774.

Beringer, J., L. Hansen, W. Wilding, et al. 1996. Factors affecting capture myopathy in white-tailed deer. J. Wildl. Manage. 60:373–380.

Berryman, J. 1992. The complexities of implementing wildlife damage management. Trans. North Am. Wildl. Nat. Resour. Conf. 57:41–50.

Berryman, J. 1994. Blurred images: and the future of wildlife damage management. Proc. Vertebr. Pest Conf. 16:2–4.

Bhat, M., R. Huffaker, and S. Lenhart. 1993. Controlling forest damage by dispersive beaver populations: centralized optimal management strategy. Ecol. Appl. 3:518–530.

Blackwell, B. and G. Bernhardt. 2004. Efficacy of aircraft landing lights in stimulating avoidance behavior in birds. J. Wildl. Manage. 68: 725–32.

Blackwell, B., T. Seamans, and B. Washburn. 2006. Use of infrared technology in wildlife surveys. Proc. Vertebr. Pest Conf. 22:467–472.

Blackwell, B. and S. Wright. 2006. Collisions of red-tailed hawks (*Buteo jamaicensis*), turkey vultures (*Cathertes aura*), and black vultures (*Coragyps atratus*) with aircraft: implications for bird strike reduction. J. Raptor Res. 40:76–80.

Bland, E. 2008. Dandelion could replace rare sources. Discovery News, Discovery Channel. http://dsc.discovery.com/news/2008/08/05/dandelion-rubber.html. Accessed 10/4/2011.

Blaustein, A. and J. Kiesecker. 2002. Complexity in conservation: lessons from the global decline of amphibian populations. Ecol. Lett. 5:597–608.

Blejwas, K., C. Williams, G. Shin, et al. 2006. Salivary DNA evidence convicts breeding male coyotes of killing sheep. J. Wildl. Manage. 70:1087–1093.

Bodenchuk, M., J. Mason, and W. Pitt. 2002. Economics of predation management in relation to agriculture, wildlife, human health and safety. Pp. 80–90 in L. Clark, J. Hone, J. Shivik, et al., eds. Human conflicts with wildlife: economic considerations. Proc. Third NWRC Spec. Symp., Natl. Wildl. Res. Cent., Fort Collins, Colo.

Bolen, E. and W. Robinson. 2002. Wildlife ecology and management. Benjamin Cummings, London. 634 pp.

Booth, T. 1994. Bird dispersal techniques. Pp. E19–E24 in S. Hygnstrom, R. Timm, and G. Larson. Prevention and control of wildlife damage. Univ. Nebr. Coop. Ext., Lincoln.

Borer, E., P. Hosseini, E. Seabloom, et al. 2007. Pathogen-induced reversal of native dominance in a grassland community. Proc. Natl. Acad. Sci. 104:5473–5478.

Boulanger, J., L. Bigler, P. Curtis, et al. 2008. Comparison of suburban vaccine distribution strategies to control raccoon rabies. J. Wildl. Dis. 44:1014–1023.

Boyle, C. 1960. Case of apparent resistance of *Rattus norvegicus* Berkenhout to anticoagulant poisons. Nature 188:517.

Bradford, W. 1856. History of Plymouth Plantation. Ed. C. Deane. Repr., Massachusetts Historical Collections, Boston. 476 pp.

Braithwait, J. 1996. Using guard animals to protect livestock. Conserv. Comm. State Mo. 14 pp.

Breck, S., N. Lance, and P. Callahan. 2006. Shocking device for protection of concentrated food sources from black bears. Wildl. Soc. Bull. 34:23–26.

Breck, S., K. Wilson, and D. Andersen. 2003. Beaver herbivory and its effect on cottonwood trees: influence of flooding along matched regulated and unregulated rivers. River Res. Applic. 19:43–58.

Britton-Simmons, K. and K. Abbott. 2008. Short- and long-term effects of disturbance and propagule pressure on biological invasion. J. Ecol. 96: 68–77.

Brock, J. 2005. Who were the first surveyors? Four surveyors of the gods: in the XVIII dynasty of Egypt–New Kingdom c. 1400 B. C. Int. Fed. Surv., Artic. Mon., March 2005. Workshop Surv., FIG Work. Week/GSDI-8 Conf., Cairo, Egypt, April 16, 2005. 15 pp.

Bromley, C., and E. Gese. 2001a. Surgical sterilization as a method of reducing coyote predation on domestic sheep. J. Wildl. Manage. 65:510–519.

Bromley, C., and E. Gese. 2001b. Effects of sterilization on territory fidelity and maintenance, pair bonds, and survival rates of free-ranging coyotes. Can. J. Zool. 79:386–392.

Brooke, M., G. Hilton, and T. Martins. 2007. Prioritizing the world's islands for vertebrate-eradication programmes. Anim. Conserv. 10(3):380–390.

Brown, K. and G. Sherley. 2002. The eradication of possums from Kapiti Island, New Zealand. Pp. 46–52 in C. Veitch and M. Clout, eds. Turning the tide: the eradication of invasive species. IUCN SSC Invasive Species Spec. Group. IUCN, Gland, Switz., and Cambridge, U.K.

Brown, P. and G. Singleton. 2002. Impacts of house mice on crops in Australia—costs and damage. Pp. 48–58 in L. Clark, J. Hone, J. Shivik, et al., eds. Human conflicts with wildlife: economic considerations. Proc. Third NWRC Spec. Symp., Natl. Wildl. Res. Cent., Fort Collins, Colo.

Bruce, H. 1959. An exteroceptive block to pregnancy in the mouse. Nature (Lond.) 184:105.

Bruggers, R., J. Brooks, R. Dolbeer, et al. 1986. Responses of pest birds to reflecting tape in agriculture. Wildl. Soc. Bull. 14:161–170.

Bruggers, R., J. Matee, J. Miskell, et al. 1981. Reduction of bird damage to field crops in Eastern Africa with methiocarb. Pest Manage. 27:230–241.

Bruinderink, G. and E. Hazebroek. 1996. Ungulate traffic collisions in Europe. Conserv. Biol. 10:1059–1067.

Bryce, R., M. Oliver, L. Davies, et al. 2010. Turning back the tide of American mink invasion at an unprecedented scale through community participation and adaptive management. Biol. Conserv. www.ptes.org/files/1315_mink_control_published_report.pdf. Accessed 7/30/11.

Bullard, R., S. Shumake, D. Campbell, et al. 1978. Preparation and evaluation of a synthetic fermented egg coyote attractant and deer repellent. J. Agric. Food Chem. 26:160–163.

Burbidge, A. and K. Morris. 2002. Introduced mammal eradications for nature conservation on western Australian islands: a review. Pp. 64–70 in C. Veitch and M. Clout, eds. Turning the tide: the eradication of invasive species.

IUCN SSC Invasive Species Spec. Group. IUCN, Gland, Switz., and Cambridge, U.K.

Burridge, M. 2005. Controlling and eradicating tick infestations on reptiles. Article No. 3, May, pp. 371–376. https://secure.vlsstore.com/Media/PublicationsArticle/PV_27_05_371.pdf. Accessed 2/21/11.

Burton, R., S. Wood, and N. Owen. 2003. Elucidation of a new oleanane glycoside from Barringtonia asiatica. ARKIVOC xiii:137–146.

Businga, N., J. Langenberg, and L. Carlson. 2007. Successful treatment of capture myopathy in three wild greater sandhill cranes (Grus canadensis tabida). J. Avian Med. Surgery 21:294–298.

Butchko, P. 1990. Predator control for the protection of endangered species in California. Proc. Vertebr. Pest Conf. 14:237–240.

Butler, J., J. Shanahan, and D. Decker. 2001. Wildlife attitudes and values: a trend analysis. HDRU Series No. 01-4, Cornell Univ., Ithaca, N.Y. 21 pp.

Cadi, A. and P. Joly. 2004. Impact of the introduction of the red-eared slider (Trachemys scripta elegans) on survival rates of the European pond turtle (Emys orbicularis). Biodivers. Conserv. 13:2511–2518.

Campbell, E., III. 1999. Barriers to movements of the brown tree snake (Boiga irregularis). Pp. 306–312 in G. Rodda, Y. Sawai, D. Chiszar, et al., eds. Problem snake management: the habu and the brown tree snake. Cornell Univ. Press, Ithaca, N.Y. 534 pp.

Campbell, K. and C. Donlan. 2005. Feral goat eradications on islands. Conserv. Biol. 19:1362–1374.

Campbell, T. and D. Long. 2009. Feral swine damage and damage management in forested ecosystems. For. Ecol. Manage. 257:2319–2326.

Canby, T. 1977. The incredible rat: lapdog of the devil. Natl. Geogr. July: 60–87.

Capizzi, D., N. Baccetti, and P. Sposimo. 2010. Prioritizing rat eradication on islands by cost and effectiveness to protect nesting seabirds. Biol. Conserv. 143:1716–1727.

Carrion, V. 2009. Monitoring and control of tilapia (Oreochromis niloticus) in the "El Junco" Lagoon, San Cristobal. Conservation and restoration of island eocosystems: control and eradication of introduced animals. www.galapagospark.org/nophprg.php?page=parque_nacional_introducidas_tilapia. Accessed 1/20/11.

Carrion, V., C. Donlan, K. Campbell, et al. 2011. Archipelago-wide island restoration in the Galápagos Islands: reducing costs of invasive mammal eradication programs and reinvasion risk. PLoS ONE 6(5): e18835. doi:10.1371/journal.pone.0018835. Accessed 8/28/11.

Catalogue of fence designs. Undated. www.environment.gov.au/biodiversity/invasive/publications/pubs/catalogue.pdf. Accessed 4/3/11.

Caudell, J., S. Shwiff, and M. Slater. 2010. Using a cost-effectiveness model to determine the applicability of OvoControl G to manage nuisance Canada geese. J. Wildl. Manage. 74(4):843–848.

Caughley, G. 1981. Overpopulation. Pp. 7–20 in P. Jewell and S. Holt, eds. Problems in management of locally abundant wild mammals. Academic Press, New York.

Caut, S., J. Casanovas, E. Virgos, et al. 2007. Rats dying for mice: modeling the competitor release effect. Aust. Ecol. 32:858–868.

Cavallo, J. 1991. Leopards and human evolution. P. 208 in J. Seidensticker and S. Lumpkin, eds. Great cats: majestic creatures of the wild. Rodale Press, Emmaus, Penn.

Cease, K. and J. Juzwik. 2001. Predominant nitidulid species (*Coleoptera: Nitidulidae*) associated with spring oak wilt mats in Minnesota. Can J. For. Res. 31:635–643.

Center, T., T. Van, M. Rayachhetry, et al. 2000. Field colonization of the melaleuca snout beetle (*Oxyops vitiosa*) in South Florida. Biol. Control 19:112–123.

Chapron, G., S. Legendre, R. Ferrière, et al. 2003. Conservation and control strategies for the wolf (*Canis lupus*) in western Europe based on demographic models. C. R. Biol. 326:575–587.

China Daily. 2003. Monkeys terrorize India workers, tourists. November 3. www.chinadaily.com.cn/en/doc/2003-11/03/content_277874.htm. Accessed 6/29/11.

Chipman, R., T. DeVault, D. Slate, et al. 2008. Non-lethal management to reduce conflicts with winter urban crow roosts in New York: 2002–2007. Proc. Vertebr. Pest Conf. 23:88–93.

Choquenot, D. and J. Hone. 2002. Using bioeconomic models to maximize benefits from vertebrate pest control: lamb predation by feral pigs. Pp. 65–79 in L. Clark, J. Hone, J. Shivik, et al., eds. Human conflicts with wildlife: economic considerations. Proc. Third NWRC Spec. Symp., Natl. Wildl. Res. Cent., Fort Collins, Colo.

Christian, J. and D. Davis. 1964. Endocrines, behavior, and population: social and endocrine factors are integrated in the regulation of growth of mammalian populations. Science 146(3651):1550–1560.

Chua, K. 2010. Risk factors, prevention and communication during Nipah virus outbreak in Malaysia. Malays. J. Pathol. 32:75–80.

Chua, K., W. Bellini, P. Rota, et al. 2000. Nipah virus: a recently emergent deadly paramyxovirus. Science 288:1432–1435.

Chua, K., B. Chua, and C. Wang. 2002. Anthropogenic deforestation, El Niño and the emergence of Nipah virus in Malaysia. Malays. J. Pathol. 24:15–21.

Cilliers, C. 1991. Biological control of water hyacinth, *Eichhornia crassipes* (*Pontederieacea*), in South Africa. Agric. Ecosyst. Environ. 37:207–217.

Cilliers, D. 2003. South African cheetah compensation fund. Carnivore Damage Prev. News 6:15–16.

CISEH. Undated. Univ. Ga. www.bugwood.org/. Accessed 2/17/2011.

Clark, L. and J. Hall. 2006. Avian influenza in wild birds: status as reservoirs, and risks to humans and agriculture. Ornithol. Monogr. 60:3–29.

Clark, L. and J. Shivik. 2002. Aerosolized essential oils and individual product compounds as brown tree snake repellents. Pest Manage. Sci. 58:775–783.

Clearwater, S., C. Hickey, and M. Martin. 2008. Overview of potential piscicides and molluscicides for controlling aquatic pest species in New Zealand. Sci. Conserv. 283, Sci. Tech. Publ., Dept. Conserv., Wellington, N.Z. 74 pp.

Cleary, E. and R. Dolbeer. 2005. Wildlife hazard management at airports, a manual for airport personnel. 2nd ed. U.S. Dep. Transp., Fed. Aviat. Admin., Off. Airport Saf. Stand., Washington, D.C.

Clergeau, P., J. Savard, G. Mennechez, et al. 1998. Bird abundance and diversity along an urban-rural gradient: a comparative study between two cities on different continents. Condor 100:413–425.

Clevenger, A., B. Chruszez, and K. Gunson. 2001. Highway mitigation fencing reduces wildlife-vehicle collisions. Wildl. Soc. Bull. 29(2):646–653.

Clevenger, A., A. Ford, and M. Sawaya. 2009. Banff wildlife crossings project: integrating science and education in restoring population connectivity across transportation corridors. Final rep. Parks Can. Agency, Radium Hot Springs, B.C. 165 pp.

Clover, M. 1954. A portable deer trap and catch-net. Calif. Fish Game 40:367–373.

Conner, M., M. Ebinger, and F. Knowlton. 2008. Evaluating coyote management strategies using a spatially explicit, individual-based, socially structured population model. Ecol. Modell. 219:234–247.

Connor, M. 1992. The red-eared slider, *Trachemys scripta elegans*. Tortuga Gaz. 28:1–3.

Conover, G., R. Simmonds, and M. Whalen, eds. 2007. Management and control plan for bighead, black, grass, and silver carps in the United States. Asian Carp Work. Group, Aquat. Nuisance Species Task Force, Washington, D.C. 223 pp.

Conover, M. 2002. Resolving human-wildlife conflicts: the science of wildlife damage management. CRC Press, Boca Raton, Flor. 440 pp.

Conover, M. and N. McCoy. 2003. Positive and negative values of blackbirds. Pp. 17–20 in G. Linz, ed. Management of North American blackbirds. Natl. Wildl. Res. Cent., Fort Collins, Colo.

Cooper, D. 1998. Road kills of animals on some New South Wales roads. Report on data collected by WIRES volunteers in 1997. Macquarie Univ., Sydney, Aust.

Cornell Univ. Undated. Living History. A bird in the hand: a survey of medieval fowling techniques. www.people.cornell.edu/. Accessed 5/6/2008.

Côté, S, T. Rooney, J.-P. Tremblay, et al. 2004. Ecological impacts of deer overabundance. Annu. Rev. Ecol. Evol. Syst. 35:113–147.

Cotton, W. 2008. Resolving conflicts between humans and the threatened Louisiana black bear. Human-Wildl. Conflicts 2(2):151–152.

Courtenay, W., D. Hensley, J. Taylor, et al. 1984. Distribution of exotic fishes in the continental United States. Pp. 41–77 in W. Courtenay and J. Stauffer, eds. Distribution, biology and management of exotic fishes. Johns Hopkins Univ. Press, Baltimore.

Courtenay, W. and J. Williams. 2004. Snakeheads (*Pisces: Channidae*)—a biological synopsis and risk assessment. U.S. Dep. Inter., U.S. Geol. Surv. Circ. 1251, 143 pp.

Courtenay, W., J. Williams, R. Britz, et al. 2004. Identity of introduced snakeheads (*Pisces, Channidae*) in Hawaii and

Madagascar, with comments on ecological concerns. Bishop Mus., Occas. Pap. No. 77, 13 pp.

Cowled, B., P. Elsworth, and S. Lapidge. 2008. Additional toxins for feral pig control: identifying and testing Achilles heels. Wildlife Res. 35(7):651–662.

Cruz, F., C. Donlan, K. Campbell, et al. 2005. Conservation action in the Galápagos: feral pig (*Sus scrofa*) eradication from Santiago Island. Biol. Conserv. 121:473–478.

Cucci, T., J. Vigne, and J. Auffray. 2005. First occurrence of the house mouse (*Mus musculus domesticus* Schwarz & Schwarz, 1943) in the Western Mediterranean: a zooarchaeological revision of subfossil occurrences. Biol. J. Linnean Soc. 84:429–445.

Cummings, J., J. Guarino, C. Knittle, et al. 1987. Decoy plantings for reducing blackbird damage to nearby commercial sunflower fields. Crop Prot. 6:56–60.

Cuthbert, R. and G. Hilton. 2004. Introduced house mice *Mus musculus*: a significant predator of threatened and endemic birds on Gough Island, South Atlantic Ocean? Biol. Conserv. 117:483–489.

Cutler, S., A. Fooks, and W. van der Poel. 2010. Public health threat of new, reemerging, and neglected zoonoses in the industrialized world. Emerging Infect. Dis. 16:1–7.

Dabritz, H., I. Gardner, M. Miller, et al. 2007. Evaluation of two *Toxoplasma gondii* serologic tests used in a serosurvey of domestic cats in California. J. Parasitol. 93:806–816.

Dabritz, H., M. Miller, E. Atwill, et al. 2007. Detection of *Toxoplasma gondii*-like oocysts in cat feces and estimates of the environmental oocyst burden. J. Am. Vet. Med. Assoc. 231:1676–1684.

Daehler, C. 1998. The taxonomic distribution of invasive angiosperm plants: ecological insights and comparison to agricultural weeds. Biol. Conserv. 84:167–180.

Dahl, R. 2005. Population equation: balancing what we need with what we have. Environ. Health Perspect. 113(9):A599–A605.

Dale, L. 2009. Personal and corporate liability in the aftermath of bird strikes: a costly consideration. Human-Wildl. Conflicts 3(2):216–225.

Darrow, P., R. Skirpstunas, S. Carlson, et al. 2008. Comparison of injuries to coyote from 3 types of cable foot-restraints. J. Wildl. Manage. 73(8):1441–1444.

Dauphiné, N. and R. Cooper. 2009. Impacts of free-ranging domestic cats (*Felis catus*) on birds in the United States: a review of recent research with conservation and management recommendations. Proc. Int. Partners Flight Conf. 4:205–219.

Davidson, W., and G. Doster. 1997. Health characteristics and white-tailed deer population density in the southeastern United States. Pp. 164–184 in W. McShea, H. Underwood, and J. Rappole, eds. The science of overabundance: deer ecology and population management. Smithsonian Books, Washington, D.C.

Davis, D. 1966. Integral animal behavior. Macmillan, New York. 118 pp.

DeCalesta, D. 1992. Impact of deer on species diversity of Allegheny hardwood stands. Proc. Northeast. Weed Sci. Soc. Abstr. 46:135.

DeCalesta, D. 1997. Deer and ecosystem management. Pp. 267–279 in W. McShea, H. Underwood, and J. Rappole, eds. The science of overabundance: deer ecology and population management. Smithsonian Books, Washington, D.C.

Decapita, M. 2000. Brown-headed cowbird control on Kirtland's warbler nesting areas in Michigan, 1972–1995. Pp. 333–341 in J. Smith. Ecology and management of cowbirds and their hosts: studies in the conservation of North American passerine birds. Univ. Texas Press, Austin.

Decker, D., T. Lauber, and W. Seimer. 2002. Human-wildlife conflict management: a practitioner's guide. Hum. Dimens. Res. Unit, Northeast Wildl. Damage Manage. Res. Outreach Coop. 48 pp.

Decker, D., D. Raik, and W. Siemer. 2004. Community-based deer management: a practitioners' guide. Northeast Wildl. Damage Manage. Res. Outreach Coop. 54 pp.

Decker, S., A. Bath, A. Simms, et al. 2008. The return of the king or bringing snails to the garden? The human dimensions of a proposed restoration of European bison (*Bison bonasus*) in Germany. Restor. Ecol. www.largeherbivore.org/assets/pdf/article-hd-stephen.pdf. Accessed 7/1/11.

Deevey, E. 1947. Life tables for natural populations of animals. Q. Rev. Biol. 22:283–314.

DEFRA. 2010. Wildlife management in England: a policy making framework for resolving human-wildlife conflicts. Dep. Environ., Food Rural Affairs, London. 38 pp.

Dejong, A. and H. Blokpoel. 1966. Hard facts about soft feathers. The bird strike hazard in Holland. Shell Aviat. News 343:2–7.

Dellamano, F. 2006. Controlling birds with netting: blueberries, cherries and grapes. N. Y. Fruit Q. 14:3–5.

Dethier, V. and E. Stellar. 1964. Animal Behavior. 2nd ed. Prentice-Hall, Englewood Cliffs, N.J. 118 pp.

DeVault, T., J. Beasley, L. Humberg, et al. 2007. Intrafield patterns of wildlife damage to corn and soybeans in northern Indiana. Human-Wildl. Conflicts 1:205–213.

Diamond, J. 1997. Guns, germs, and steel: the fates of human societies. W. W. Norton. New York.

Diamond, S., R. Giles, R. Kirkpatrick, et al. 2000. Hard mast production before and after the chestnut blight. South. J. Appl. For. 24:196–201.

Di Castri, F. 1990. On invading species and invaded ecosystems: the interplay of historical chance and biological necessity. Pp. 3–16 in F. Di Castri, A. Hansen, and M. Debussche, eds. Biological Invasions in Europe and the Mediterranean Basin. Kluwer Acad. Publ. Dordrecht, Neth.

Ding, J., R. Mack, P. Lu, et al. 2008. China's booming economy is sparking and accelerating biological invasions. BioScience 58:317–324.

Distefano, E. 2005. Human-wildlife conflict worldwide: collection of case studies, analysis of management strategies and good practices. SARD Initiative Rep. FAO, Rome, Italy. 34 pp.

Dolbeer, R. 1994. Blackbirds. Pp. E25–E32 in S. Hygnstrom, R. Timm, and G. Larson. Prevention and control of wildlife damage. Univ. Nebr. Coop. Ext., Lincoln.

Dolbeer, R. 1998. Population dynamics: the foundation of wildlife damage management for the 21st century. Proc. Vertebr. Pest Conf. 18:2–11.

Dolbeer, R., J. Belant, and J. Sillings. 1993. Shooting gulls reduces strikes with aircraft at John F. Kennedy International Airport. Wildl. Soc. Bull. 21:442–450.

Dolbeer, R. and R. Chipman. 1999. Shooting gulls to reduce strikes with aircraft at John F. Kennedy International Airport, 1991–1998. Spec. rep. Port Authority N. Y. and N. J., U.S. Dep. Agric., Natl. Wildl. Res. Cent., Sandusky, Ohio.

Dolbeer, R., N. Holler, and D. Hawthorne. 1994a. Identification and assessment of wildlife damage: an overview. Pp. A1–A18 in S. Hygnstrom, R. Timm, and G. Larson. Prevention and control of wildlife damage. Univ. Nebr. Coop. Ext., Lincoln.

Dolbeer, R., N. Holler, and D. Hawthorne. 1994b. Identification and control of wildlife damage. Pp. 474–506 in T. Bookhout, ed. Research and management techniques for wildlife and habitats. Wildl. Soc., Bethesda, Mary.

Dolbeer, R., A. Stickley, and P. Woronecki. 1979. Starling (*Sturnus vulgaris*) damage to sprouting wheat in Tennessee and Kentucky, USA. Prot. Ecol. 1:159–169.

Dolbeer, R., P. Woronecki, and R. Bruggers. 1986. Reflecting tapes repel blackbirds from millet, sunflowers, and sweet corn. Wildl. Soc. Bull. 14:418–425.

Dolbeer, R. and S. Wright. 2009. Safety management systems: how useful will the FAA National Wildlife Strike Database be? Human-Wildl. Conflicts 3(2):167–178.

Dolbeer, R., S. Wright, and E. Cleary. 2000. Ranking the hazard level of wildlife species to aviation. Wildl. Soc. Bull. 28:372–378.

Dollinger, A. 2000. Egyptian vermin. www.reshafim.org.il /ad/egypt/timelines/topics/pests.htm. Accessed 10/20/10.

Donlan, C., H. Greene, J. Berger, et al. 2005. Re-wilding North America. Nature 436:913–914.

Donlan, C., P. Martin, and G. Roemer. 2007. Lessons from land: present and past signs of ecological decay and the overture to the earth's sixth mass extinction. Pages 14–26 in J. Estes, D. DeMaster, D. Doak, et al., eds. Whales, whaling, and ocean ecosystems. Univ. Calif. Press, Berkeley.

Dove, C., R. Snow, M. Rochford, et al. 2011. Birds consumed by the invasive Burmese python (*Python molurus bivittatus*) in Everglades National Park, Florida, USA. Wilson J. Ornithol. 123:126–131.

Dunn, E., C. Francis, P. Blancher, et al. 2005. Enhancing the scientific value of the Christmas Bird Count. Auk 122:338–346.

Dye, T. 2008. Understanding public policy. 12th ed. Prentice Hall, Upper Saddle River, N.J. 368 pp.

Eason, C., K. Fagerstone, J. Eisemann, et al. 2010. A review of existing and potential New World and Australian vertebrate pesticides with a rationale for linking use patterns to registration requirements. Int. J. Pest Manage. 56:109–125.

Eisemann, J., B. Petersen, and K. Fagerstone. 2003. Efficacy of zinc phosphide for controlling Norway rats, roof rats, house mice, *Peromyscus spp.*, prairie dogs and ground squirrels: literature review (1942–2000). Proc. Wildl. Damage Manage. Conf. 10:335–349.

Eisen, L. and R. Eisen. 2011. Using geographic information systems and decision support systems for the prediction, prevention, and control of vector-borne diseases. Annu. Rev. Entomol. 56:41–61.

Ellis, C. and D. Schneider. 2000. Predicting the spread of zebra mussels (*Dreissena polymorpha*) in Illinois lakes and streams. Int. Conf. Integr. GIS Environ. Model (GIS/EM4). Alberta, Can., Sept. 2–8. www.colorado.edu/research/cires /banff/pubpapers/237/. Accessed 7/27/11.

Emerton, L. 1999. Balancing the opportunity costs of wildlife conservation for communities around Lake Mburo National Park, Uganda. Evaluating Eden Project and Community Conservation Research Project. Discuss. Pap. No. 5. Univ. Manchester, Univ. Zimbabwe, Univ. Cambridge, and Afr. Wildl. Found. 27 pp.

Engeman, R. 2002. Economic considerations of damage assessment. Pp. 36–41 in L. Clark, J. Hone, J. Shivik, et al., eds. 2002. Human conflicts with wildlife: economic considerations. Proc. Third NWRC Spec. Symp., Natl. Wildl. Res. Cent., Fort Collins, Colo.

Engeman, R. and L. Allen. 2000. Overview of a passive tracking index for monitoring wild canids and associated species. Integr. Pest Manage. Rev. 5:197–203.

Engeman, R. and D. Campbell. 1991. Pocket gopher reoccupation of burrow systems following population reduction. Crop Prot. 18:523–525.

Engeman, R., B. Constantin, S. Shwiff, et al. 2007. Adapative and economic management methods for feral hog control in Florida. Human-Wildl. Conflicts 1(2):178–185.

Engeman, R. and M. Linnell. 1998. Trapping strategies for deterring the spread of brown tree snake from Guam. Pac. Conserv. Biol. 4:348–353.

Engeman, R., M. Pipas, K. Gruver, et al. 2000. Monitoring coyote population changes with a passive activity index. Wildl. Res. 27:553–557.

Engeman, R., S. Shwiff, B. Constantin, et al. 2002. An economic analysis of predator removal approaches for protecting marine turtle nests at Hobe Sound National Wildlife Refuge. Ecol. Econ. 42:469–478.

Engeman, R., S. Shwiff, H. Smith, et al. 2002. Monetary valuation methods for economic analysis of the benefit-costs of protecting rare wildlife species from predators. Integr. Pest Manage. Rev. 7:139–144.

Engeman, R. and G. Witmer. 2000. IPM strategies: indexing difficult to monitor populations of pest species. Proc. Vertebr. Pest Conf. 19:183–189.

Eve, J. 1981. Management implications of disease. Pp. 413–423 in W. Davidson, F. Hayes, V. Nettles, et al., eds. Diseases and parasites of white-tailed deer. Misc. Publ. No. 7. Tall Timber Res. Stn., Tallahassee, Fla.

Everest, J., J. Miller, D. Ball, et al. 1999. Kudzu in Alabama. Ala. Coop. Ext. Syst. www.aces.edu/pubs/docs/A/ANR -0065/. Accessed 2/21/11.

Fagerstone, K. 2003. Mitigating impacts of terrestrial invasive species. Pp. 1–3 in D. Pimental, ed. Encyclopedia of pest management. 3rd quarterly online update. Marcel Dekker, New York.

Fagerstone, K., L. Miller, J. Eisemann, et al. 2008. Registration of wildlife contraceptives in the United States of America, with OvoControl and GonaCon immunocontraceptive vaccines as examples. Wildl. Res. 35:586–592.

Fagerstone, K., L. Miller, G. Killian, et al. 2010. Review of issues concerning the use of reproductive inhibitors, with particular emphasis on resolving human-wildlife conflicts in North America. Integr. Zool. 1:15–30.

Fagerstone, K. and E. Schafer. 1998. Status of APHIS vertebrate pesticides and drugs. Proc. Vertebr. Pest Conf. 18:319–324.

Fairaizl, S. and W. Pfeifer. 1987. The lure crop alternative. Proc. Great Plains Wildl. Damage Control Workshop 8:163–168.

Fall, M. and D. Schneider. 1968. Pest birds and modern architecture. Proc. Bird Control Semin. 4:129–135.

FAO, United Nations. 1997. Code of conduct for the import and release of exotic biological control agents. Biocontrol News Inf. 18:119N–124N.

Farag, S., Y. Kimura, H. Ito, et al. 2009. New isoflavone glycosides from *Iris spuria L.* (Calizona) cultivated in Egypt. J. Nat. Med. 63(1):91–95.

Farri, T. and R. Boroffice. 1999. An overview on the status and control of water hyacinth in Nigeria. Pp. 18–24 in M. Hill, M. Julien, and T. Center, eds. Proceedings of the first IOBC global working group meeting for the biological and integrated control of water hyacinth. Weeds Res. Div., ARC, S. Afr.

Fausch, K. and R. White. 1981. Competition between brook trout (*Salvelinus fontinalis*) and brown trout (*Salmo trutta*) for positions in a Michigan Stream. Can. J. Fish. Aquat. Sci. 38:1220–1227.

Feldhamer, G., L. Drickamer, S. Vessey, et al. 2007. Mammalogy. Johns Hopkins Univ. Press, Baltimore. 643 pp.

Fernandez, A., R. Richardson, D. Tschirley, et al. 2009. Wildlife conservation in Zambia: impacts on rural household welfare. Work. Pap. No. 41, Food Secur. Res. Proj., Lusaka, Zambia. 19 pp.

Ferraz, K., M. Lechevalier, H. Couto, et al. 2003. Damage caused by capybaras in a corn field. Sci. Agric., Piracicaba, Braz. www.scielo.br/scielo.php?script=sci_arttext& pid=S0103-90162003000100029&lng=en&nrm=iso. doi: 10.1590/S0103-90162003000100029. Accessed 10/13/2011.

Ficetola, G., C. Coïc, M. Detaint, et al. 2007. Pattern of distribution of the American bullfrog *Rana catesbeiana* in Europe. Biol. Invasions. DOI 10.1007/s10530-006-9080-y.

Ficetola, G., W. Thuiller, and E. Padoa-Schioppa. 2009. From introduction to establishment of alien species: bioclimatic differences between presence and reproduction localities in the slider turtle. Divers. Distrib. 15:108–116.

Fine, P. 2002. The invasibility of tropical forests by exotic plants. J. Trop. Ecol. 18:687–705.

Finlayson, B., R. Schnick, R. Cailteux, et al. 2000. Rotenone use in fisheries management: administrative and technical guidelines manual. Am. Fish. Soc., Bethesda, Mary. 214 pp.

Fischer, J. and P. Dunlevy. 2010. Eradicating rats on Lehua Island, Hawaii, with the help of GIS and GPS. ArcNews online. www.esri.com/news/arcnews/spring10articles /lehua-island-hawaii.html. Accessed 8/26/11.

Fischer, J. and D. Lindenmayer. 2000. An assessment of published results of animal relocations. Biol. Conserv. 96(1):1–11.

Fisher, M., T. Garner, and S. Walker. 2009. Global emergence of *Batrachochytrium dendrobatidis* and amphibian chytridiomycosis in space, time, and host. Annu. Rev. Microbiol. 63:291–310.

Fitzwater, W. 1972. Barrier fencing in wildlife management. Proc. Vertebr. Pest Conf. 5:49–55.

Fleishman, E., N. McDonal, R. Mac Nally, et al. 2003. Effects of floristics, physiognomy and non-native vegetation on riparian communities in a Mojave Desert watershed. J. Anim. Ecol. 72:484–490.

Fleming, P., L. Allen, S. Lapidge, et al. 2006. A strategic approach to mitigating the impacts of wild canids: proposed activities of the Invasive Animals Cooperative Research Centre. Aust. J. Exp. Agric. 46:753–762.

Flowers, R. 1986. Supplemental feeding of black bears in tree-damaged areas of western Washington. Pp. 147–148 in Proc. Anim. Damage Manage. Pac. Northwest For. Symp., Washington State Univ., Pullman, Wash.

Flueck, W., J. Smith-Flueck, and C. Naumann. 2003. The current distribution of red deer (*Cervus elaphus*) in America. Z. Jagdwiss. 49:112–119.

Frank, L., D. Simpson, and R. Woodroffe. 2003. Foot snares: an effective method of capturing African lions. Wildl. Soc. Bull. 31(1):309–314.

Franklin, W., and K. Powell. 1994. Llamas: a part of integrated sheep protection. Iowa State Univ., Univ. Ext. Pm-1527, Ames, Iowa. 12 pp.

Fraser, D. and E. Thomas. 1982. Moose-vehicle accidents in Ontario: relation to highway salt. Wildl. Soc. Bull. 10:261–265.

Freed, L., R. Cann, M. Goff, et al. 2005. Increase in avian malaria at upper elevations in Hawaii. Condor 107:753–764.

Frenkel, J., J. Dubey, and N. Miller. 1970. *Toxoplasmosis gondii* in cats: fecal stages identified as coccidian oocysts. Science 167:893–896.

Frost, P. and I. Bond. 2008. The CAMPFIRE programme in Zimbabwe: payments for wildlife services. Ecol. Econ. 65:776–787.

Fu, Z. 1997. Rabies and rabies research: past, present and future. Vaccine 15:S20–S24.

Fuller, P. 2011a. *Channa marulius*. USGS Nonindigenous Aquatic Species Database. Gainesville, Fla. http://nas.er .usgs.gov/queries/FactSheet.aspx?speciesID=2266 Revision Date: 12/5/2008. Accessed 2/26/11.

Fuller, P. 2011b. *Ictalurus furcatus*. USGS Nonindigenous Aquatic Species Database, Gainesville, Fla. (http://nas.er .usgs.gov/queries/FactSheet.aspx?speciesID=740 Revis. Date: 5/5/2010.) Accessed 2/27/11.

Fuller, P. 2011c. *Oncorhynchus mykiss*. USGS Nonindigenous Aquatic Species Database. Gainesville, Fla. http://nas.er .usgs.gov/queries/FactSheet.aspx?speciesID=910 Revis. Date: 2/3/2011. Accessed 2/27/11.

Fuller, P., E. Maynard, and D. Raikow. 2011. *Morone americana*. USGS Nonindigenous Aquatic Species Database. Gainesville, Fla. http://nas.er.usgs.gov/queries

/FactSheet.aspx?speciesID=777 Revis. Date: 9/29/2006. Accessed 2/27/11.

Galef, B. and P. Henderson. 1972. Mother's milk: a determinant of the feeding preference of weaning rat pups. J. Comp. Physiol. Psychol. 78:213–219.

Garcia, J., W. Hankins, and K. Rusiniak. 1974. Behavioral regulation of the milieu interne in man and rat. Science 185: 824–831.

Gatt, M. Undated. Strategy for preventing and managing human-wildlife conflicts in Ontario. www.scribd.com/doc /43307904/Human-Wildlife-Conflict. Accessed 7/24/11.

Gehring, T., K. VerCauteren, and J. Landry. 2010. Livestock protection dogs in the 21st century: is an ancient tool relevant to modern conservation challenges? BioScience 60:299–308.

Gehrt, S. 2007. Ecology of coyotes in urban landscapes. Proc. Wildl. Damage Manage. Conf. 12:303–311.

Gese, E. 2004. Survey and census techniques for canids. Pp. 273–279 in C. Sillero-Zubiri, M. Hoffman, and D. Macdonald, eds. Canids: foxes, wolves, jackals, and dogs. IUCN/SSC Canid Spec. Group. Gland, Switz., and Cambridge, U.K.

Gigliotti, L., D. Shroufe, and S. Gurtin. 2009. The changing culture of wildlife management. Pp. 75–89 in M. Manfredo, J. Vaske, P. Brown, et al., eds. Wildlife and society: the science of human dimensions. Island Press, Washington, D.C.

Gildorf, J., S. Hygnstrom, K. VerCauteren, et al. 2004. Evaluation of a deer-activated bioacoustic frightening device for reducing deer damage in cornfields. Wildl. Soc. Bull. 32:515–523.

GISD. 2010. Sturnus vulgaris. www.issg.org/database/ welcome/disclaimer.asp. Accessed 1/26/2011.

Glahn, J. and B. Blackwell. 2000. Using the Desman and Dissuader laser to disperse double-crested cormorants and other birds. Unpubl. doc. Wildl. Serv. 5 pp.

Glahn, J., G. Ellis, P. Fioranelli, et al. 2001. Evaluation of moderate and low-powered lasers for dispersing double-crested cormorants from their night roosts. Proc. East. Wildl. Damage Manage. Conf. 9:34–45.

Glahn, J. and D. King. 2004. Bird depredation. Chapter 16 in C. Tucker and J. Hargreaves, eds. Biology and culture of channel catfish. Elsevier, Amsterdam, Netherlands. 676 pp.

Glahn, J., D. Reinhold, and P. Smith. 1999. Wading bird depredations on channel catfish (Ictalurus punctatus) in Northwest Mississippi. J. World Aquacult. Soc. 30:107–114.

Global Footprint Network. 2010. Climate change is not the problem. www.footprintnetwork.org/images/uploads /2010_Annual_Report.pdf. Accessed 8/12/11. 57 pp.

Goldberg, T. 2002. Largemouth bass virus: an emerging problem for warm-water fisheries. Am. Fish. Soc. Symp. 31:411–416.

Gosselink, T., T. Van Deelen, R. Warner, et al. 2003. Temporal habitat partitioning and spatial use of coyotes and red foxes in east-central Illinois. J. Wildl. Manage. 67:90–103.

Goswami, V., M. Madhusudan, and K. Karanth. 2007. Application of photographic capture-recapture modeling to estimate demographic parameters for male Asian elephants. Anim. Conserv. 10:391–399.

Gough, P. and J. Beyer. 1981. Bird-vectored diseases. Proc. Great Plains Wildl. Damage Control Workshop 5:260–272.

Gould, F. 2007. Broadening the application of evolutionarily based genetic pest management. Evolution 62(2):500–510.

Graham, M., B. Notter, W. Adams, et al. 2010. Patterns of crop-raiding by elephants, Loxodonta africana, in Laikipia, Kenya, and the management of human-elephant conflict. Syst. Biodiversity 8(4):435–445.

Grant, E., D. Philipp, K. Inendino, et al. 2003. Effects of temperature on susceptibility of largemouth bass to largemouth bass virus. J. Aquat. Anim. Health 15:215–220.

Greaves, J. 1986. Managing resistance to rodenticides. Pp. 236–244 in Pesticide resistance: strategies and tactics for management. Natl. Acad. Press, Washington, D.C.

Gregory, N., L. Milne, A. Rhodes, et al. 1998. Effect of potassium cyanide on behavior and time to death in possums. N. Z. Vet. J. 46:60–64.

Grizzle, J., I. Altinok, W. Fraser, et al. 2002. First isolation of largemouth bass virus. Dis. Aquat. Org. 50:233–235.

Gustavson, C. 1976. Prey-lithium aversion. Coyotes and wolves. Behav. Biol. 17:61–72.

Haas, R., M. Thomas, and G. Towns. 2003. An assessment of the potential use of Gambusia for mosquito control in Michigan. Mich. Dep. Nat. Res., Fish. Tech. Rep. 2003–2.

Hagle, S., K. Gibson, and S. Tunnock. 2003. Field guide to diseases and insect pests of northern and central Rocky Mountain conifers. Rep. Num. R1-03-08. USDA, For. Serv., State and Private For., North. Reg. Intermt. Reg., Missoula, Mont. 197 pp.

Haim, A., U. Shanas, O. Brandes, et al. 2007. Suggesting the use of integrated methods for vole population management in alfalfa fields. Integr. Zool. 2:184–190.

Handegard, L. 1988. Using aircraft for controlling blackbird/sunflower depredations. Proc. Vertebr. Pest Conf. 13:293–294.

Haney, J. 2007. Wildlife compensation schemes from around the world: an annotated bibliography. Conserv. Sci. Econ. Program, Defenders Wildl. www.defenders.org /resources/publications/programs_and_policy/science _and_economics/wildlife_compensation_schemes_from _around_the_world.pdf. Accessed 5/23/11.

Haney, J., T. Kroeger, F. Casey, et al. 2007. Wilderness discount on livestock compensation costs for imperiled gray wolf Canis lupus. USDA For. Serv. Proc. RMRS-P 49:1–11.

Hansen, M. 2010. The Asian carp threat to the Great Lakes. Letter dated February 9. www.glfc.org/fishmgmt/Hansen _testimony_aisancarp.pdf. Accessed 2/27/11. 8 pp.

Hardin, G. 1968. The tragedy of the commons. Science 162:1243–1248.

Hardy, A., J. Fuller, M. Huijser, et al. 2007. Evaluation of wildlife crossing structures and fencing on U. S. Highway 93 Evaro to Polson Phase 1: preconstruction data collection and finalization of evaluation plan. Final Rep.

FHWA/MT-06-008/1744-1 Dep. Transp., State Mont., Fed. Highw. Adm., U. S. Dep. Transp. 213 pp.

Harlock Jackson. 2006. Australian animal welfare strategy: review of existing animal welfare arrangements for the companion animals working group. Fitzroy, Vic., Aust. 67 pp.

Hart, D. and W. Sussman. 2005. Man the hunted: primates, predators, and human evolution. Westview Press, Boulder, Colo. 312 pp.

Havelaar, A. 2007. Methodological choices for calculating the disease burden and cost-of-illness of foodborne zoonoses in European countries. MedVetNet EU. http://ctsu.nottingham.ac.uk/idea2008/docs/Report_07-002%20Cost%20of%20Illness.pdf. Accessed 7/18/11. 16 pp.

Hawthorne, D. 2004. The history of federal and cooperative animal damage control. Sheep Goat Res. J. 19:13–15.

Heinrich, J. and S. Craven. 1990. Evaluation of three damage abatement techniques for Canada Geese. Wildl. Soc. Bull. 18:405–410.

Helle, E. and K. Kauhala. 1993. Age structure, mortality, and sex ratio of the raccoon dog in Finland. J. Mammal. 74:936–942.

Higgins, C. and W. Mitsch. 2001. The role of muskrats (Ondatra zibethicus) as ecosystem engineers in created freshwater marshes. Annu. Rep., Olentangy River Wetland Res. Park, Columbus, Ohio, pp. 81–86.

Hille, C., F. Osborn, and A. Plumptre. 2002. Human-wildlife conflict: identifying the problem and possible solutions. Wildl. Conserv. Soc., New York. 137 pp.

Hoare, R. 2001. A decision support system for managing human-elephant conflict situations in Africa. Afr. Elephant Spec. Group, Species Survival Comm., IUCN. 103 pp.

Hodge, G., ed. 1991. Pocket guide to the humane control of wildlife in cities & towns. Falcon Press for Humane Soc. USA. 112 pp.

Hoffmann, A., B. Montgomery, J. Popovici, et al. 2011. Successful establishment of Wolbachia in Aedes populations to suppress dengue transmission. Nature 476:454–457.

Holcomb, L. 1976. Experimental use of Av-Alarm for repelling Quelea from rice in Somalia. Proc. Bird Control Semin. 7:275–278.

Holdenrieder, O. and T. Sieber. 2007. First record of Disula destructiva in Switzerland and preliminary inoculation experiments on native European Cornus species. Pp. 51–56 in H. Evans and T. Oszako, eds. Alien invasive species and international trade. For. Res. Inst., Warsaw, Poland. 179 pp.

Hopfenberg, R. 2003. Human carrying capacity is determined by food availability. Popul. Environ. 25(2):109–117.

Howald, G., C. Donlan, J. Galvan, et al. 2007. Invasive rodent eradication on islands. Conserv. Biol. 21:1258–1268.

Howard, W. 1994. Rattlesnakes. Pp. F21–F26 in S. Hygnstrom, R. Timm, and G. Larson. Prevention and control of wildlife damage. Univ. Nebr. Coop. Ext., Lincoln.

Howells, R. and G. Garrett. 1992. Status of some exotic sport fishes in Texas waters. Tex. J. Sci. 44:317–324.

Huijser, M., J. Duffield, A. Clevenger, et al. 2009. Cost-benefit analyses of mitigation measures aimed at reducing collisions with large ungulates in the United States and Canada; a decision support tool. Ecol. Soc. 14(2):15.

Huijser, M., P. McGowen, J. Fuller, et al. 2007. Wildlife-vehicle collision reduction study. Rep. to Congr., U.S. Dep. Transp., Fed. Highw. Admin, Washington D.C.

Hunt, R. and A. McDougall. 2005. Managing coyotes in Australia: are we prepared? Proc. NSW Pest Anim. Control Conf. 3. http://205.186.139.30/wp-content/uploads/2010/03/Hunt2005c.pdf. Accessed 4/22/11. 5 pp.

Hutson, C., K. Lee, J. Abel, et al. 2007. Monkeypox zoonotic associations: insights from laboratory evaluation of animals associated with the multi-state US outbreak. Am. J. Trop. Med. Hyg. 76:757–768.

Hygnstrom, S. and S. Craven. 1988. Electric fences and commercial repellents for reducing deer damage in cornfields. Wildl. Soc. Bull. 16:291–296.

Hygnstrom, S., R. Timm, and G. Larson, eds. 1994. Prevention and control of wildlife damage. Univ. Nebr. Coop. Ext., Lincoln.

Hygnstrom, S. and K. VerCauteren. 2000. Cost-effectiveness of five burrow fumigants for managing black-tailed prairie dogs. Int. Biodeterior. Biodegrad. 45:159–168.

IdahoStatesman.com. 2011. After Idaho gets wolves delisted, Congress takes aim at Endangered Species Act. www.idahostatesman.com/2011/06/26/1704236/endangered-species-act-under-fire.html. Accessed 7/11/11.

Ingold, D. 1994. Influence of nest-site competition between European starlings and woodpeckers. Wilson Bull. 106:227–241.

Ingold, D. 1998. The influence of starlings on flicker reproduction when both naturally excavated cavities and artificial nest boxes are available. Wilson Bull. 110:218–225.

Innes, J., B. Warburton, D. Williams, et al. 1995. Large scale poisoning of ship rats (Rattus rattus) in indigenous forests of the North Island, New Zealand. New Zeal. J. Ecol. 19(1):5–17.

Inslerman, R., J. Miller, D. Baker, et al. 2006. Baiting and supplemental feeding of game wildlife species. The Wildlife Soc. Techn. Rev. 06-1, Bethesda, Mary. 56 pp.

IPCC. 2007. Summary for policymakers. In S. Solomon, D. Qin, M. Manning, et al., eds. Climate change 2007: the physical science basis. Contribution of Working Group I to the Fourth Assessment Report of the Intergovernmental Panel on Climate Change. Cambridge Univ. Press, Cambridge, U.K.

Islam, Z. and M. Hossain. 2003. Response of rice plants to rat damage at the reproductive phase. IRRN 28:45–46.

Jackson, W. and A. Ashton. 1986. Case histories of anti-coagulant resistance. Pp. 355–370 in Pesticide resistance: strategies and tactics for management. Natl. Acad. Press, Washington, D.C.

Jackson, W. and R. Marsh, eds. 1977. Test methods for vertebrate pest control and management materials. STP625-EB. Am. Soc. Test. Mater. 258 pp.

Jacobson, C. 2008. Wildlife conservation and management in the 21st century: understanding challenges for institutional transformation. Ph.D. thesis, Cornell Univ., Ithaca, N.Y. 147 pp.

Jacobson, E., P. Ginn, J. Troutman, et al. 2005. West Nile virus infection in farmed American alligators (*Alligator mississippiensis*) in Florida. J. Wildl. Dis. 41:96–106.

Jadhav, D. 2010. Piscicidal plants used by Bhil tribe of Ratlam district (M. P.) India. J. Econ. Taxon. Bot. 34(4):757–759.

Jansen, C. 1974. Behavior patterns observed in coyote-sheep interactions. M.S. thesis, Colo. State Univ., Ft. Collins. 57 pp.

Jenkins, M., S. Jose, and P. White. 2007. Impacts of an exotic disease and vegetation change on foliar calcium cycling in Appalachian forests. Ecol. Appl. 17:869–881.

Ji, W. 2009. A review of the potential of fertility control to manage brushtail possums in New Zealand. Human-Wildl Conflicts 31:20–29.

Jimenez, M. 2003. Progress on water hyacinth (*Eichhornia crassipes*) management. In R. Labrada, ed. Weed management for developing countries, Addendum a. FAO Plant Prod. Prot. Pap. 120.

Johnson, C., J. Isaac, and D. Fisher. 2007. Rarity of top predator triggers continent-wide collapse of mammal prey: dingoes and marsupials in Australia. P. R. Soc. B 274:341–346.

Johnson, C. and S. Wroe. 2003. Causes of extinction of vertebrates during the Holocene of mainland Australia: arrival of the dingo, or human impact? Holocene 13:941–948.

Johnson, D. and P. Stiling. 1998. Distribution and dispersal of *Cactoblastis cactorum* (Lepidoptera: Pyralidae), an exotic Opuntia-feeding moth, in Florida. Fla. Entomol. 81:12–22.

Jojola, S., G. Witmer, and D. Nolte. 2005. Nutria: an invasive rodent pest or valued resource? Proc. Wildl. Damage Manage. Conf. 11:120–126.

Jones, H. 2010. Prognosis for ecosystem recovery following rodent eradication and seabird restoration in an island archipelago. Ecol. Appl. 20:1204–1216.

Jones, H., B. Tershy, E. Zavaleta, et al. 2008. Severity of the effects of invasive rats on seabirds: a global review. Conserv. Biol. 22:16–26.

Jones, J. and J. Dubey. 2010. Waterborne toxoplasmosis—recent developments. Exp. Parasitol. 124:10–25.

Josselyn, J. 1674. New England's rarities discovered. Applewood Books, Bedford, Mass. 118 pp.

Juckett, G. and J. Hancox. 2002. Venomous snakebites in the United States: management review and update. Am. Fam. Physician 65:1367–1374.

Justia US Law. Undated. 184 F.2d 616: Sickman et al. v. United States (two cases). Ryal et al. v. United States. http://law .justia.com/cases/federal/appellate-courts/F2/184/616 /304065/. Accessed 7/16/11.

Kadlec, R., J. Pries, and H. Mustard. 2006. Muskrats (*Ondatra zibethicus*) in treated wetlands. Ecol. Eng. 29:143–153.

Kangaroo Advisory Committee. 1997. Living with Eastern Grey Kangaroos in the A.C.T.—Public Land Report 3. Aust. Cap. Territory Gov., Canberra, Aust.

Karanth, K. and J. Nichols. 1998. Estimation of tiger densities in India using photographic captures and recaptures. Ecol. 79:2852–2862.

Kateregga, E. and T. Sterner. 2007. Indicators for an invasive species: water hyacinths in Lake Victoria. Ecol. Indic. 7:362–370.

Kats, L. and R. Ferrer. 2003. Alien predators and amphibian declines: review of two decades of science and the transition to conservation. Divers. Distrib. 9:99–110.

Kauffman, C. 2007. Histoplasmosis: a clinical and laboratory update. Clin. Microbiol. Rev. 21:115–132.

Kauffman, M., J. Brodie, and E. Jules. 2010. Are wolves saving Yellowstone's aspen? A landscape-level test of a behaviorally mediated trophic cascade. Ecol. 91:2742–2755.

Kaufman, L. 1992. Catastrophic change in species-rich freshwater ecosystems: the lessons of Lake Victoria. BioScience 42: 846–858.

Kay, C. 2007. Were native people keystone predators? A continuous-time analysis of wildlife observations made by Lewis and Clark in 1804–1806. Can. Field-Nat. 121:1–16.

Kay, S. and S. Hoyle. 2001. Mail order, the Internet, and invasive aquatic weeds. J. Aquat. Plant Manage. 39:88–91.

Keitt, B., K. Campbell, A. Saunders, et al. 2011. The global islands invasive vertebrate eradication database: a tool to improve and facilitate restoration of island ecosystems. Eradication Manage. IUCN, Gland, Switz.

Kellert, S. 1979. Public attitudes toward critical wildlife and natural habitat issues. U.S. Fish Wildl. Serv. Rep. Washington, D.C.

Kellert, S. 1980. Activities of the American public relating to animals, phase 2. U.S. Dept. of Interior, Fish Wildl. Svc. 178 pp.

Kellert, S. 1991. Japanese perceptions of wildlife. Conserv. Biol. 5(3):297–308.

Kellert, S. 1993. The biological basis for human values of nature. Pp. 42–69 in S. Kellert and E. Wilson, eds. The biophilia hypothesis. Island Press / Shearwater Books, Washington, D.C.

Kellert, S. 2007. Biophilia, children and restoring connections to nature in the modern built environment. www .childrenandnature.org/reports/9_2006/PPTs/kellert .pdf. Accessed 6/13/11.

Kellert, S. and E. O. Wilson. 1993. Biophilia hypothesis. Island Press, Washington, D.C. 492 pp.

Kessler, C. 2002. Eradication of feral goats and pigs and consequences for other biota on Sarigan Island, Commonwealth of the Northern Mariana Islands. Pp. 132–140 in C. Veitch and M. Clout, eds. Turning the tide: the eradication of invasive species. Occas. Pap. No. 27, IUCN Species Survival Comm.

Kikillus, K., K. Hare, and S. Hartley. 2009. Minimizing false-negatives when predicting the potential distribution of an invasive species: a bioclimatic envelope for the red-eared slider at global and regional scales. Anim. Conserv. 13, Suppl. 1:5–15.

Kilpatrick, H., S. Spohr, and K. Lima. 2001. Effects of population reductions on home range size of white-tailed deer in high densities. Can. J. Zool. 79:949–954.

King, D. 2005. Interactions between the American white pelican and aquaculture in the southeastern United States: an overview. Waterbirds 28, Spec. Publ. 1:83–86.

King, D. and D. Anderson. 2005. Recent population status of the American white pelican: a continental perspective. Waterbirds 28, Spec. Publ. 1:48–54.

King, D., L. Twigg, and J. Gardner. 1989. Tolerance to sodium monofluoroacetate in dasyurids from western Australia. Aust. Wildl. Res. 16:131–140.

King, L. 2011. Beehive fence construction manual. Elephants Bees Proj. Save the Elephants, Kenya. 17 pp.

King, L., I. Douglas-Hamilton, and F. Vollrath. 2011. Beehive fences as effective deterrents for crop-raiding elephants: field trials in northern Kenya. Afr. J. Ecol. doi: 10.111/j.1 365-2028.2011.01275.x. Accessed 8/20/11.

King, L., A. Lawrence, I. Douglas-Hamilton, et al. 2009. Beehive fence deters crop-raiding elephants. Afr. J. Ecol. 47:131–137.

Kluckhohn, C. 1951. Values and value orientations in the theory of action. Pp. 388–433 in T. Parsons and E. Shils. Toward a general theory of action. Harvard Univ. Press, Cambridge, Mass.

Knight, J. 1994. Mountain lion. Pp. C93–C99 in S. Hygnstrom, R. Timm, and G. Larson, eds. Prevention and control of wildlife damage. Univ. Nebr. Coop. Ext., Lincoln.

Knowlton, F. 1992. Reflections on an onerous deed. Probe, December, p. 1.

Koehn, J. 2004. Carp (Cyprinus carpio) as a powerful invader in Australian waterways. Freshwater Biol. 49:882–894.

Koehn, J., A. Brumley, and P. Gehrke. 2000. Managing the impacts of carp. Bur. Rural Sci., Dep. Agric., Fish. For.–Aust., Canberra. 249 pp.

Koenig, W. 2003. European starlings and their effect on cavity-nesting birds. Conserv. Biol. 17:1134–1140.

Kowalczyk, R., A. Zalewski, B. Jędrzejewska, et al. 2009. Reproduction and mortality of invasive raccoon dogs (Nyctereutes procyonoides) in the Bialowieza Primeval Forest (eastern Poland). Ann. Zool. Fenn. 46:291–301.

Kreeger, T. 1997. Overview of delivery systems for the administration of contraceptives to wildlife. Pp. 29–48 in T. Kreeger, ed. Contraception in wildlife management. Tech. Bull. No. 1853. Anim. Plant Health Insp. Serv., USDA, Washington, D.C.

Krutilla, J. 1967. Conservation reconsidered. Am. Econ. Rev. 57:777–786.

Lack, D. 1954. The natural regulation of animal numbers. Oxford Univ. Press, Oxford, U.K. 343 pp.

Langley, R. 2005. Alligator attacks on humans in the United States. Wilderness Environ. Med. 16:119–124.

Langley, R. 2010. Adverse encounters with alligators in the United States: an update. Wilderness Environ. Med. 21:156–163.

Lapidge, S., D. Dall, J. Dawes, et al. 2005. Starlicide—the benefits, risks and industry need for DRC-1339 in Australia. Proc. Australas. Vertebr. Pest Conf. 13:235–238.

Lapidge, S. and C. Eason. 2010. Pharmacokinetics and methaemoglobin reductase activity as determinants of species susceptibility and non-target risks from sodium nitrite manufactured feral pig baits. Report for the Australian Government Department of the Environment, Water, Heritage and the Arts. Canberra, Australia. 18 pp.

Larkin, R., T. Van Deelen, R. Sabick, et al. 2003. Electronic signaling for prompt removal of an animal from a trap. Wildl. Soc. Bull. 31:392–398.

Latschar, J. 2009. Deer management program 2009. Press Release, Natl. Park Serv. http://civilwarinteractive.com /Newswire/?p=4260. Accessed 10/4/2011.

Lehrman, D. 1964. The reproductive behavior of ring doves. Sci. Am. 211:48–54.

Lennon, R., J. Hunn, and R. Schnick. 1971. Reclamation of ponds, lakes, and streams with fish toxicants: a review. FAO Fisheries Tech. Papers, T100. Rome. 99 pp.

Leopold, A. 1933. Game management. Univ. Wis. Press, Madison. 481 pp.

Leprieur, F., O. Beauchard, S. Blanchet, et al. 2008. Fish invasions in the world's river systems: when natural processes are blurred by human activities. PLoS Biol, Feb 6.Accessed 1/16/2011.

Lever, C. 2003. Naturalized reptiles and amphibians of the world. Oxford Univ. Press, Oxford, U.K. 318 pp.

Levins, R. 1969. Some demographic and genetic consequences of environmental heterogeneity for biological control. Bull. Entomol. Soc. Am. 15:237–240.

Li, K. and O. DeMasi. 2009. It's a coyote eat deer feed tick world: a deterministic model of predator-prey interaction in the northeast. Snapshots in Res., vol 11, Undergrad. Sci. Eng. Rev., Rice Univ. 9 pp.

Liddick, D. 2006. Eco-terrorism: radical environmental and animal liberation movements. Praeger, Westport, Conn. 189 pp.

Lindgren, P., T. Sullivan, and D. Crump, 1995. Review of synthetic predator odor semiochemicals as repellents for wildlife management in the Pacific Northwest. Pp. 217–230 in USDA Natl. Wildl. Res. Cent. Symp., Natl. Wildl. Res. Cent. Repellent Conf.

Linhart, S., G. Dasch, R. Johnson, et al. 1992. Electronic frightening devices for reducing coyote predation on domestic sheep: efficacy under range conditions and operational use. Proc. Vertebr. Pest Conf. 15:386–392.

Linhart, S. and F. Knowlton. 1975. Determining the relative abundance of coyotes by scent station lines. Wildl. Soc. Bull. 3:119–124.

Link, R. 2005. Living with wildlife: muskrats. Washington Dep. Fish Wildl. http://wdfw.wa.gov/living/muskrats. html. Accessed 6/4/11.

Linnell, J., E. Nilsen, U. Lande, et al. 2005. Zoning as a means of mitigating conflicts with large carnivores: principles and reality. Pp. 162–175 in R. Woodroffe, S. Thirgood, and A. Rabinowitz, eds. People and wildlife: conflict or coexistence? Cambridge Univ. Press, Cambridge, U.K. 497 pp.

Linz, G. and H. Homan. 2011. Use of glyphosate for managing invasive cattail (Typha spp.) to disperse blackbird (Icteridae) roosts. Crop Prot. 30(2):98–104.

Linz, G., D. Schaaf, P. Mastrangelo, et al. 2004. Wildlife conservation sunflower plots as dual-purpose wildlife management strategy. Proc. Vertebr. Pest Conf. 21:291–294.

Lips, K., F. Brem, R. Brenes, et al. 2006. Emerging infectious disease and the loss of biodiversity in a neotropical amphibian community. Proc. Natl. Acad. Sci. USA 103:3165–3170.

Littauer, G. 1990. Avian predators. Frightening techniques for reducing bird damage at aquaculture facilities. Publ. No. 401, South. Reg. Aquacult. Cent. 4 pp.

Lockett, M. 1998. The effect of rotenone on fishes and its use as a sampling technique: a survey. Z. Fischkd. 5:13–45.

Lockwood, J., P. Cassey, and T. Blackburn. 2005. The role of propagule pressure in explaining species invasions. Trends Ecol. Evol. 20:223–228.

Long, D., T. Campbell, and G. Massei. 2010. Evaluation of swine-specific feeder systems. Rangelands 32:8–13.

Long, K. and A. Robley. 2004. Cost effective feral animal exclusion fencing for areas of high conservation value in Australia. Dep. Sustainability Environ., Heidelberg, Melbourne, Aust. 54 pp.

Longcore, T., C. Rich, and L. Sullivan. 2009. Critical assessment of claims regarding management of feral cats by trap-neuter-return. Conserv. Biol. www.audubond ebspark.org/chapter_assets/Longcoreetal2009ConBio .pdf. Accessed 7/16/11.

Loope, L., O. Hamann, and C. Stone. 1988. Comparative conservation biology of oceanic archipelagoes. Bioscience 38:272–282.

Lotka, A. 1925. Elements of physical biology. Williams & Wilkins, Baltimore.

Louda, S., and P. Stiling. 2004. The double-edged sword of biological control in conservation and restoration. Conserv. Biol. 18:50–53.

Lowe, S., M. Browne, S. Boudejelas, et al. 2000. 100 of the world's worst alien species: a selection from the global invasive species database. Invasive Species Spec. Group, Species Survival Comm., World Conserv. Union. 12 pp.

Lowney, M. 1999. Damage by black and turkey vultures in Virginia, 1990–1996. Wildl. Soc. Bull. 27:715–719.

Luby, S., M. Rahman, M. Hossain, et al. 2006. Foodborne-transmission of Nipah virus, Bangladesh. Emerg. Infect. Dis. 12:1888–1894.

Lucas, G. and H. Synge. 1978. The IUCN plant red data book. World Conserv. Union, Morges, Switz.

MacArthur, R. and E. Wilson. 1967. The theory of island biogeography. Princeton Univ. Press, Princeton, N.J.

MacInnes, C., S. Smith, R. Tinline, et al. 2001. Elimination of rabies from red foxes in eastern Ontario. J. Wildl. Dis. 37:119–132.

Mack, R., D. Simberloff, W. Lonsdale, et al. 2000. Biotic invasions: causes, epidemiology, global consequences, and control. Ecol. Appl. 10:689–710.

Madden, F. 2004. Creating coexistence between humans and wildlife: global perspectives on local efforts to address human-wildlife conflict. Hum. Dimens. Wildl. 9:247–257.

Mahy, B. and C. Brown. 2000. Emerging zoonoses: crossing the species barrier. Rev. Sci. Tech. 19:33–40.

Malthus, R. 2008. An essay on the principle of population. Oxford Univ. Press, Oxford, U.K. (orig. pub. 1798). 208 pp.

Manfredo, M. 2008. Who cares about wildlife? Springer Science + Business Media, New York. 228 pp.

Manfredo, M. and A. Bright. 2008. Attitudes and study of human dimensions of wildlife. Pp. 75–109 in M.

Manfredo. Who cares about wildlife? Springer Science + Business Media, New York. 228 pp.

Manfredo, M. and T. Teel. 2008. Integrating concepts: demonstration of a multilevel model for exploring the rise of mutualism value orientations in post-industrial society. Chapter 8 in M. Manfredo. Who Cares About Wildlife? Springer Science + Business Media, LLC.

Manfredo, M., T. Teel, and H. Zinn. 2009. Understanding global values toward wildlife. Chapter 3 in M. Manfredo, J. Vaske, P. Brown, et al., eds. Wildlife and society: the science of human dimensions. Island Press, Washington, D.C. 350 pp.

Marcus, J., J. Dinan, R. Johnson, et al. 2007. Directing nest site selection of least terns and piping plovers. Waterbirds 30:251–258.

Marker, L. and S. Sivamani. 2009. Policy for human-leopard conflict management in India. CAT News 50:23–26.

Marsh, R., A. Koehler, and T. Salmon. 1990. Exclusionary methods and materials to protect plants from pests—a review. Proc. Vertebr. Pest Conf. 14:174–180.

Martin, J., R. Raid, and L. Branch. 2007. Quantifying damage potential of three rodent species on sugarcane. J. Am. Soc. Sugar Cane Technol. 27:48–54.

Martin, P. 1971. Prehistoric overkill. Pp. 612–624 in T. Detwyler, ed. Man's impact on environment. McGraw-Hill, New York. 731 pp.

Martin, P. 2005. Twilight of the mammoths: ice age extinctions and the rewilding of America (organisms and environments). Univ. Calif. Press, Berkeley. 269 pp.

Martorello, D., T. Eason, and M. Pelton. 2001. A sighting technique using cameras to estimate population size of black bears. Wildl. Soc. Bull. 29:560–567.

Maslow, A. 1954. Motivation and personality. Harper & Row, New York.

Mason, J., ed. 1997. Repellents in wildlife management. Proc. Second DWRC Spec. Symp., Natl. Wildl. Res. Cent., Fort Collins, Colo.

Mason, J., M. Avery, J. Glahn, et al. 1991. Evaluation of methyl anthranilate and starch-plated dimethyl anthranilate as bird repellent feed additives. J. Wildl. Manage. 55:182–187.

Mason, J., L. Clark, and P. Shah. 1992. Taxonomic differences between birds and mammals in their responses to chemical irritants. Pp. 311–317 in R. Doty and D. Müller-Schwarze, eds. Chemical signals in vertebrates, vol 6. Plenum Press, New York.

Mastro, L., M. Conover, and S. Frey. 2008. Deer-vehicle collision prevention techniques. Human-Wildl. Conflicts 2:80–92.

Mauldin, R. and P. Savarie. 2010. Acetaminophen as an oral toxicant for Nile monitor lizards (*Varanus niloticus*) and Burmese pythons (*Python molurus bivittatus*). Wildl. Res. 37:215–222.

Mayo, J., T. Straka, and D. Leonard. 2003. The cost of slowing the spread of the gypsy moth (*Lepidoptera: Lymantriidae*). J. Econ. Entomol. 96:1448–1454.

McAfee, W. 1966. Lahontan cutthroat trout. Pp. 225–231 in A. Calhoun, ed. Inland Fish. Manage. Calif. Dep. Fish Game, Sacramento.

McBeath, D. 1941. Whitetail traps and tags. Mich. Conserv. 10:6–7, 11.

McCoy, N. 2002. Economic tools for managing impacts of urban Canada geese. Pp. 117–122 in L. Clark, J. Hone, J. Shivik, et al., eds. 2002. Human conflicts with wildlife: economic considerations. Proc. Third NWRC Spec. Symp., Natl. Wildl. Res. Cent., Fort Collins, Colo.

McCullough, D. 1997. Irruptive behavior in ungulates. Pp. 69–98 in W. McShea, H. Underwood, and J. Rappole, eds. The science of overabundance: deer ecology and population management. Smithsonian Books, Washington, D.C.

McDonald, K. 1990. *Rheobatrachus* Liem and *Taudactylus* Straughan & Lee (*Anura: Leptodactylidae*) in Eungella National Park, Queensland: distribution and decline. Trans. R. Soc. South Aust. 114:187–194.

McDowall, R. 1990. New Zealand freshwater fishes: a natural history and guide. Heinemann Reed, Auckland, N.Z.

McFadyen, R. 2000. Successes in biological control of weeds. Pp. 3–14 in N. Spencer, ed. Proceedings X International Symposium on Biological Control of Weeds. Mont. State Univ., Bozeman.

McGlynn, T. 1999. The worldwide transfer of ants: geographical distribution and ecological invasions. J. Biogeogr. 26:535–548.

McIlroy, J. 1994. Susceptibility of target and non-target animals to 1080. Pp. 90–95 in Seawright, A. and C. Eason, eds. Proc. Sci. Workshop on 1080, Royal Soc. New Zealand, Misc. Series 28. Wellington.

McKinney, M. 2002. Urbanization, biodiversity, and conservation. BioScience 52:883–890.

McMichael, A. 2004. Environmental and social influences on emerging infectious diseases: past, present and future. Philos. Trans. R. Soc. Lond. Ser. B 359:1049–1058.

McMurtry, D. 2002. Controlling conflicts with urban Canada geese in Missouri. Mo. Dep. Conserv. PLS 051, USDA Wildl. Serv. 25pp.

McNay, M. 2002. A case history of wolf-human encounters in Alaska and Canada. Alaska Dep. Fish Game Wildl. Tech. Bull. 13. 52 pp.

Merrill, J., E. Cooch, and P. Curtis. 2006. Managing an overabundant deer population by sterilization: effects of immigration, stochasticity and the capture process. J. Wildl. Manage. 70:268–277.

Meyer, J-Y and J. Florence. 1996. Tahiti's native flora endangered by the invasion of *Miconia calvescens*. J. Biogeogr. 23:775–781.

Millennium Ecosystem Assessment. 2005. Ecosystems and human well-being: General synthesis. Island Press, Washington, D.C. 155 pp.

Miller, J. 1994. Muskrats. Pp. B61–B69 in S. Hygnstrom, R. Timm, and G. Larson, eds. Prevention and control of wildlife damage. Univ. Nebr. Coop. Ext., Lincoln.

Miller, J. 2007. Evolution of the field of wildlife damage management in the United States and future challenges. Human-Wildl. Conflicts 1(1):13–20.

Miller, J. and G. Yarrow. 1994. Beavers. Pp. B1–B11 in S. Hygnstrom, R. Timm, and G. Larson, eds. Prevention and control of wildlife damage. Univ. Nebr. Coop. Ext., Lincoln.

Miller, M., P. Conrad, W. Miller, et al. 2008. Type X *Toxoplasma gondii* in a wild mussel and terrestrial carnivores from coastal California: new linkages between terrestrial mammals, runoff and toxoplasmosis of sea otters. Int. J. Parasitol. 38:1319–1328.

Moloney, P., and J. Hearne. 2009. The population dynamics of converting properties from cattle to kangaroo production. 18th World IMACS/MODSIM Congress, Cairns, Aust. http://mssanz.org.au/modsim09. Accessed 10/13/2011.

Montag, J. 2003. Compensation and predator conservation: limitations of compensation. In C. Angst, J. Landry, J. Linnell, et al., eds. Carnivore Prev. News 6:2–6.

Morey, D. 1994. The early evolution of the domestic dog. Am. Sci. 82:336–347.

Morey, P., E. Gese, and S. Gehrt. 2007. Spatial and temporal variation in the diet of coyotes in the Chicago metropolitan area. Am. Midl. Nat. 158:147–161.

Morrison, P. and R. Allcorn. 2006. The effectiveness of different methods to deter large gulls (*Larus spp.*) from competing with nesting terns (*Sterna spp.*) on Coquet Island RSPB reserve, Northumberland, Engl. Conserv. Evidence 3:84–87.

Moser, B. and G. Witmer. 2000. The effects of elk and cattle foraging on the vegetation, birds, and small mammals of the Bridge Creek Wildlife Area, Oregon. Int. Biodeterior. Biodegrad. 45:151–157.

Muchapondwa, E. 2003. The economics of community-based wildlife conservation in Zimbabwe. Ph.D. Thesis, Göttenborg Univ., Swed. 221 pp.

Mungall, E. 2001. Exotics. Pp. 736–764 in S. Demarais and P. Krausman, eds. Ecology and management of large mammals in North America. Prentice Hall, Upper Saddle River, N.J.

Muñoz-Igualada, J., J. Shivik, F. Domínguez, et al. 2010. Traditional and new cable restraint systems to capture fox in central Spain. J. Wildl. Manage. 74(1):181–187.

Murie, A. 1944. The wolves of Mount McKinley. Fauna Ser. No. 5, U.S. Natl. Park Serv. 238 pp.

Murphy, E., C. Eason, S. Hix, et al. 2007. Developing a new toxin for potential control of feral cats, stoats and wild dogs in New Zealand. Pp. 469–473 in G. Witmer, W. Pitt, and K. Fagerstone., eds. Managing Vertebrate Invasive Species. Proc. Int. Symp. USDA/APHIS/WS, Natl. Wildl. Res. Ctr., Fort Collins, Colorado.

Musiani, M., C. Mamo, L. Boitani, et al. 2003. Wolf depredation trends and the use of barriers to protect livestock in western North America. Conserv. Biol. 17:1–10.

Nagano, N., S. Oana, Y. Nagano, et al. 2006. A severe *Salmonella enterica* serotype Paratyphi B infection in a child related to a pet turtle, *Trachemys scripta elegans*. Jpn. J. Infect. Dis. 59:132–134.

Nash, P., C. Furcolow, K. Bynum, et al. 2007. 20, 25–diazacholesterol as an oral contraceptive for black-tailed prairie dog population management. Human-Wildl. Conflicts 1(1):60–67.

NASS. 2010. Sheep and goats death loss. www.mohairusa .com/misc/Sheep_and_Goats_Death_Loss.pdf. Accessed 5/22/11.

Nation, T. 2007. The influence of flowering dogwood (*Cornus florida*) on land snail diversity in a southern mixed hardwood forest. Am. Midl. Nat. 157:137–148.

National Geographic. 2009. Atlas of the human journey. https://genographic.nationalgeographic.com/genographic/lan/en/atlas.html. Accessed 4/23/2009.

NBII. Undated. Wildlife diseases node. http://wildlifedisease.nbii.gov/. Accessed 3/12/11.

Nelson, K. 2005. Cooperative rabies management program national report 2005. USDA–Wildl. Serv., Concord, N.H. 137 pp.

Nelson, L. 2009. Medieval history lectures: the great famine (1315–1317) and the black death (1346–1351). www.vlib.us/medieval/lectures/. Accessed 5/17/09.

Nico, L. 1999. *Cyprinus carpio*. Nonindigenous Aquat. Species Database, Gainesville, Fla. http://nas.er.usgs.gov/queries/FactSheet.aspx?speciesID=4. Accessed 1/16/2011.

Nico, L. 2005. *Hypophthalmichthys molitrix*. USGS Nonindigenous Aquat. Species Database, Gainesville, Fla. http://nas.er.usgs.gov/queries/FactSheet.asp?speciesID=549.

Nico, L. and P. Fuller. 2011. *Cyprinella lutrensis*. USGS Nonindigenous Aquat. Species Database, Gainesville, Fla. http://nas.er.usgs.gov/queries/FactSheet.aspx?speciesID=518. Revision Date: 4/19/2010.

Nicolaus, L., J. Casell, R. Carlson, et al. 1983. Taste-aversion conditioning of crows to control predation on eggs. Science 220:212–214.

Ninni, A. 1865. Sulla mortalita dei gambari (*Astacus fluviatilis* L.) nel veneto e piu particalarmenta nella provincia trevigiana. Alti Inst. Veneto Ser. III 10:1203–1209.

Nogales, M., A. Martín, B. Tershey, et al. 2004. A review of feral cat eradication on islands. Conserv. Biol. 18(20):310–319.

Nolte, D. and M. Dykzeul. 2002. Wildlife impacts on forest resources. Pp. 163–168 in L. Clark, ed. Human conflicts with wildlife: economic considerations. USDA/APHIS. Fort Collins, Colo.

NOVA. 2012. www.pbs.org/wgbh/nova/.

Nugent, B., K. Gagne, and M. Dillingham. 2008. Managing gulls to reduce fecal coliform bacteria in a municipal drinking water source. Proc. Vertebr. Pest Conf. 23:26–30.

Nussey, D., L. Kruuk, A. Morris, et al. 2007. Environmental conditions in early life influence aging rates in a wild population of red deer. Curr. Biol. 17:R1000–R1001.

NWRC. Undated. Acetaminophen for control of brown tree snake. www.aphis.usda.gov/wildlife_damage/nwrc/registration/content/56228-34%20Acetaminophen%20Label%2007-11SPECIMEN.pdf. Accessed 10/27/2010.

Nyahongo, J. and E. Røskaft. 2011. Perception of people towards lions and other wildlife killing humans, around Selous Game Reserve, Tanzania. Int. J. Biodiversity Conserv. 3(4):110–115.

O'Dowd, D., P. Green, and P. Lake. 2003. Invasional 'meltdown' on an oceanic island. Ecol. Lett. 6:812–817.

Odum, E. 1971. Fundamentals of ecology. 3rd ed. Saunders, Philadelphia. 574 pp.

O'Gara, B. and D. Getz. 1986. Capturing golden eagles using a helicopter and net gun. Wildl. Soc. Bull. 14:400–402.

Ogden, D. 1971. How national policy is made. http://ageconsearch.umn.edu/bitstream/17268/1/ar710005.pdf. Accessed 7/7/11.

Ogutu-Ohwayo, R. 2004. Management of the Nile perch, *Lates niloticus*, fishery in Lake Victoria in light of the changes in its life history characteristics. Afr. J. Ecol. 42: 306–314.

O'Hare, J., J. Eisemann, K. Fagerstone, et al. 2007. Use of alpha-chloralose by USDA Wildlife Services to immobilize birds. Proc. Wildl. Damage Manage. Conf. 12:103–113.

Okarma, H. and W. Jędrzejewski. 1997. Livetrapping wolves with nets. Wildl. Soc. Bull. 25:78–82.

Olden, J., J. McCarthy, J. Maxted, et al. 2006. The rapid spread of rusty crayfish (*Orconectes rusticus*) with observations on native crayfish declines in Wisconsin (USA) over the past 130 years. Biol. Invasions 8:1621–1628.

Olsen, B., V. Munster, A. Wallensten, et al. 2006. Global patterns of influenza A virus in wild birds. Science 312:384–388.

Oppel, S., B. Beaven, M. Bolton, et al. 2010. Eradication of invasive mammals on islands inhabited by humans and domestic animals. Conserv. Biol. 25(2):232–240.

Oregon Wild. 2010. Lawsuit filed to stop federal, state-sanctioned killing of endangered wolves. www.oregonwild.org/about/press-room/press-releases/lawsuit-filed-to-stop-federal-state-sanctioned-killing-of-endangered-wolves. Accessed 7/10/11.

Osborn, F. and C. Hill. 2005. Techniques to reduce crop loss: human and technical dimensions in Africa. Pp. 72–85 in R. Woodroffe, S. Thirgood, and A. Rabinowitz, eds. People and wildlife: conflict or coexistence? Cambridge Univ. Press, Cambridge, U.K.

Osgood, D. and J. Zieman. 1998. The influence of subsurface hydrology on nutrient supply and smooth cordgrass (*Spartina alterniflora*) production in a developing barrier island marsh. Estuaries 21(4B):767–783.

Oswalt, C. and S. Oswalt. 2010. Documentation of significant losses in *Cornus florida* L. populations throughout the Appalachian ecoregion. Int. J. For. Res. 2010, Artic. ID 401951, 10 pp.

Oswalt, W. 1999. Eskimos and explorers. 2d ed. Univ. Nebr. Press, Lincoln. 341 pp.

Padilla, D. and S. Williams. 2004. Beyond ballast water: aquarium and ornamental trades as sources of invasive species in aquatic ecosystems. Front. Ecol. Environ. 2:131–138.

Paine, R. 1966. Food web complexity and species diversity. Am. Nat. 100:65–75.

Pallister, J., D. Halliday, A. Robinson, et al. 2011. Assessment of virally vectored autoimmunity as a biocontrol strategy for cane toads. PLoS ONE 6(1): e14576. doi:10.1371/journal.pone.0014576. Accessed 8/26/11.

Palmateer, S. 1987. Current and future status of rodenticides and predacides. Proc. Great Plains Wildl. Damage Control Workshop 8: 16–17.

Panagiotakopulu, E. 2004. Pharaonic Egypt and the origins of plague. J. Biogeogr. 31:269–275.

Parkes, J., D. Ramsey, N. Macdonald, et al. 2010. Rapid eradication of feral pigs (*Sus scrofa*) from Santa Cruz Island, California. Biol. Conserv. 143(3):634–641.

Parrish, D. and F. Margraf. 1994. Spatial and temporal patterns of food use by white perch and yellow perch in Lake Erie. J. Freshwater Ecol. 9:29–35.

Parrish, H. 1966. Incidence of treated snakebites in the United States. Public Health Rep. 81:269–276.

Patel, R. and T. Burke. 2009. Urbanization—an emerging humanitarian disaster. N. Engl. J. Med. 361:741–743.

Pauchard, A. and K. Shea. 2006. Integrating the study of non-native plant invasions across spatial scales. Biol. Invasions 8:399–413.

Pearson, D. and R. Fletcher. 2008. Mitigating exotic impacts: restoring deer mouse populations elevated by an exotic food subsidy. Ecol. Appl. 18:321–334.

Peh, K. and N. Sodhi. 2002. Characteristics of nocturnal roosts of house crows in Singapore. J. Wildl. Manage. 66:1128–1133.

Peterjohn, B. and J. Sauer. 1993. North American Bird Breeding Survey annual summary 1990–1991. Bird Popul. 1:52–67.

Phillips, I., R. Vinebrooke, and M. Turner. 2009. Ecosystem consequences of potential range expansions of *Orconectes virilis* and *Orconectes rusticus* crayfish in Canada—a review. Environ. Rev. 17:235–248.

Pigg, J., J. Stahl, M. Ambler, et al. 1993. Two potential sources of exotic fish in Oklahoma. Proc. Okla. Acad. Sci. 73:67.

Pimentel, D., S. McNair, J. Janecka, et al. 2001. Economic and environmental threats of alien plant, animal, and microbe invasions. Agric. Ecosyst. Environ. 84:1–20.

Pimentel, D., M. Whitecraft, Z. Scott, et al. 2010. Will limited land, water, and energy control human population numbers in the future? www.progressivesforimmigrationreform.org/David_Pimentel_Human_Ecology_Class_2009_paper.pdf. Accessed 8/12/11.

Pimentel, D., R. Zuniga, and D. Morrison. 2005. Update on the environmental and economic costs associated with alien-invasive species in the United States. Ecol. Econ. 52:273–288.

Pitt, W., P. Box, and F. Knowlton. 2003. An individual-based model of canid populations: modeling territoriality and social structure. Ecol. Modell. 166:109–121.

Pitt, W., L. Driscoll, and R. Sugihara. 2011. Efficacy of rodenticide baits for the control of three invasive rodent species in Hawaii. Arch. Environ. Contam. Toxicol. 60:533–542.

Pitt, W., D. Vice, and M. Pitzler. 2005. Challenges of invasive reptiles and amphibians. Proc. Wildl. Damage Manage. Conf. 11:112–119.

Plice, L. and T. Balgooyen. 1999. A remotely operated trap for American kestrels using nestboxes. J. Field Ornithol. 70:158–162.

Pochop, P., J. Cummings, and R. Engeman. 2001. Field evaluation of a visual barrier to discourage gull nesting. Pac. Conserv. Biol. 7:143–145.

Porta, J. 1658. 15th booke of natural magick—of fishing, fowling and hunting. Basic Books, New York. 409 pp.

Portney, P. Undated. The concise encyclopedia of economics: Benefit-cost anaylsis. www.econlib.org/library/Enc/BenefitCostAnalysis.html. Accessed 5/27/11.

Pounds, J., M. Bustamante, L. Coloma, et al. 2006. Widespread amphibian extinctions from epidemic disease driven by global warming. Nature 439:161–167.

Pounds, J., M. Fogden, J. Savage, et al. 1997. Tests of null models for amphibian declines on a tropical mountain. Conserv. Biol. 11:1307–1322.

Pourrut, X., B. Kumulungui, T. Wittmann, et al. 2005. The natural history of the Ebola virus in Africa. Microbes Infect. 7:1005–1014.

Powell, R. and G. Proulx. 2003. Trapping and marking terrestrial mammals for research: integrating ethics, performance criteria, techniques, and common sense. ILAR J. 44:259–276.

Power, A., R. Walker, K. Payne, et al. 2004. First occurrence of the nonindigenous green mussel (*Perna viridis*) in coastal Georgia, United States. J. Shellfish Res. 23:741–744.

Prugh, L., C. Stoner, C. Epps, et al. 2009. The rise of the mesopredator. BioScience 59: 779–791.

Pruss, S., N. Cool, R. Hudson, et al. 2002. Evaluation of a modified neck snare to live-capture coyotes. Wildl. Soc. Bull. 30(2):508–516.

Pyšek, P. and D. Richardson. 2006. The biogeography of naturalization in alien plants. J. Biogeogr. 33:2040–2050.

Raffaelli, D. and H. Moller. 2000. Manipulative field experiments in animal ecology: do they promise more than they can deliver? Adv. Ecol. Res. 30:299–338.

Rammell, C. and P. Fleming. 1978. Compound 1080 properties and use of sodium monofluoroacetate in New Zealand. Minist. Agric. Fish., Wellington, N.Z.

Ratcliffe, N., I. Mitchell, K. Varnham, et al. 2009. How to prioritize rat management for the benefit of petrels: a case study of the UK, Channel Islands and Isle of Man. Ibis 151:699–708.

Rauzon, M. 2007. Island restoration: exploring the past, anticipating the future. Mar. Ornithol. 35:97–107.

Raven, P. and G. Johnson. 1992. Biology. 3rd ed. Mosby–Year Book, St. Louis.

Rayner, T. and R. Creese. 2006. A review of rotenone use for the control of non-indigenous fish in Australian fresh waters, and an attempted eradication of the noxious fish, *Phalloceros caudimaculatus*. N. Z. J. Mar. Freshwater Res. 40:477–486.

Reddy, P., D. Gorelick, C. Brasher, et al. 1970. Progressive disseminated histoplasmosis as seen in adults. Am. J. Med. 48:629–636.

Redpath, S. and S. Thirgood. 2009. Hen harriers and red grouse: moving towards consensus? J. Appl. Ecol. 46:961–963.

Reed, K., J. Meece, J. Henkel, et al. 2003. Birds, migration and emerging zoonoses: West Nile virus, Lyme disease, influenza A and enteropathogens. Clin. Med. Res. 1:5–12.

Reeves, C. 1992. Egyptian medicine. Osprey Publ., London. 72 pp.

Reidinger, R. 1997. Recent studies on flavor aversion learning in wildlife damage management. Pp. 101–120 in J. Mason, ed. Repellents in wildlife management: proceedings of a symposium. Natl. Wildl. Res. Cent., Fort Collins, Colo.

Reidinger, R., G. Beauchamp, and M. Barth. 1982. Conditioned aversion to a taste perceived while grooming in rats. Physiol. Behav. 28:715–723.

Reidinger, R., J. Libay, and A. Kolz. 1985. Field trial of an electrical barrier for protecting rice fields from rat damage. Philipp. Agric. 68:169–179.

Reidinger, R. and J. Mason. 1983. Exploitable characteristics of neophobia and food aversions for improvements in rodent and bird control. Pp. 20–39 in D. Kaukeinen, ed. Vertebrate pest control and management materials, fourth symp., ASTM STP 817. Am. Soc. Test. Mater., Philadelphia.

Rhyan, J., K. Aune, B. Hood, et al. 1995. Bovine tuberculosis in a free ranging mule deer (*Odocoileus hemionus*) from Montana. J. Wildl. Dis. 31:432–435.

Rice, P. 2009. Recent finds: paleoanthropology, part 2. Gen. Anthropol. 16:12–16.

Richardson, W. and T. West. 2000. Serious bird strike accidents to military aircraft: updated list and summary. Proc. Int. Bird Strike Comm. 25:67–97 (WP SA1).

Rigg, R. 2001. Livestock guarding dogs: their current use world wide. IUCN/SSC Canid Spec. Group Occas. Pap. No. 1. 133 pp.

Riley, S. and D. Decker. 2000. Wildlife stakeholder acceptance capacity for cougars in Montana. Wildl. Soc. Bull. 28:931–939.

Rogers, L. 2009. Does diversionary feeding create nuisance bears and jeopardize public safety? Proc. West. Black Bear Workshop 10. www.bearstudy.org/website/images /stories/Publications/diversionary_feeding_of_black _bears_9_june_2009.pdf. Accessed 5/5/11.

Rollins, L., A. Woolnough, and W. Sherwin. 2006. Population genetic tools for pest management: a review. Wildl. Res. 33(4):251–261.

Romney, A., S. Weller, and W. Batchelder. 1986. Culture as consensus: a theory of culture and informant accuracy. Amer. Anthropol. 88(2):313–338.

Ronconi, R., Z. Swaim, H. Lane, et al. 2010. Modified hoop-net techniques for capturing birds at sea and comparison with other capture methods. Mar. Ornithol. 38:23–29.

Rondeau, D. and E. Bulte. 2007. Wildlife damage and agriculture: a dynamic analysis of compensation schemes. Am. J. Agric. Econ. 89:490–507.

Rooney, T. and D. Waller. 2003. Direct and indirect effects of white-tailed deer in forest ecosystems. For. Ecol. Manage. 181:165–176.

Root, J., R. Puskas, J. W. Fischer, et al. 2009. Landscape genetics of raccoons (*Procyon lotor*) associated with ridges and valleys of Pennsylvania—implications for oral rabies vaccination programs. Vector-Borne Zoonotic Dis. 9:583–588.

Roy, S., G. Smith, and J. Russell. 2009. The eradication of invasive mammal species: can adaptive resource management fill the gaps in our knowledge? Human-Wildl. Conflicts 3(1):30–40.

Royer, N. Undated. History of fancy mice. www.afrma.org /rminfo4b.htm. Accessed 1/25/11.

Ruhe, W. 2005. Bird avoidance models vs. real time bird strike warning systems—a comparison. Int. Bird Strike Comm. IBSC27/WPIX-3. 5 pp. www.int-birdstrike.org /Athens_Papers/IBSC27%20WPIX-3.pdf. Accessed 7/28/11.

Rupprecht, C., J. Smith, M. Fekadu, et al. 1995. The ascension of wildlife rabies: a cause for public health concern or intervention? Emerging Infect. Dis. 1:107–114.

Russell, F. 1980. Snake venom poisoning in the United States. Ann. Rev. Med. 31:247–259.

Russell, F., D. Zippin, and N. Fowler. 2001. Effect of white-tailed deer (*Odocoileus virginianus*) on plants, plant populations and communities: a review. Am. Midl. Nat. 146(1):1–26.

Rutberg, A., ed. 2005. Humane wildlife solutions: the role of immunocontraception. Humane Soc. Press, Washington, D.C. 107 pp.

Salmony, S. 2006. The human population: accepting species limits. Environ. Health Perspect. 114(1):A17–A18.

Sampson, S. 2005. Dietary overlap between two Asian carp and three native filter feeding fishes of the Illinois and Mississippi rivers. Master's thesis, Univ. Ill., Urbana.

Sargeant, A., R. Greenwood, M. Sovada, et al. 1993. Distribution and abundance of predators that affect duck production—Prairie Pothole region. Resour. Publ. 194, U.S. Fish Wildl. Serv., Washington, D.C.

Saul, E. 1967. Birds and aircraft: a problem at Auckland's new international airport. J. R. Aeronaut. Soc. 71:366–375.

Savarie, P., J. Shivik, G. White, et al. 2001. Use of acetamino-phen for large-scale control of brown tree snakes. J. Wildl. Manage. 65:356–365.

Savarie, P., D. Vice, L. Bangerter, et al. 2004. Operational field evaluation of a plastic bulb reservoir as a tranquilizer trap device for delivering propiopromazine hydrochloride to feral dogs, coyotes, and gray wolves. Proc. Vertebr. Pest Conf. 21:64–69.

Schaeffer, J. and F. Margraf. 1987. Predation on fish eggs by white perch (*Morone americana*) in western Lake Erie. Environ. Biol. Fishes 18:77–80.

Schaller, M. 2002. Evaluation of wildlife damage to forests in Germany. Pp. 123–126 in L. Clark, J. Hone, J. Shivik, et al., eds. 2002. Human conflicts with wildlife: economic considerations. Proc. Third NWRC Spec. Symp., Natl. Wildl. Res. Cent., Fort Collins, Colo.

Schaller, M. 2007. Forests and wildlife management in Germany—a mini-review. Eurasian J. For. Res. 10(1):59–70.

Schemnitz, S. 1996. Capturing and handling wild animals. Pp. 106–124 in T. Bookhout, ed. Research and manage-ment techniques for wildlife and habitats. Wildl. Soc., Bethesda, Mary.

Schramm, H., J. Grizzle, L. Hanson, et al. 2004. Improving survival of tournament-caught bass and the effects of tournament handling on largemouth bass virus disease. Completion rep., Sport Fish Restor. 134 pp.

Schuhmann, P. and K. Schwabe. 2002. Fundamentals of economic principles of wildlife management. Pp. 1–16 in L. Clark, J. Hone, J. Shivik, et al., eds. 2002. Human conflicts with wildlife: economic considerations. Proc. Third NWRC Spec. Symp., Natl. Wildl. Res. Cent., Fort Collins, Colo.

Seamans, T., S. Barras, and G. Bernhardt. 2007. Evaluation of two perch deterrents for starlings, blackbirds, and pigeons. Int. J. Pest Manage. 53:45–51.

Seamans, T. and K. VerCauteren. 2006. Evaluation of ElectroBraid as a white-tailed deer barrier. Wildl. Soc. Bull. 34:8–15.

Selye, H. 1936. A syndrome produced by diverse nocuous agents. Nature 138:32.

Sengl, J., J. Spencer, and Z. Bowers. 2008. Developing standard operating procedures for wildlife damage management activities in urban and suburban areas in southern Nevada. Proc. Vertebr. Pest Conf. 23:201–205.

Sequin, E., M. Jaeger, P. Brussard, et al. 2003. Wariness of coyotes to camera traps relative to social status and territory boundaries. Can. J. Zool. 81:2015–2025.

Serbesoff-King, K. 2003. Melaleuca in Florida: a literature review on the taxonomy, distribution, biology, ecology, economic importance and control measures. J. Aquat. Plant Manage. 41:98–112.

Shanmuganathan, T., J. Pallister, S. Doody, et al. 2010. Biological control of the cane toad in Australia: a review. Anim. Conserv. 13, Suppl. 1:16–23.

Sherley, G., ed. 2000. Invasive species in the Pacific: a technical review and draft regional strategy. South Pacific Regional Environment Programme, Apia, Samoa. 190 pp.

Shivik, J., D. Martin, M. Pipas, et al. 2005. Initial comparison: jaws, cables, and cage-traps to capture coyotes. Wildl. Soc. Bull. 33:1375–1383.

Shivik, J., P. Savarie, and L. Clark. 2002. Aerial delivery of baits to brown tree snakes. Wildl. Soc. Bull. 30:1062–1067.

Shope, R. 1992. Impacts of global climate change on human health: Spread of infectious disease. Pp. 363–370 in S. Majumdar, L. Kalkstein, B. Yarnal, et al., eds. Global climate change: Implications, challenges and mitigation measures. Penn. Acad. Sci., Easton.

Shwiff, S. 2004. Economics in wildlife damage management studies: common problems and solutions. Proc. Vertebr. Pest Conf. 21:346–349.

Shwiff, S. and R. Merrell. 2004. Coyote predation management: an economic analysis of increased antelope recruitment and cattle production in south central Wyoming. Sheep Goat Res. J. 19:29–33.

Shwiff, S. and R. Sterner. 2002. An economic framework for benefit-cost analysis in wildlife damage studies. Proc. Vertebr. Pest Conf. 20:340–344.

Sicard, B., W. Diarra, and H. Cooper. 1999. Ecophysiology and chronobiology applied to rodent pest management in semi-arid agricultural areas in sub-Saharan west Africa. Pp. 409–440 in G. Singleton, L. Hinds, H. Leirs, et al., eds. Ecologically-based management of rodent pests. Aust. Centre Int. Agric. Res., Canberra.

Sillero-Zubiri, C. and D. Switzer. 2004. Management of canids near people. Pp. 257–266 in C. Sillero-Zubiri, M. Hoffmann, and D. Macdonald, eds. Canids: foxes, wolves, jackals, and dogs. Status survey and conservation action plan. 2d edition. IUCN Canid Spec. Group, Gland, Switz., and Cambridge, U.K.

Simberloff, D. 2005. The politics of assessing risk for biological invasions: the USA as a case study. Trends Ecol. Evol. 20(5):216–222.

Sims, B. 1995. Predator politics in Texas. Pp. 141–142 in Proc. Symp. Tex. Sheep Goat Raisers Assoc. http://digital commons.unl.edu/cgi/viewcontent.cgi?article=1007& context=coyotesw&sei-redir=1#search=%22wildlife %20damage%20politics%20texas%22. Accessed 7/22/11.

Singh, S. and A. Singh. 2009. Toxic effect of *Euphorbia pulcherima* plant to fingerlings of *Labeo rohita* (Hamilton) in different culturing conditions. World J. Fish Mar. Sci. 1(4):324–329.

Singleton, G. 2009. Is there a role for genetic pest management in rodent pest management? Abstr. GPM Conf., N.C. State Univ., Raleigh. www.ncsu.edu/project/gpm/. Accessed 8/26/11.

Singleton, G., P. Brown, and J. Jacob. 2004. Ecologically-based rodent management: its effectiveness in cropping systems in South-East Asia. NJAS 52:163–171.

Singleton, G., L. Hinds, H. Leirs, et al., eds. 1999. Ecologically-based management of rodent pests. Aust. Cent. Int. Agric. Res., Canberra. 492 pp.

Singleton, G., Sudarmaji, and S. Suriapermana. 1998. An experimental field study to evaluate a trap barrier system and fumigation for controlling the rice field rat, *Rattus argentiventer*, in rice crops in West Java. Crop Prot. 17:55–64.

Skonhoft, A. Undated. The cost, or benefit, of predation. An analysis of the recent wolf re-colonization in Scandinavia. Preliminary and uncompleted. 18 pp.

Slate, D., R. Chipman, C. Rupprecht, et al. 2002. Oral rabies vaccination: a national perspective on program development and implementation. Proc. Vertebr. Pest Conf. 20:232–240.

Slate, D., C. Rupprecht, J. Rooney, et al. 2005. Status of oral rabies vaccination in wild carnivores in the United States. Virus Res. 111:68–76.

Smith, K., A. Dobson, F. McKenzie, et al. 2005. Ecological theory to enhance infectious disease control and public health policy. Front. Ecol. Environ. 3:29–37.

Smith, M. 2004. Capturing problematic urban Canada geese in Reno, Nevada: goose roundups vs. use of alpha-chloralose. Proc. Vertebr. Pest Conf. 21:97–100.

Smith, T., S. Herrero, T. DeBruyn, et al. 2008. Efficacy of bear deterrent spray in Alaska. J. Wildl. Manage. 72:640–645.

Soria, M., M. Gardener, and A. Tye. 2002. Eradication of potentially invasive plants with limited distributions in the Galápagos Islands. Pp. 287–292 in C. Veitch and M. Clout, eds. Turning the tide: the eradication of invasive species. Occas. Pap. No. 27, IUCN Species Survival Comm.

Spinage, C. 2003. Cattle plague: a history. Kluwer Acad. / Plenum Publ., New York. 770 pp.

Spurr, E. and J. Coleman. 2005a. Cost-effectiveness of bird repellents for crop protection. Proc. Australas. Vertebr. Pest Conf. 13:227–233.

Spurr, E. and J. Coleman. 2005b. Review of Canada goose population trends, damage, and control in New Zealand. Manaaki Whenua Press, Lincoln, N.Z. 32 pp.

Sterner, R. 1994. Zinc phosphide: implications of optimal foraging theory and particle-dose analyses to efficacy, acceptance, bait shyness, and non-target hazards. Proc. Vertebr. Pest Conf. 16:152–159.

Sterner, R. 2002. Spreadsheets, response surfaces and intervention decisions in wildlife damage management. Pp. 42–46 in L. Clark, J. Hone, J. Shivik, et al., eds. Human conflicts with wildlife: economic considerations. Proc. Third NWRC Spec. Symp., Natl. Wildl. Res. Cent., Fort Collins, Colo.

Sterner, R. 2008. The IPM paradigm: vertebrates, economics and uncertainty. Proc. Vertebr. Pest Conf. 23:194–200.

Sterner, R. 2009. The economics of threatened species conservation: a review and analysis. Chapter 8 in J. Aronoff, ed. Handbook of nature conservation. Nova Sci. Publ. 479 pp.

Sterner, R. and K. Crane. 2000. Sheep-predation behaviors of wild-caught, confined coyotes: some historical data. Proc. Vertebr. Pest Conf. 19:325–330.

Sterner, R. and K. Tope. 2002. Repellents: projections of direct benefit-cost surfaces. Proc. Vertebr. Pest Conf. 20:319–325.

Steuber, J., M. Pitzler, and J. Oldenburg. 1995. Protecting juvenile salmonids from gull predation using wire exclusion below hydroelectric dams. Proc. Great Plains Wildl. Damage Control Workshop 12:38–41.

Stevens, G., J. Rogue, R. Weber, et al. 2000. Evaluation of a radar-activated, demand-performance bird hazing system. Int. Biodeterior. Biodegrad. 45:129–137.

Stickley, A. and J. King. 1995. Long-term trial of an inflatable effigy scare device for repelling cormorants from catfish ponds. Proc. East. Wildl. Damage Control Conf. 6:89–92.

Stinzing, A. and K. Lang. 2003. Dogwood anthracnose. Erster Fund von Discula destructiva an Cornus florida in Deutschland. Nachrichtenbl. Deut. Pflanzenschutzd. 55:1–5.

Stoddart, D. and P. Smith. 1986. Recognition of odor-induced bias in the live-trapping of Apodemus sylvaticus. Oikos 46:194–199.

Stone, C. and S. Anderson. 1988. Introduced animals in Hawaii's natural areas. Proc. Vertebr. Pest Conf. 13:134–140.

Storm, D., C. Nielsen, E. Schauber, et al. 2007. Deer-human conflict and hunter access in an exurban landscape. Human-Wildl. Conflicts 1(1):53–59.

Strayer, D. 1991. Projected distribution of the zebra mussel (Dreissena polymorpha) in North America. Can. J. Fish Aquat. Sci. 48:1389–1395.

Strayer, D. 2009. Twenty years of zebra mussels: lessons from the mollusk that made headlines. Front Ecol. Environ. 7:135–141.

Stuart, I. and A. Conallin. 2009. The Williams carp separation cage: new innovations and a commercial trial. Pp. 108–112 in J. Pritchard, ed. Proc. Native Fish Forum, Murray-Darling Basin Auth., Canberra, Australia.

Stuart, I., A. Williams, J. McKenzie, et al. 2006. Managing a migratory pest species: a selective trap for common carp. N. Amer. J. Fisheries Manage. 26:888–893.

Stuart, J. 2000. Additional notes on native and non-native turtles of the Rio Grande Drainage Basin, New Mexico. Bull. Chicago Herpetol. Soc. 35:229–235.

Sullivan, T. and W. Klenner. 1993. Influence of diversionary food on red squirrel populations and damage to crop trees in young lodgepole pine forests. Ecol. Appl. 3:708–718.

Sullivan, T., D. Sullivan, and E. Hogue. 2001. Influence of diversionary food on vole (Microtus montanus and Microtus longicaudus) populations and feeding damage to coniferous tree seedlings. Crop Prot. 20:103–112.

Sutton, W., D. Larson, and L. Jarvis. 2004. A new approach for assessing the costs of living with wildlife in developing countries. DEA Res. Discuss. Pap. No. 69, Minist. Environ. Tourism, Windhoek, Namibia. 27 pp.

Takahashi, M. 2009. Overview of the structure and the challenges of Japanese wildlife law and policy. Biol. Conserv. 142(9):1958–1964.

Takami, T., T. Yoshihara, Y. Miyakoshi, et al. 2002. Replacement of the white-spotted char Salvelinus leucomaenis by brown trout Salmo trutta in a branch of the Chitose River, Hokkaido. Nippon Suisan Gakkaishi 68:24–28.

Talbot, P. 1912. In the shadow of the bush. George H. Doran, New York. 642 pp.

Tan, S. 1990. The watchtower casts no shadow: nonliability of federal and state governments for property damage inflicted by wildlife. http://heinonline.org/HOL/LandingPage?collection=journals&handle=hein.journals/ucollr61&div=24&id=&page=. Accessed 7/10/11.

Tantardini, A., M. Calvi, B. Cavagna, et al. 2004. Primo rinvenimento in Italia di antracnosi causata da Discula destructiva su Cornus florida e C. nuttallii. Inf. Fitopatol. 12:44–47.

Taylor, B. and R. Irwin. 2004. Linking economic activities to the distribution of exotic plants. Proc. Natl. Acad. Sci. 101:17725–17730.

Taylor, J., W. Courtenay, and J. McCann. 1984. Known impact of exotic fishes in the continental United States. Pp. 322–373 in W. Courtenay and J. Stauffer, eds. Distribution, biology, and management of exotic fish. Johns Hopkins Univ. Press, Baltimore.

Taylor, J. and B. Dorr. 2003. Double-crested cormorant impacts to commercial and natural resources. Proc. Wildl. Damage Manage. Conf. 10:43–51.

Teel, T., R. Krannich, and R. Schmidt. 2002. Utah stakeholders' attitudes toward selected cougar and black bear management practices. Wildl. Soc. Bull. 30:2–15.

Terry, L. 1984. Deterring waterfowl from water features: a wire grid system to deter waterfowl from using ponds on airports. Unpubl. Rep., Denver Wildl. Res. Cent. 19 pp.

Texas Agricultural Extension Service. 1991. A Matter of Perspective. Video.

Thakuri, S. 2009. Population dynamics modeling and distribution of key mammals in relation to habitat and anthropic variables. Master's thesis. Tribhuvan Univ. Regd. No. 5-1-33-568-98. 97 pp.

Thiele, J., G. Linz, H. Homan, et al. 2012. Developing an effective management plan for starlings roosting in downtown Omaha, Nebraska. Pp. 87–90 in S. Frey, ed. Proc. 14th Wildl. Damage Manage. Conf., Nebraska City.

Thirgood, S. and S. Redpath. 2005. Hen harriers and red grouse: the ecology of a conflict. Pp. 192–208 in R. Woodroffe, S. Thirgood, and A. Rabinowitz, eds. People and wildlife: conflict or coexistence? Cambridge Univ. Press, Cambridge, UK.

Thirgood, S. and S. Redpath. 2008. Hen harriers and red grouse: science, politics and human-wildlife conflict. J. Appl. Ecol. 45:1550–1554.

Thomas, C., A. Cameron, R. Green, et al. 2004. Extinction risk from climate change. Nature 427:145–148.

Thompson, D., R. Stuckey, and E. Thompson. 1987. Spread, impact, and control of purple loosestrife (*Lythrum salicaria*) in North American Wetlands. 55 pp. www .npwrc.usgs.gov/resource/plants/loosstrf/index.htm Version 04Jun1999. Accessed 2/21/11.

Thompson, R., S. Kutz, and A. Smith. 2009. Parasitic zoonoses and wildlife: emerging issues. Int. J. Environ. Res. Public Health 2:678–693.

Thompson, R., G. Mitchell, and R. Burns. 1972. Vampire bat control by systemic treatment of livestock with an anticoagulant. Science 177:806–808.

Thompson, S., J. Jonkel, and P. Sowka. 2009. Practical electric fencing resource guide: controlling predators. Living with Wildl. Found., Inc. www.lwwf.org/electric_fence _2009.pdf. Accessed 4/4/11.

Thornton, C. and M. Quinn. 2009. Coexisting with cougars: public perceptions, attitudes, and awareness of cougars on the urban-rural fringe of Calgary, Alberta, Canada. Human-Wildl. Conflicts 3(2):282–295.

Thorpe, J. 2003. Fatalities and destroyed civil aircraft due to bird strikes 1912–2002. Proc. Int. Bird Strike Comm. 26:85–113.

Thouless, C. and J. Sakwa. 1995. Shocking elephants: fences and crop raiders in Laikipia District, Kenya. Biol. Conserv. 72(1):99–107.

Tidemann, C. 2005. Common Indian myna. Aust. Natl. Univ. http://fennerschool-associated.anu.edu.au//myna /problem. html Accessed 1/21/11.

Tierkel, E. 1975. Canine Rabies. Pp. 123–137 in G. Baer, ed. The natural history of rabies. Vol. 2. Acad. Press, New York. 454 pp.

Tilghman, N. 1989. Impacts of white-tailed deer on forest regeneration in northwestern Pennsylvania. J. Wildl. Manage. 53:524–532.

Tillman, E., J. Humphrey, and M. Avery. 2002. Use of vulture carcasses and effigies to reduce vulture damage to property and agriculture. Proc. Vertebr. Pest Conf. 20:123–128.

Tindall, S., C. Ralph, and M. Clout. 2007. Changes in bird abundance following common myna control on a New Zealand island. Pac. Conserv. Biol. 13:202–212.

Tischaefer, R. and C. Tischaefer. Undated. The egg trap company. www.theeggtrapcompany.com/index.htm. Accessed 8/18/11.

Todd, T. 1986. Occurrence of white bass–white perch hybrids in Lake Erie. Copeia 1986:196–199.

Tourenq, C., S. Aulagnier, L. Durieux, et al. 2001. Identifying rice fields at risk from damage by the greater flamingo. J. Appl. Ecol. 38:170–179.

Towns, D., I. Atkinson, and C. Daugherty. 2006. Have the harmful effects of introduced rats on islands been exaggerated? Biol. Invasions 8(4):863–891.

Townsend, C., M. Begon, and J. Harper. 2003. Essentials of ecology. 2nd ed. Blackwell Science, Massachusetts. 544 pp.

Trefethen, J. 1975. An American crusade for wildlife. Winchester Press, New York.

Treves, A. 2008. The human dimensions of conflicts with wildlife around protected areas. Pp. 214–228 in M. Manfredo, J. Vaske, P. Brown, et al., eds. Wildlife and Society: the Science of Human Dimensions. Island Press, Washington, D.C. 350 pp.

Treves, A., R. Jurewicz, L. Naughton-Treves, et al. 2009. The price of tolerance: wolf damage payments after recovery. Biodivers. Conserv. www.nelson.wisc.edu/people/treves /Pubs/2.pdf. Accessed 7/30/11.

Treves, A., R. Wallace, L. Naughton-Treves, et al. 2006. Co-managing human-wildlife conflicts: a review. Hum. Dimensions Wildl. 11:383–396.

Troll, C. 1939. Luftbildplan und ökologische Bodenforschung (Aerial photography and ecological studies of the earth). Zeitschrift der Gesellschaft für Erdkunde zu Berlin 7/8: 241–298.

Turner, C., T. Center, D. Burrows, et al. 1998. Ecology and management of *Melaleuca quinquenervia*, an invader of wetlands in Florida, USA. Wetlands Ecol. Manage. 5:165–178.

Turner, M., R. Gardner, and R. O'Neill. 2001. Landscape ecology in theory and practice: pattern and process. Springer, New York. 401 pp.

Twohey, M., J. Heinrich, J. Seelye, et al. 2003. The sterile-male-release technique in Great Lakes sea lamprey management. J. Great Lakes Res. 29, Suppl. 1:410–423.

TWS. 2007. Final TWS Position Statement: baiting and supplemental feeding of game wildlife species. http:// joomla.wildlife.org/documents/positionstatements/42 -Baiting%20and%20Feeding.pdf. Accessed 5/2/11.

Umpqua Watersheds. 2004. Help save black bears from the timber industry. www.umpqua-watersheds.org/archive /local/bears.html#lawsuit. Accessed 7/10/11.

Unestam, T. 1976. Defense reactions in and susceptibility of Australian and New Guinean freshwater crayfish to European-crayfish-plague fungus. Aust. J. Exp. Biol. Med. Sci. 53: 349–359.

Unestam, T. and D. Weiss. 1970. The host-parasite relationship between freshwater crayfish and the crayfish disease fungus *Aphanomyces astaci*: responses to infection by a susceptible and a resistant species. J. Gen. Microbiol. 60:77–90.

United Nations. 2001. World population monitoring 2001. Popul. Div., Dep. Econ. Soc. Aff. www.un.org/esa /population/publications/wpm/wpm2001.pdf. Accessed 8/23/11.

United Nations. 2009. World urbanization prospects: the 2009 revision. Popul. Div., Dep Econ. Soc. Aff. http://esa .un.org/unpd/wup/Documents/WUP2009_Highlights _Final.pdf. Accessed 8/15/11.

USDA. 2002. Wildlife services: helping producers manage predation. Program Aid No. 1722, Anim. Plant Health Insp. Serv. 7 pp.

USEPA. Undated. What is a pesticide? www.epa.gov /pesticides/about/index.htm#what_pesticide. Accessed 4/22/11.

van Aarde, R. and T. Jackson. 2007. Megaparks for meta-populations: addressing the causes of locally high elephant numbers in southern Africa. Biol. Conserv. 134:289–297.

Van der Lee, S. and L. Boot. 1955. Spontaneous pseudopreg-nancy in mice. Acta Physiol. Pharmacol. Neer. 4:442.

Vantassel, S. 2008. Ethics of wildlife control in humanized landscapes: a response. Proc. Vertebr. Pest Conf. 23:294–300.

Vantassel, S., K. Powell, and T. Miller. 2008. Survey of changes to cable-trap regulations in the United States during 1980–2007. Univ. Nebr., Lincoln. 24 pp. http:// digitalcommons.unl.edu/icwdm/. Accessed 8/16/11.

Veitch, C. 2001. The eradication of feral cats (Felis catus) from Little Barrier Island, New Zealand. N. Z. J. Zool. 28:1–12.

Veitch, C., and M. Clout, eds. 2002. Turning the tide: the eradication of invasive species. World Conserv. Union, Gland, Switz.

VerCauteren, K. 2003. The deer boom: discussions on population growth and range expansion of the white-tailed deer. Pp. 15–20 in G. Hisey and K. Hisey, eds. Bowhunting records of North American whitetail deer. Pope and Young Club, Chatfield, Minn.

VerCauteren, K., S. Hygnstrom, R. Timm, et al. 2002. Development of a model to assess rodent control in swine facilities. Pp. 59–64 in L. Clark, J. Hone, J. Shivik, et al., eds. Human conflicts with wildlife: economic consider-ations. Proc. Third NWRC, Spec. Symp., Natl. Wildl. Res. Cent., Fort Collins, Colo.

VerCauteren, K., M. Lavelle, and S. Hygnstrom. 2006a. Fences and deer-damage management: a review of designs and efficacy. Wildl. Soc. Bull. 34:191–200.

VerCauteren, K., M. Lavelle, and S. Hygnstrom. 2006b. A simulation model for determining cost-effectiveness of fences for reducing deer damage. Wildl. Soc. Bull. 34(1):16–22.

VerCauteren, K., M. Lavelle, and S. Moyles. 2003. Coyote-activated frightening devices for reducing sheep predation on open range. Proc. Wildl. Damage Manage. Conf. 10:146–151.

VerCauteren, K., M. Lavelle, and G. Phillips. 2008. Livestock protection dogs for deterring deer from cattle and feed. J. Wildl. Manage. 72:1443–1448.

VerCauteren, K., M. Pipas, P. Peterson, et al. 2003. Stored-crop loss due to deer consumption. Wildl. Soc. Bull. 31:578–582.

VerCauteren, K., N. Seward, D. Hirchert, et al. 2005. Dogs for reducing wildlife damage to organic crops. Proc. Wildl. Damage Manage. Conf. 11:286–293.

Verhulst, P. 1838. Notice sur la loi que la population poursuit dans son accroissement. Corresp. Math. Phys. 10:113–121.

Vilà, M., C. Basnou, P. Pyšek, et al. 2010. How well do we understand the impacts of alien species on ecosystem services? A pan-European, cross-taxa assessment. Front. Ecol. Environ. 8(3):135–144.

Virginia Tech. 2005. Virginia Tech. Center for Human-Wildlife Conflict Research statement of mission, goals, and objectives. www.humanwildlife.org/documents/ White%20paper.pdf. Accessed 7/25/11.

Vitasek, J. 2004. A review of rabies elimination in Europe. Vet. Med. 49(5):171–185.

Volterra, V. 1926. Variazioni e fluttuazioni del numero d'individui in specie animali conviventi. Mem. R. Accad. Naz. dei Lincei. Ser. VI, vol. 2.

Vuillaume, P., M. Aubert, J. Demerson, et al. 1997. Vaccina-tion des renards contre la rage par depot d'appats vaccinaux a 'entrée des terriers. Ann. Med. Vet. 141:55–62.

Wager, R. and P. Jackson. 1993. The action plan for Austra-lian freshwater fishes. www.environment.gov.au /biodiversity/threatened/publications/action/fish/index .html. Accessed 1/26/11.

Walton, M. and C. Field. 1989. Use of donkeys to guard sheep and goats in Texas. Proc. East. Wildl. Damage Control Conf. 4:87–94.

Ward, A., K. VerCauteren, W. Walter, et al. 2009. Options for the control of disease 3: targeting the environment. Chapter 8 in R. Delahay, G. Smith and M. Hutchings, eds. Management of Disease in Wild Mammals. Springer, New York.

Ward, J. and S. Williams. 2010. Effectiveness of deer repellents in Connecticut. Human-Wildl. Interact. 4:56–66.

Warren, R. 2011. Deer overabundance in the USA: recent advances in population control. Anim. Prod. Sci. 51:259–266.

WeedSOFT. Undated. http://weedsoft.unl.edu/. Accessed 7/27/11.

Weisbrod, B. 1964. Collective-consumption services of individual-consumption goods. Q. J. Econ. 78:471–477.

Welsh, R. and D. Muller-Schwarze. 1989. Experimental habitat scenting inhibits colonization by beaver, Castor canadensis. J. Chem. Ecol. 15:887–893.

Werner, S., J. Carlson, S.Tupper, et al. 2009. Threshold concentrations of an anthranquinone-based repellent for Canada geese, red-winged blackbirds, and ring-necked pheasants. Appl. Anim. Behav. Sci. 121:190–196.

Werner, S. and F. Provenza. 2011. Reconciling sensory cues and varied consequences of avian repellents. Physiol. Behav. 102:158–163.

West, B., A. Cooper, and J. Armstrong. 2009. Managing wild pigs: A technical guide. Human-Wildl. Interact. Monogr. 1:1–55.

Westman, K. and R. Savolainen. 2001. Long term study of competition between two co-occurring crayfish species, the native Astacus astacus L and the introduced Pacifasta-cus leniusculus Dana, in a Finnish Lake. Bull. Francais de la Peche et de la Pisciculture 361:613–627.

Wheat, L., T. Slama, H. Eitzen, et al. 1981. A large urban outbreak of histoplasmosis: clinical features. Ann. Intern. Med. 94:331–337.

Whitten, W. 1956. Modification of the oestrous cycle of the mouse by external stimuli associated with the male. J. Endocrinol. 13:399–404.

WHO. Undated a. Drug-resistant Salmonella. www.who.int /mediacentre/factsheets/fs139/en. Accessed 3/18/11.

WHO. Undated b. Emerging zoonoses. www.who.int /zoonoses/emerging_zoonoses/en/. Accessed 3/13/11.

WHO. Undated c. Influenza A (H1N1)—update 49. www .who.int/csr/don/2009_06_15/en/. Accessed 3/26/11.

WHO. Undated d. Rabies. www.who.int/mediacentre /factsheets/fs099/en/. Accessed 5/8/11.

Whyte, I. 2004. The ecological basis of the Kruger National Park's new elephant management policy and expected outcomes after implementation. Pachyderm 36:99–108.

Williams, A. and J. Wells. 2004. Characteristics of vehicle-animal crashes in which vehicle occupants are killed. Insur. Inst. Highw. Safety, Arlington, Va.

Williams, J. 1974. The effect of artificial rat damage on coconut yields in Fiji. Int. J. Pest Manage. 20(3):275–282.

Williams, T. 1999. The terrible turtle trade. Audubon 101:44, 46–48, 50–51.

Williams, T. 2009. Felines fatales. Audubon, Sept/Oct. www .scottchurchdirect.com/docs_ted/felines-fatales.pdf. Accessed 7/16/11.

Wilson, A. 2005. Recent advances in the control of oak wilt in the United States. Plant Pathol. J. 4:177–191.

Wilson, D., F. Cole, J. Nichols, et al. 1996. Measuring and monitoring biological diversity: standard methods for mammals. Smithsonian Inst. Press, Washington, D.C. 409 pp.

Wilson, E. 1984. Biophilia: the human bond with other species. Harvard Univ. Press, Cambridge, Mass.

Wilson, E. 1993. Biophilia and the conservation ethic. Pp. 31–41 in S. Kellert and E. Wilson, eds. The biophilia hypothesis. Island Press / Shearwater Books, Washington, D.C. 484 pp.

Wireless traps. Undated. www.theeggtrapcompany.com /index.htm. Accessed 8/18/11.

Witmer, G. 2007. The ecology of vertebrate pests and integrated pest management (IPM). Pp. 393–410 in M. Kogan and P. Jepson, eds. Perspectives in ecological theory and integrated pest management. Cambridge Univ. Press, Cambridge, U.K.

Witmer, G., F. Boyd, and Z. Hillis-Starr. 2007. The successful eradication of introduced roof rats (Rattus rattus) from Buck Island using diphacinone, followed by an irruption of house mice (Mus musculus). Wildl. Res. 34:108–115.

Witmer, G., B. Constantin, and F. Boyd. 2005. Feral and introduced carnivores: issues and challenges. Proc. Wildl. Damage Manage. Conf. 11:90–101.

Witmer, G. and K. VerCauteren. 2001. Understanding vole problems in direct seeding—strategies for management. Pages 104–110 in Veseth, R., ed. Proc. Northwest Direct Seed Cropping Syst. Conf.

Witte, F., T. Goldschmidt, P. Goudswaard, et al. 1992. Species extinction and concomitant ecological changes in Lake Victoria. Neth. J. Zool. 42:214–232.

Wittenberg, R. and M. Cock, eds. 2001. Invasive alien species: a toolkit of best prevention and management practices. CAB Int., Wallingford, U.K., xvii–228.

Wood, W. 1865. Wood's New England's prospect. (Orig. pub. 1634.) http://books.google.com/books ?id=ZWoFAAAAQAAJ&printsec=frontcover&

source=gbs_ge_summary_r&cad=0#v=onepage&q &f=false. 131 pp.

Woodroffe, R., S. Thirgood, and A. Rabinowitz. 2005a. The future of coexistence: resolving human-wildlife conflicts in a changing world. Pp. 388–405 in R. Woodroffe, S. Thirgood, and A. Rabinowitz, eds. People and wildlife: conflict or coexistence? Cambridge Univ. Press, Cambridge, U.K. 497 pp.

Woodroffe, R., S. Thirgood, and A. Rabinowitz. 2005b. The impact of human-wildlife conflict on natural systems. Pp. 1–12 in R. Woodroffe, S. Thirgood, and A. Rabinowitz, eds. People and wildlife: conflict or coexistence? Cambridge Univ. Press, Cambridge, U.K. 497 pp.

Woodward, A. and D. David. 1994. Alligators. Pp. F1–F6 in S. Hygnstrom, R. Timm, and G. Larson, eds. Prevention and control of wildlife damage. Univ. Nebr. Coop. Ext, Lincoln.

Woolhouse, M. and S. Gowtage-Sequeria. 2005. Host range and emerging and reemerging pathogens. Emerging Infect. Dis. 11:1842–1847.

Woolhouse, M., D. Haydon, and R. Antia. 2005. Emerging pathogens: the epidemiology and evolution of species jumps. Trends Ecol. Evol. 20:238–244.

World Wildlife Fund. 2010. Living planet report 2010: biodiversity, biocapacity, and development. 117 pp.

Woronecki, P., R. Dolbeer, and T. Seamans. 1990. Use of alpha-chloralose to remove waterfowl from nuisance and damage situations. Proc. Vertebr. Pest Conf. 14:343–349.

Wright, E. 1982. Bird problems and their solutions in Britain. Proc. Vertebr. Pest Conf. 10:186–189.

Wyckoff, A., S. Henke, T. Campbell, et al. 2009. Feral swine contact with domestic swine: a serologic survey and assessment of potential for disease transmission. J. Wildl. Dis. 45(2):422–429.

Wywialowski, A. 1991. Implications of the animal rights movement for wildlife damage management. Proc. Great Plains Wildl. Damage Conf. 10:28–32.

Yashon, J. 1994. Bird strike deterrence and threat management at Ben Gurion International Airport, Israel. Bird Strike Comm. Europe Proc., Work. Pap. 22:317–320.

Yoder, J. 2002. Damage abatement and compensation programs as incentives for wildlife management on private land. Pp. 17–28 in L. Clark, J. Hone, J. Shivik, et al., eds. Human conflicts with wildlife: economic considerations. Proc. Third NWRC Spec. Symp., Natl. Wildl. Res. Cent., Fort Collins, Colo.

Yoder, J., C. Harral, and M. Beach. 2010. Giardiasis surveillance—United States, 2006–2008. Morbidity and Mortality Weekly Report, Surveillance Summaries 59:15–25.

Young, J. and C. Clements. 2009. Cheatgrass: fire and forage on the range. Univ. Nevada Press, Reno. 348 pp.

Zavaleta, E., R. Hobbs, and H. Mooney. 2001. Viewing invasive species removal in a whole ecosystem context. Trends Ecol. Evol. 16:454–459.

Ziegltrum, G. 2004. Efficacy of black bear supplemental feeding to reduce conifer damage in western Washington. J. Wildl. Manage. 68:470–474.

Ziegltrum, G. 2008. Commentary: impacts of the black bear supplemental feeding program on ecology in western Washington. Human-Wildl. Conflicts 2(2):153–159.

Ziegltrum, G. and D. Nolte. 1995. Black bear damage management in Washington State. Proc. East. Wildl. Damage Manage. Conf. 7:104–107.

Zimmermann, H., V. Moran, and J. Hoffmann. 2000. The renowned cactus moth, *Cactoblastis cactorum*: its natural history and threat to native *Opuntia* floras in Mexico and the United States of America. Divers. Distrib. 6:259–269.

Zinn, H. and M. Manfredo. 2000. An experimental test of rational and emotional appeals about a recreation issue. Leisure Sci. 22:183–194.

Zinn, H. and C. Pierce. 2002. Values, gender, and concern about potentially dangerous wildlife. Environ. Behav. 34(2):239–256.

Zinsstag, J., E. Schelling, F. Roth, et al. 2007. Human benefits of animal interventions for zoonosis control. Emerging Infect. Dis. 13(4):527–531.

Zuckerman, L. and R. Behnke. 1986. Introduced fishes in the San Luis Valley, Colorado. Pp. 435–452 in R. Stroud, ed. Fish culture in fisheries management. Am. Fish. Soc., Bethesda, Mary.

Index